T0260052

AFTER EUNUCHS

After Eunuchs

SCIENCE, MEDICINE, AND
THE TRANSFORMATION OF SEX
IN MODERN CHINA

Howard Chiang

Columbia University Press

New York

Columbia University Press gratefully acknowledges the generous support for
this book provided by Publisher's Circle member Dorothy Ko.

Columbia University Press
Publishers Since 1893
New York   Chichester, West Sussex
cup.columbia.edu
Copyright © 2018 Columbia University Press
All rights reserved

Library of Congress Cataloging-in-Publication Data
Names: Chiang, Howard, 1983– author.
Title: After eunuchs : science, medicine, and the transformations of sex in modern China /
Howard Chiang.
Description: New York : Columbia University Press, [2018] | Includes bibliographical
references and index.
Identifiers: LCCN 2017055402 | ISBN 9780231185783 (cloth : alk. paper) |
ISBN 9780231546331 (e-book)
Subjects: LCSH: Gender identity—China. | Sex role—China.
Classification: LCC HQ1075.5.C6 C45 2018 | DDC 305.30951—dc23
LC record available at https://lccn.loc.gov/2017055402

Columbia University Press books are printed on permanent and durable acid-free paper.
Printed in the United States of America

Cover image: "Social Manners in Secure," 2015, Raintree Chan. Used with permission
of Raintree Chan.

For Hao-Te

Contents

Illustrations

Acknowledgments

Over the years I have incurred many debts, undoubtedly many more than what I will be able to acknowledge here. My most heartfelt thanks go to my PhD advisors, Benjamin Elman and Angela Creager. The reading seminar that I took with Ben on the history of East Asian science marked a point of no return in my journey as a scholar. At that juncture of my intellectual curiosity, his erudition proved to be a most fitting shoulder for my obsession with pluralist interpretations of the past—far more and richer than what I have been able to absorb. Similarly, the unparalleled support that I received from Angela made the History of Science Program at Princeton a comfortable second home. Although I ultimately changed the geographical focus of my research, Angela's versatility—as revealed in academic contexts and otherwise—guided me through difficult times and made possible exciting moments in my doctoral study. Even though they granted me a great deal of liberty in the process of completing my dissertation, they always knew when and where to push me, when and where to encourage me, and how to make me come to terms with the shortcomings of my writing. Their wise counsel, intellectual freedom, and unwavering support will continue to inspire me in the future. I will consider it a success if I can achieve even a fraction of what they have accomplished. This book takes an important first step in that direction.

Outside Princeton, the person that I have come to think very highly of as a mentor, role model, collaborator, and friend is Ari Larissa Heinrich.

Given our shared interest in the history of science, modern Chinese cultural studies, global queer studies, and critical theories of the body, we have had the opportunity to collaborate on many occasions, and in all of them Ari has impressed me time and again with the depth of his knowledge and his discerning approach to raising questions and solving problems. Although he often presents himself as a newcomer to many of our collaborative projects, I consider his modesty and elegant style of responding to criticisms a cornerstone of his scholarly integrity, something that I will continue to emulate in the years to come. And I am grateful to Margot Canaday, Sean Hsiang-lin Lei, Helen Hok-Sze Leung, and Susan Stryker for their generous advice on research and professional development. I am especially honored to have taught for Margot's course on the history of gender and sexuality in modern America.

While at Princeton, I benefited tremendously from the advice and scholarly training of many faculty members, including Graham Burnett, Margot Canaday, Janet Chen, Angela Creager, Ben Elman, Michael Gordin, Molly Greene, Liz Lunbeck, Michael Mahoney, Susan Naquin, Paul Starr, Emily Thompson, Helen Tilley, and other participants in the History of Science Program Seminar, Women and Gender Graduate Student Colloquium, and Contemporary China Colloquium. I thank for their support and friendship Sare Aricanli, Tuna Artun, Daniel Burton-Rose, John DiMoia, Chunmei Du, Carolyn Eisert, Matt Ellis, Yulia Frumer, Benjamin Gross, Nathan Ha, Tomoko Kitagawa, Victoria Lee, Maribel Morey, Karam Nachar, Anthony Pedro, Margaret Schotte, Wayne Soon, Daniel Trambaiolo, Brigid Vance, Shellen Wu, Doogab Yi, Evan Young, and Xinxian Zheng. Tireless critics unwilling to settle for easy answers, Karam, Sare, Daniel, Nathan, and Brigid provided perceptive comments on different parts of my early work.

The seed of the intellectual impulse for writing this book, however, was really planted during my undergraduate years at the University of Southern California. Above anyone else, Lois Banner opened my eyes to the historical study of gender, sexuality, the body, sexology, and queer subcultures. Our endless, stimulating conversations inspired me to pursue a career as a professional historian. Brooke Carlson was the first person who taught me how to write well in college. Dianna Blaine, too, helped me sharpen my composition skills, but her classes also played an important role in shaping my interest in gender studies and feminist theory. Karen Zivi's lectures convinced me the power of the theories and writings of Michel

Foucault. The graduate seminars that I took with Carla Kaplan on feminist theory and Joseph Hawkins on LGBT studies continue to influence my approach to research in these areas. After graduating from USC, I became more intellectually mature under the tutelage of Alice Kessler-Harris and Rebecca Jordan-Young at Columbia University.

Many individuals deserve acknowledgment for their unwavering support and for taking substantial time from their work to improve mine. Celina Hung, Lily Wong, and Alvin Wong provided extensive feedback on the entire book manuscript. I thank them for their thoughtful and swift replies to my often last-minute requests. Andrew Jones, Haiyan Lee, Carlos Rojas, and Joan Scott gave useful comments on an earlier draft of chapter 1. Tani Barlow, David Luesink, and Yi-Li Wu offered extremely helpful suggestions for improving chapter 2. Hongwei Bao, Wenqing Kang, Thomas Laqueur, Petrus Liu, and Giovanni Vitiello pushed me to clarify the critical interventions of chapter 3. Francesca Bray, Angelina Chin, Huei-chu Chu, Charlotte Furth, Derek Hird, Izumi Nakayama, Tong Lam, Angela Leung, Megen Steffen, and Hsiu-yun Wang read or heard various drafts of chapter 5 and made astute suggestions on how I might improve it.

Numerous friends and colleagues encouraged me, advised me, challenged me, and listened to me graciously as I obsessed about my work. I am grateful for my conversations with Bridie Andrews, Daniel Asen, Andrea Bachner, Maxine Berg, Beverly Bossler, Francesca Bray, Hsiu-fen Chen, Hsiao-wen Cheng, Ta-wei Chi, Angelina Chin, Rebecca Earle, David Eng, Harriet Evans, Judith Farquhar, Ross Forman, Veronika Fuechtner, Takashi Fujitani, Anne Gerritsen, Anna Hajkova, Marta Hanson, Heinrietta Harrison, Douglas Haynes, Todd Henry, James Hevia, TJ Hinrichs, Sarah Hodges, Walter Hsu, Kuang-chi Hung, Quinn Javers, Ryan Jones, Eric Karchmer, Michael Keevak, Kyu Hyun Kim, Joyman Lee, Yu-lin Lee, Sean Lei, Angela Ki Che Leung, Helen Leung, Xiaodong Lin, David Luesink, Joanne Meyerowitz, Rana Mitter, Carla Nappi, Sarah Richardson, Giorgio Riello, Gayle Salamon, Tze-lan Sang, Volker Scheid, Laura Schwartz, Joan Scott, Fabien Simonis, Claudia Stein, Susan Stryker, Mathew Thomson, Charles Walton, James Welker, Harry Wu, Yi-Li Wu, and Shirley Ye.

I am especially indebted to Hongwei Bao, Jia Yinghua, Travis Kong, Angela Leung, Carla Nappi, Peter Szto, Wen-Ji Wang, and Yi-Li Wu for sharing with me their unpublished manuscripts or items from their personal collections. I have also benefited from conversations with the following individuals at various stages of this project: Steven Angelides, Heike Bauer,

Brian Bernards, Jens Damm, Donna Drucker, Prasenjit Duara, Sara Friedman, Daiwie Fu, Charlotte Furth, David Halperin, Chie Ikeya, Andrea Janku, Travis Kong, Dorothy Ko, Eugenia Lean, Ping-Hui Liao, Vivienne Lo, Tamara Loos, Fran Martin, Casey Miller, John Moffett, Afsaneh Najmabadi, Rayna Rapp, Leila J. Rupp, Hugh Shapiro, Grace Shen, Shu-mei Shih, Megan Sinnott, Matthew Sommer, Liza Steele, Marc Stein, Michiko Suzuki, Mirana Szeto, E. K. Tan, Chien-hsin Tsai, Jing Tsu, Rubie Watson, David Der-wei Wang, Yin Wang, Chia-Ling Wu, Jeu-Jeng Yuann, and Zhongmin Zhang.

I am extremely grateful for the encouragement of Anne Routon and Caelyn Cobb at Columbia University Press, whose support gave this book greater scholarly appeal and coherence, as well as the marvelous assistance of Miriam Grossman. The detailed comments and suggestions of the two anonymous reviewers were incredibly insightful and helpful. Their feedback made this book intellectually richer in the process.

The research and writing for this book was made possible by the generous financial assistance of various programs at Princeton University, including the Graduate School's University Fellowship and the History Department's Shelby Cullom Davis Stipend from 2006 to 2011. Research grants from the Department of History, the East Asian Studies Program, and the Princeton Institute for International and Regional Studies enabled me the conduct research in China and Taiwan in 2007, 2008, and 2010. The Institute of Modern History at Academia Sinica in Taipei kindly provided me an office for my visiting research fellowship in 2008–2009 and the Academia Sinica Fellowship for Doctoral Candidate in the Humanities and Social Sciences in 2011–2012. In the spring of 2010, an Andrew W. Mellon Foundation Research Fellowship supported my stay at the Needham Research Institute in Cambridge, UK, and at the same time the Wellcome Trust Centre for the History of Medicine at University College London provided me an office space for my visiting student affiliation. A PhD Writing Completion Fellowship from the East Asian Studies Program at Princeton supported my work in the fall of 2011. Gratitude is also due to the Universities' China Committee in London; the Centre for the History of Medicine, the Department of History, the Global History and Culture Centre, the Humanities Research Centre, and the Humanities Research Fund at the University of Warwick; and the History Department and the Faculty of Arts at the University of Waterloo for funding this project during its final stages.

All translations from foreign sources are my own except where noted.

Earlier versions of parts of chapters 1, 3, and, 5 originally appeared in "Epistemic Modernity and the Emergence of Homosexuality in China," *Gender and History* 22, no. 3 (2010); "How China Became a 'Castrated Civilization' and Eunuchs a 'Third Sex,'" in *Transgender China*, edited by Howard Chiang (Palgrave Macmillan, 2012); "Data of Desire: Translating (Homo) Sexology in Republican China," in *Sexology and Translation: Cultural and Scientific Encounters Across the Modern World, 1880–1930*, edited by Heike Bauer (Temple University Press, 2015); "Gender Transformations in Sinophone Taiwan," *positions: asia critique* 25, no. 3 (2017), republished by permission of the copyright holder, Duke University Press; and "Christine Goes to China: Xie Jianshun and the Discourse of Sex Change in Cold War Taiwan," in *Gender, Health, and History in Modern East Asia*, edited by Angela Ki Che Leung and Izumi Nakyama (Hong Kong University Press, 2017).

Finally, I am thankful for the steadfast support of my parents, extended family, and nonacademic friends. But I owe the greatest debt to Hao-Te Shih. In the past decade, he has put up with my demand for solitary moments, loved me through good times and bad, and shared with me the pleasures of everyday life. It is to him that I dedicate this book.

AFTER EUNUCHS

Introduction

Toward a Genealogy of Sex

T hroughout Chinese history, one gender-liminal figure stood out as exceptional: the eunuch.[1] According to some medical writers, babies born with ambiguous genitalia were sent to the imperial palace as "natural eunuchs."[2] More often, a normal boy was transformed into a eunuch through castration surgery at a relatively young age. This group of castrated men, who typically began their careers as low-ranking servants inside the palace, wielded enormous political power during distinct epochs of Chinese history. The eunuchs of the Ming dynasty (1368–1644), next to those of the Eastern Han (25–220) and the Tang (618–907), are perhaps the most notorious and well-studied cohort for having exerted colossal control over the state.[3] After the fall of the Ming, Manchu emperors throughout the Qing dynasty (1644–1911) imposed greater constraints on the activities of eunuchs and curtailed eunuch employment.[4] Court officials, literati, and other cultural elites often attacked the eunuch system by stressing the corruptive behaviors and the absence of scholarly credentials of these castrated men.

Interestingly, early critiques of castration in Chinese discourses tended to neglect the question of sexual identity altogether. Destiny was not anatomy. Before the twentieth century, men and women in China rarely pleaded their social division based on biological "facts" alone.[5] It is true that physicians had long differentiated bodies on the basis of gender since the early period. For example, they intervened in childbirth and

[1]

pregnancy-related problems and sometimes prescribed separate drugs for women.[6] But their thinking mostly built on an androgynous medical model that historian Charlotte Furth calls the "Yellow Emperor's body," which was rooted as much in the conceptualization of bodily and spatial location as in the yin–yang cosmology.[7] It was only in the Song dynasty (960–1279) that an independent department for treating women's disease emerged in full force with the maturation of *fuke* (婦科, women's medicine or gynecology).[8] In Europe, doctors and philosophers shifted their views of men and women from two versions of a single-sexed body to incommensurable opposites only by the time of the Enlightenment.[9]

Despite the salience of gender hierarchy to Confucian patriarchal norms, or precisely in order to valorize the latter's cultural assertion, Chinese physicians did not identify the seed of sexual difference in the corporeal flesh until the nineteenth century. Early modern *fuke* experts, all of whom were men, drew attention to the physical symptoms of bodily process, such as blood depletion, rather than systematic understandings of an isolated organ.[10] In the 1980s, by contrast, the birth and expansion of *nanke* (男科, men's medicine) underscored the importance of genital physiology and the social and psychological basis of sexuality.[11] Bridging the gap from *fuke* to *nanke*, this book seeks to understand the process through which these novel psychobiological understandings of sex emerged in modern Chinese culture. One of my main goals is to locate the changing meanings of gender, sexuality, and the body within a growing global hegemony of Western biomedicine in the nineteenth and twentieth centuries.

From Eunuchs to Transsexuals

There is little reason to assume that sex had always already been an arresting concept for those most erudite in the study of nature. Li Shizhen (李時珍, 1518–1593), the author of *Bencao gangmu* (本草綱目), a book that epitomized materia medica in the Ming dynasty and continues to be cited as the most authoritative and comprehensive encyclopedia in Chinese medicine today, posited a spectrum of human reproductive anomalies with five "nonmales" (非男, *feinan*) and five "nonfemales" (非女, *feinü*).[12] The defective nonmales were the natural eunuch, the bullock, the leaky, the coward, and the changeling and could not become fathers; the deficient nonfemales were incapable of biological mothering and included the corkscrew, the striped, the drum, the horned, and the pulse.[13] Being the most systematic classification of "hermaphroditism" in late imperial China, Li's typology defined the boundaries of sexual normativity in terms of reproductive capabilities alone.[14] Modern biomedical understandings of genital anatomy, endocrine secretions, chromosomes, and sexual psychology—traits that seem so universal for the natural definitions of sex—fell beyond its ordering of knowledge. Although some *fuke* doctors later placed a more liberal emphasis on the womb and the breast in diagnosing female-specific ailments, they rarely held the independent workings of these organs responsible for the unusual development of gender expression.[15]

Around the turn of the twentieth century, the concept of sex slowly entered the Chinese lexicon.[16] Already in the Self-Strengthening movement (1861–1895), when the urban center of Chinese culture and society relocated from the heartland to the shore, missionary doctors dedicated themselves to introducing Western-style medicine, including the establishment of new asylums and the translation of modern anatomical knowledge.[17] Their contributions stamped the first sustained effort to redefine Chinese understandings of sexual difference in terms of Western reproductive anatomy. The gradual spread of Western biomedical epistemology from elite medical circles to vernacular culture reached a crescendo in the 1920s.[18] Learning from their Euro-American colleagues, Chinese biologists promoted a vision of sex dimorphism, which construed the bodily morphology and function of the two sexes as opposite, complementary, and fundamentally different. Their writings endowed the concept of *xing* (性, sex) with an integral feature of visuality by foregrounding epistemic connections between what they called "primary," "secondary," and "tertiary" sexual characteristics. They extended these connections to all life forms across the human/nonhuman divide, and they tried to explain hermaphroditism with genetic theories of sex determination. Over time the visual evidence of bioscience recast existing boundaries and polarities of gender in a new normative light.[19]

The bioscientific naturalization of gender coincided with the collapse of the Qing imperium as well as the unprecedented success of the feminist and education reform movements.[20] Against this political backdrop, the demise of eunuchism paralleled certain rhetorical features of the anti-footbinding discourse such as the exposing and display of bare bodies.[21] Three voices contributed to the making of an "archive" documenting the methods of Chinese castration, a repository of evidence that was decisively lacking before the late nineteenth century: that of Western spectators, eunuchs themselves, and members of the last imperial family. An anti-eunuch sentiment arose out of the photographic, textual, and oral records these voices left behind, and as the gender identity of eunuchs was evaluated anew in the modern era through the lens of Western biomedicine, China's association with the metaphor of a "castrated civilization" intensified over time. The period between the 1870s and the 1930s thus constituted a transitional phase when the castrated male body—joining women's bound feet and the leper's crippled body—seemed out of sync with the Chinese body politic at large.[22] Those decades witnessed the development of a growing interest in eunuchs' sexual identity, something also absent in earlier critiques of castration. The hegemony of Western biomedicine, in other words, cultivated the increasingly common association of eunuchs with the nominal label of the "third sex." The demise of eunuchism indexed the birth of the concept of sex itself.

And the meaning of sex soon began to change. At the dawn of the century, the word "*xing*" carried visual connotations of male and female biology. In scientific and popular formulations, women and men were simply understood, respectively, as human equivalents of the *ci* (雌, female) and *xiong* (雄, male) types of lower organisms. Most observers adhered to a biological determinism. In various efforts to delineate different components of sex, they considered reproductive anatomy, morphological characteristics, and sexual chromosomes on different levels of visual representation. But in the aftermath of the New Culture movement (1915–1919), iconoclastic intellectuals such as Zhang Jingsheng (張競生, 1888–1970) and Pan Guangdan (潘光旦, 1899–1967) contended that the hidden nature of human erotic preference could also be discovered and known in a scientific way. Sex, they argued, was not merely an observable trait, but it was something to be desired as well. These May Fourth public intellectuals participated in a new concerted effort, though not without dissent from their interlocutors, to emulate European sexological science. Their translation of Western sexological texts, concepts, methodologies, and styles of reasoning provided the crucial historical conditions under which, and the means through which, sexuality emerged as an object of empirical knowledge. The disciplinary formation of Chinese sexology in the May Fourth era, therefore, added a new element of carnality to the scientific meaning of sex.

By the second third of the century, the vocabulary of sex had expanded to encompass an intrinsic magnitude of malleability. The idea of hormones provided Chinese sex researchers, tabloid writers, popular authors, and social commentators a new scientific basis for discussing gender and the human body. Beginning in the mid-1920s, they appropriated from Western endocrinologists the theory of universal bisexuality, which posited that everyone was partly male and partly female. This chemical and quantitative definition of sex was supported by findings arising from selected laboratories in Europe, especially in Vienna, where famous animal sex reversal experiments were conducted. In the early twentieth century, the intriguing results of these experiments reached a global community of scientists.[23] In the United States, biologists imagined sex as a "plastic dichotomy," which gave them the liberty to claim both the fluidity and the rigidity of sexual binarism to varying degrees.[24] Psychologists, psychoanalysts, and other sex researchers, too, debated on the validity of the universal bisexual condition as they unpacked the distinctions between homosexuality, bisexuality, and transsexuality.[25] As Chinese scientists entertained the possibility of sex transformation based on these foreign ideas and experimental findings, they referred to indigenous examples of reproductive anomalies—such as eunuchs and hermaphrodites—as points of reference and, most importantly, redescribed these old phenomena in the new language of biological sex. Meanwhile, in the mid-1930s the explosion of journalistic sensationalism on Yao Jinping (姚錦屏, b. 1915), a woman from Tianjin who allegedly claimed to have turned into a man overnight, greatly amplified people's awareness of the possibility of human sex alteration. By the 1940s these epistemological shifts found some of their loudest pronouncements in the depiction of sex change in popular fictions.

As scientists and doctors sought to pin down the technical definitions of sex, non-experts took a more serious interest in broadening its social valence. Emerging from the domains of biology, sex psychology, and endocrinology, the multiple interpretations of sex saturated the Chinese cultural agenda in the Republican period. The anti-footbinding movement and the demise of castration had already acquainted the public with images of "natural" male and female bodies. The new idea of romantic love had begun to push people to break from conventional arranged marriages and to form nuclear families.[26] Popular versions of Freud and other sexologists bolstered the recognition of psychosexual development as the cornerstone of individual subjectivity.[27] Narratives of male and female same-sex relations called up complex associations from ideologies of proper and imprudent gender orientation.[28] Similarly, stories on prostitutes surged and reappeared with conflicting messages about decent sexual behavior in the popular press.[29] And the mass media had made sex and its possible transmutation a mainstay of visual culture. In broad outline and narrow, Chinese society had "sexualized" during the first half of the twentieth century.

When the boundaries of sex no longer appeared as impermeable as they once had seemed, the Chinese-speaking community met its first transsexual, Xie Jianshun (謝尖順, b. 1918). In 1953, four years after Mao Zedong's political regime took over mainland China and the Republican state was forced to relocate its base, the success of native doctors in converting a man into a woman made news headlines in Taiwan. Enthusiasts frequently labeled Xie the "Chinese Christine." This was an allusion to the contemporaneous American ex-G.I. transsexual celebrity, Christine Jorgensen (1926–1989), who had traveled to Denmark for her sex reassignment surgery in the spring of 1950 and became a worldwide household name immediately afterward due to her self-fashioned glamorous, feminine look.[30] Within a week, the characterization of Xie in the Taiwanese press changed from an average citizen whose ambiguous sex provoked uncertainty and national anxiety to a transsexual cultural icon whose fate would indisputably contribute to the global staging of Taiwan in the neo-imperial shadow of the United States. The saga of Xie Jianshun and other sex-change stories illustrate how the Republican government regained sovereignty in Taiwan by inheriting a Western biomedical epistemology of sex from Republican-era scientific globalism—a medical worldview especially conducive to the prevailing American model of health care in the early Cold War era.[31] In other words, the reciprocal relationship between medical scientific knowledge and the transformation of the body—in terms of both corporeal and geopolitical arrangements—culminated, historically, in the conditions under which transsexuality emerged first and foremost in places like Taiwan across the postcolonial Pacific Rim, which was geographically and culturally situated at the overlapping margins of Chineseness and transpacific U.S. hegemony.

From eunuchs to transsexuals, this book revises the view that China "opened up" to the global circulation of sexual ideas and practices only after the economic reforms of the late 1970s.[32] Building on a growing body of revisionist historical scholarship, the bulk of the narrative highlights the 1920s as an earlier, more pivotal turning point in the modern definitions of Chinese sexual identity and desire.[33] However, in contrast to the view propounded by such historians as Frank Dikötter, who maintains that modernizing elites of the early Republican period failed to introduce new and foreign ideas of sexual variations but only sexuality qua "heterogenitality," this study, with its focus on marginal sexualities, provides ample evidence on the contrary.[34] Drawing on scientific publications, medical journals, newspaper clippings, popular magazines, tabloid presses, scholarly textbooks, fictional and periodical literatures, and other previously untapped sources, this study portrays the decades between empire and communism as a globally significant, rather than catastrophic, interlude in China's modern history.[35] The evolving discourse of same-sex desire and the biologization of gender norms constituted two epistemological ruptures that complicated the shifting correlations of sex, gender, and sexuality in the Republican period.[36] The extensive media attention on sex change in postwar Taiwan marked a culminating episode of these earlier developments.

Weaving together intellectual, social, and cultural history, this book aims to accomplish three goals: it shows how sexual knowledge became a crucial element in the formulation of Chinese modernity; it highlights the role of the body as a catalyst in the mutual transformations of Chinese nationalism and the social significance of sex; and, grounded in the visual and conceptual analysis of sexual science, it establishes a genealogical relationship between the demise of eunuchism and the emergence of transsexuality in China.[37] This genealogy maps the underexplored history of China's modern geobody onto the more focused history of the biomedicalized human body.[38] The coevolution of "China" and "sex," two seemingly immutable constructs, sheds light on the gradual displacement of colonial modernity by Sinophone articulations in the course of the twentieth century.[39]

From Etymology to Epistemology

For a book that centers on the history of sex, it is not immediately intuitive why such a history is explored through the prisms of science and transformation. Although tracing the etymology of the Chinese word for sex presents itself as a logical task, this book implements a slightly different approach and makes an auxiliary methodological claim for the value of epistemological inquiry. That is, a central motif that threads the following chapters is the treatment of sex as an epistemic concept and not just a word.

Readers who are familiar with the work of Michel Foucault would quickly notice the affinity between his genealogical interrogation and the turn from etymology to epistemology underlying this book's methodological ambition.[40] The rest of this introduction explains *why* and *how* this study adopts an analytical approach in science studies called historical epistemology.[41]

A cursory review of Chinese books published in the first decade of the twentieth century reveals that "*xing*," the modern Chinese word for sex, did not appear in any of the titles. This was true even for books with a thematic focus on reproductive medicine.[42] Strictly speaking, *xing* did not mean sex before the twentieth century, and feminist critics have argued for the incompatibility of translating "gender" into *xingbie* (性別) in the late Qing and early Republican periods.[43] For example, for the many definitions of *xing* found in the *New China Character Dictionary* (2004)—such as "natural instincts," "inherent tendencies," "disposition," "temperament," "the nature of something (or of someone)," and "life," among others—there is a plethora of corresponding sources in classical Chinese in which the word "*xing*" was used.[44] They include Gaozi's well-known expression, from the Confucian text *Mencius*, that the appetite for food and the appetite for sex together constitute human nature: *shi se xing ye* (食色性也). Note that, in this expression, what *xing* connotes is natural instinct and not sex, which is represented by the word "*se*." The "Gaozi" chapter of *Mencius* also supplies the source for the meaning of *xing* as natural disposition or temperament: "When Heaven is about to confer a great office on anyone, it first exercises his mind with suffering . . . it stimulates his mind, hardens his nature [*xing*], and supplies his incompetence."[45] Similarly, Buddhist texts use *xing* to refer to the nature of something, thereby contrasting it to *xiang* (像), the superficial appearance of all things.[46] The usage of *xing* to mean life can be found in *The Chronicles of Zuo* (左傳, *Zuozhuan*, fourth century BCE): "The people enjoy their lives [*xing*], and there are no enemies or thieves" or "New palaces are reared . . . the strength of people is taxed to an exhausting degree . . . the people feel that their lives [*xing*] are not worth preserving."[47] When it finally arrives at a definition of *xing* pertaining to sex, the *New China Character Dictionary* remains silent on a corresponding classical source. As historian Leon Rocha has observed, "If one takes for granted that dictionaries attempt to record usages of a certain word in common currency, then *xing* until the twentieth century continued to signify what Heaven had decreed; *xing* named an unsexed, ungendered concept of innate human nature or essence."[48]

On the level of etymological investigation, Rocha offers two main explanations for how *xing* came to mean sex by the 1920s. First, he traces the origin of this association to Ye Dehui's (1864–1927) "Preface to *The Classic of the Plain Girl*," which appeared in 1907. According to Rocha and Jai Ben-ray, the first mentioning of *xing* qua sex appeared in the following sentence: "The spirit of the study of sex [性學, *xingxue*], how could the pedantic Confucian scholars possibly be able to see its essence?"[49] It is important to note that the sexual designation of *xing* first appeared in a phrase, *xingxue*, that would later become the standard translation of terms such as sexology, sex science, sexual sciences, and sex research. We return to the disciplinary formation of sexology in the third chapter of this book.[50] For now, suffice it to conclude from this first sexual designation of *xing* that it crucially depended on a scientific, naturalized understanding of sex that radically diverges from its earlier obscene, negative, or moralistic connotations, as found in such traditional expressions as *yin* (淫, lewd), *se* (色, lust), and *pi* (癖, obsession). In the 1910s a few Chinese vernacular publications on women and feminism began to equate *xing* with sex, either to refer to the relationship between men and women or to mean sexual desire.[51]

Another important etymological origin of the modern definition of *xing* is the Japanese-mediated translation of the English word "sex." In this etymological route, the 1920s represents a watershed moment, "as *sei*-sex was institutionalised in Japanese reference works for the first time."[52] Many Japanese dictionaries published from the 1890s to the 1910s, for instance, contained no record of *sei*-sex; the term is missing from Otsuki Fumihiko's *Sea of Words—Genkai* (1891), Owada Tateki's *Nihon da jiten* (1897), Shozaburo Kanazawa's *Forest of Words—Jirin* (1907), Shigeno Yasutsugu's *Sanseido kanwa dajiten* (1910), and Matsui Kanji and Uedo Kazutoshi's *Fuzanbo dainihon kokugo jiten* (1915). By 1927 the equation of *sei* with sex was most powerfully articulated by Ochiai Naobumi and Haga Yaichi in their *Fountain of Words—Gensen*: "*Sei*. English: *sex*, the differences in the psychological and physical qualities of men and women."[53] However, as Rocha has carefully pointed out, a few scattered Japanese–English dictionaries published before the 1920s did record *sei*-sex, including Shibata Shoukichi and Koyasu Takashi's *Eiwa jii* (1873), the Japanese translation of *Webster's*

Unabridged Dictionary (1888) by Tanahashi Ichiro and Frank Warrington Eastlake, and Kanda Naibu's *Mohan shin eiwa daijiten* (1911) and *Shuchin konsaisu eiwa jiten* (1922). Therefore, Rocha concludes that

> the pattern that emerges is that, until the twentieth century, the character called *xing* in Chinese was used in Japanese to also signify nature, life and so forth, and from the 1870s to 1880s, the *kanji* was used to signify sex and this new usage became more popular in the 1920s, displacing older words such as *iro* (the Japanese equivalent of the Chinese *se*). This corroborates Furukawa Makoto's finding that *sei* (as sex) became a fashionable word in the 1920s. . . . We could venture the hypothesis that the Japanese used the *kanji* called *xing* in Chinese to translate "sex" and "sexuality" *before* the Chinese.[54]

The strange career of the translation of "sex" into "*xing*," above all, exemplifies what translation theorist Lydia Liu has called "return graphic loans": "the Japanese used *kanji* (Chinese characters) to translate European terms, and the neologisms were then imported back into the Chinese language."[55] Specifically, these "return graphic loans" were "imported back into Chinese with a radical change in meaning."[56] Therefore, given all the meanings of "*xing*" other than "sex" before the twentieth century, the established equivalence of ideas about sexual anatomy, sexual behavior, sexual desire, and so on with the Chinese word "*xing*" must be understood as a product of these early twentieth-century neologistic constructions.

The translational trajectory of sex, as mediated via the Japanese language, reflects the broader historical patterns of the translation of other words that are central to our study, including, most importantly, "science." In classical Chinese, *kexue* (科學) means "studies for the civil examinations"; after the Japanese appropriation of the term (*kagaku*), however, it became "science" in twentieth-century China. This history of mediated translation had direct bearing on the meaning and practice of knowledge production. Before the 1920s, in places such as the Jiangnan Arsenal and the Commercial Press in Shanghai, scientific thinkers, men of letters, missionaries, and philologist activists all participated in the "unruly" practices of translating science and compiling and rewriting foreign texts. In the ad hoc practices of the Jiangnan Arsenal, the *chouren* (疇人) tradition of the eighteenth century and the *gezhi* (格致) learning of the nineteenth century were reappropriated in a similar way to refer to "an open spectrum of learning established through a translative relationship between different systems of knowledge."[57] In the transcompilation activities of the Commercial Press, as evident in the production of *Botanical Nomenclature* (植物學大辭典, *Zhiwuxue da cidian*, 1908–1917) and *Nomenclature of Zoology* (動物學大辭典, *Dongwuxue da cidian*, 1921), the editors combined the classification systems of Carl Linnaeus (1707–1778) and Michel Adanson (1727–1806) with the empirical executions of the classical philological tradition.[58] As historian Meng Yue puts it, "These encyclopedic dictionaries functioned in opposition to the complete 'takeover' of Meiji word usage and terminology by preserving, in philological ways, alternative histories of universal concepts and negotiating space for what was about to turn into 'local knowledge.'"[59] In other words, the participants of these transculturation activities espoused a universality of knowledge production by vindicating the translatability of discrepant systems of knowledge.

The Japanese-mediated translation of "science" into *kexue* in the early twentieth century reduced the epistemological and practical potentials of *chouren* and *gezhi*. This mediated process of translation equated both words with "Chinese learning" and positioned them as such in opposition to "Western learning." Conservative officials, such as Weng Tonghe (翁同龢, 1830–1904) and Zhang Zhidong (張之洞, 1837–1909), and reformists, such as Liang Qichao (梁啟超, 1873–1929), mapped the new concept of *kexue* onto Western learning and juxtaposed it as a hierarchically superior form of knowledge against Chinese learning, *chouren*, and *gezhi*. In the early twentieth century, *chouren* and *gezhi* would lose all of the "uncontrollable proliferation of textualities" that characterized their nineteenth-century scientific praxes.[60] Similarly, by the time when a younger generation of intellectuals, such as Hu Shih (胡適, 1891–1962) and Wang Yunwu (王雲五, 1888–1979), replaced the earlier group of philologist editors at the Commercial Press, the early yet more creative phase of the press's history, which Meng Yue designates as an era of "semiotic modernity," came to an end.[61] Therefore, even though the chronology of this book begins precisely at the concluding moments of this semiotic modernity, it is far from an attempt to flatten out the rich and prolific elements of earlier scientific practice.

According to Wang Hui, by the time the mainstream intellectuals of the New Culture movement, such as Chen Duxiu (陳獨秀, 1879–1942), Hu

Shih, Wu Zhihui (吳稚暉, 1865–1953), and Ding Wenjiang (丁文江, 1887–1936), among others, adopted the *kexue* translation of science, their efforts both reflected and produced a fundamental rearrangement of modern Chinese thought in the early Republican period: "a scientific 'worldview based on axiomatic principles' (公理世界觀, *gongli shijieguan*) reformed and replaced the traditional 'worldview based on heavenly principles' (天理世界觀, *tianli shijieguan*)."[62] The articulation of issues of civic impartiality, social legitimacy, and individual rights drew on the new abstract scientific vocabularies of objectivity, validity, rationality, and so on. If the birth of the "modern Chinese nation" was directly connected to the rise of a new "modern scientific worldview" in China, this study begins to make that connection even more explicit to its epistemological undergirding of the concept of sex. Put differently, my goal is to shift our attention away from the familiar competing discourses of nationalism between the constitutionalists (e.g., Liang Qichao) and revolutionaries (e.g., 章炳麟, Zhang Binling 1868–1936) around the turn of the twentieth century. As many historians have pointed out, their respective visions of national citizenship and anti-Manchu Han nationalism mainly revolved around the questions of race and ethnicity.[63] This book therefore aims to reorient our analytical spotlight on the largely overlooked, yet no less salient, issue of sex.

In order to "map" the conceptual cartography of *xing*, I focus on some of its key epistemic nodal points around which its modern definition of sex coalesced. An etymological investigation, such as the one executed by Rocha, certainly has its heuristic value. But words have a life of their own outside pure linguistic boundaries.[64] The way they are *used* to make certain statements meaningful suggests that their history—and historicity— far exceeds the etymological realm. It might be useful, for instance, to consider the social significance of key terms as stemming from their operation not only as words with specific definitions but as concepts whose comprehensibility depends largely on the context of discussion and knowledge production. Here is where I depart from Raymond Williams' "keyword" approach and adopt the method of historical epistemology developed by such philosophers of science as Arnold Davidson, Ian Hacking, Lorraine Daston, and Hans-Jörg Rheinberger.[65] Insofar as certain translated terms, such as "science," "race," and "sex," played a central role in elite and vernacular Chinese discourses of nation formation, their conceptual contingency and historicity are as important as, if not more so than, their etymological origins in giving rise to their cultural propagation.[66] It is on this conceptual register of words—their epistemological conditions—that constitutes the focus of my genealogical analysis of sex.

By epistemology, I follow the French tradition, best exemplified by the work of Gaston Bachelard, Georges Canguilhem, and Foucault, to reflect on "the historical conditions *under* which, and the means *with* which, things are made into objects of knowledge."[67] This definition of epistemology differs from one typical of the classical tradition, which tends to be preoccupied with the question of what it is that makes knowledge scientific. In other words, the historical epistemology approach I adopt in this book does not invoke a theory of knowledge that inquires into a presumed scientific basis of the structures and nature of knowledge but is primarily concerned with "the process of generating scientific knowledge and the ways in which it is initiated and maintained."[68] Rheinberger characterizes this shift in terms of "a transformation of the problem situation": "A reflection on the relationship between concept and object from *the point of view of the knowing subject* was gradually replaced by a reflection of the relationship between object and concept that started from *the object to be known*."[69]

With this historical epistemology approach, this study proceeds from treating *"xing"* as a concept and "sex" as its corollary scientific object to be known. One should avoid the assumption that the nature of the historical relationship of sex to science was fundamentally fixed so that an undisguised view of *xing* (as sex) was merely waiting to be acquired by Chinese scientists. This would fall under the classical epistemologist tradition. A more useful point of departure, which is the one employed here, is to start from the opposite of that very assumption: by looking at the conditions that *had to be established* in order for *xing* qua sex to become "objects of empirical knowledge under historically variable conditions."[70]

My larger thesis is that the modern formulation of *xing* hinges on the rise of new structures of knowledge in the early twentieth century. These structures of knowledge formed an "epistemic nexus" around which the visual realm of life, the subjectivity of human desire, and the malleability of the body became intertwined in the concept of sex. The historical process whereby *xing* grew into the conceptual equivalent of sex reflects a broader underlying transformation in its epistemological designation of human nature: from the rock-solid essence of things into a mutable ontological referent. This transformation triggered a concurrent evolution in which the bodily arbiter of gender (e.g., uterus and testis) shifted from being the outgrowth of cosmological forces to being the very determinant of sexual difference. Rooted in the effort of scientists who popularized ideas in biology, sexology, and endocrinology and who reached out to a wider public in a piecemeal fashion, *xing* acquired new elements of visibility, carnality, and elasticity in coming to mean "sex" in the twentieth century. Each of these three historical-epistemological conditions are examined independently in the intervening chapters of the book, but first our analysis must attend to the global geopolitical context that paved the way for the emergence of sex itself.

CHAPTER I

China Castrated

Before Transsexuality

In 1976 the Chinese physician Chen Cunren (陳存仁, 1908–1990) learned by word of mouth that Singapore was a metropolis where beautiful, castrated cross-dressers gathered. During his visit his friends explained that these glamorous castrated men were actually prostitutes. Both local residents and tourists considered these cross-gendered sex workers, especially those with their male genitals removed, more "precocious" than ordinary prostitutes, and they therefore "cost a fortune." One day Chen and his friends waited in the area of the city where they were told that these castrates would normally show up late in the evening. He left early, though, due to fatigue. As such, he did not bear witness to these spectacular divas. A photo later shown to him by his friend, however, confirmed their presence, which Chen found both puzzling and intriguing: puzzling because it did not occur to him that a considerable number of men (apparently "up to three hundred") would castrate themselves voluntarily prior to entering the sex industry; intriguing because this proved that the practice of castration had lingered and remained a rather genuine experience despite the dismantling of the Qing imperial institution.[1]

Chen's discussion of this Southeast Asian transgender phenomenon came at the end of a lengthy article that he wrote on Chinese eunuchs for the Hong Kong-based *Dacheng* (大成) magazine in 1977.[2] An authority in

Chinese medicine, Chen founded the journal *Health News* (健康報, *Jian-kang bao*) in 1928 and was one of the five delegates who traveled to Nanjing to protest the Nationalist government's attempt to abolish Chinese medicine in 1929.[3] It might seem strange that a respected physician writing about castration in China would link it to the phenomenon of sex reassignment in contemporary Singapore, but Chen had correctly identified a corporeal practice, as the precedent to modern transsexual surgery, that had become an important object of criticism worldwide since the nineteenth century. In the eyes of both the Western imperialists and the Chinese, castration, like footbinding, represented a powerful symbol of backwardness, oppression, despotism, and national shame.[4] Before scientists were keen to introduce foreign ideas about sex transformation, one of the processes whereby the image of China as intrinsically weak gained global currency was the cultural demise of eunuchism.[5]

Prior to the nineteenth century, eunuchism drew conflicting opinions in Chinese society. The most popular perception of eunuchs stemmed from its condemnation by scholar officials. This well-known distaste for eunuchs has a history that goes back to the origins of the practice of castration itself in ancient China. The earliest documentation of castration can be traced to the inscriptions of characters resembling a knife and the male reproductive organ in the oracle bones from the late Shang dynasty (1600–1046 BC).[6] The word "eunuch" (閹人, *yanren*) first appeared in the Western Zhou period (1046–771 BC), since castration (宮刑, *gongxing*) was listed as one of the five major forms of slavery punishment (五刑, *wuxing*) up to the second century BC.[7] Initially, castrated slaves belonged to a relatively low social stratum, being auxiliary servants without any political ambition. However, as the number of slaves grew with the centralization of monarchism in the Warring States (453–221 BC) and Qin periods (221–206 BC), their role became increasingly imbued with political meaning.

By the Han dynasty (206 BC–AD 220), the social space that eunuchs occupied in the court had moved closer to the emperor than ever before. The interest of Han emperors in working with eunuchs to circumvent the power of extended royal relatives gave eunuchs an unprecedented degree of economic, military, and political freedom. The later Han period was henceforth also known as the First Epoch of the Rise of Eunuchs (第一次宦官時代, *diyici huanguan shidai*).[8] Similarly, the Second Epoch of the Rise of Eunuchs in the Tang dynasty (618–907) and the Third Epoch of the Rise of Eunuchs in the Ming dynasty (1368–1644) represented crucial episodes during which the intervention of eunuchs in governmental affairs reached its peak.[9] These decisive turning points also curtailed the potential threat and mobility of eunuchs in the immediate subsequent dynasties, such as the Song (960–1279) and the Qing (1644–1911).[10] While many Confucian elites urged the emperor to reduce the size of the harem in response to the political ascendance of eunuchs, they never advocated the total elimination of the eunuchs institution.[11]

A contrary theme in Chinese history casts these controversial agents in a more positive light. This interpretation construes the subjects of castration neither as dynastic evils nor as the rivals of Confucian statesmen but as heroic figures offsetting the negative connotations of castration. The two most famous examples are Sima Qian (司馬遷) and Zheng He (鄭和). Strictly speaking, the great historian Sima Qian (145 or 135–86 BC) was not a servant of the imperial court, but he was sentenced to castration by Emperor Wu of Han for defending the military officer Li Ling (with the implication being that Sima was indirectly attacking Emperor Wu's brother-in-law). Best known for his monumental work, *Records of the Grand Historian* (史記, *Shiji*), which covers two millennia of history from the Yellow Emperor to his time, Sima has since been heralded as "the father of Chinese historiography." He is often credited for formulating an innovative biographical approach to historical writing that became the blueprint for the remainder of the official *Twenty-Four Histories*.[12]

Besides Sima, another individual who embodied values that contradicted the negative portrayal of castration was the Ming dynasty Muslim eunuch Zheng He (1371–1433). Zheng is often remembered for leading the naval expeditions commissioned by the Yongle Emperor between 1405 and 1433. The aim of these expeditions was to establish a Chinese presence and impose imperial control over the Indian Ocean trade, thereby strengthening the Chinese tributary system. Among the places that Zheng's fleets visited were Brunei, Thailand, Southeast Asia, India, the Horn of Africa, and Arabia. As a result, Zheng's expedition significantly expanded the Ming dynasty's global maritime awareness. In 1904 the reformer Liang Qichao wrote a biography that recuperated Zheng and his accomplishments from the oblivion.[13]

Despite these diverging attitudes, it is indisputable that eunuchs had played an important role in the history of imperial China. And yet scholars have paid relatively little attention to the actual measures of Chinese castration and have remained considerably silent on its history.[14] Studies of

Chinese eunuchs tend to center on the extent of their involvement in the political arena. Indeed, this has been the predominant focus of Chinese historiography ever since the institutional lives of eunuchs were systematically documented in 1769 in *The History of the Palace* (*Guochao gongshi*), a project commissioned by the Qianlong Emperor.[15] Conventional wisdom prefers to explain the corrupt activities of Chinese eunuchs as the result of their internalized anger and frustration with their lost manhood.[16] Literary critic Gary Taylor, who invites us to view the eunuch as "not a defective man but an improved one," might have a point here.[17] Apart from the court officials, eunuchs were after all the only "men" whom the emperor and his family trusted. A revisionist history of eunuchs must therefore resist the temptation of narrating through the lens of their political life alone, in addition to letting go of emasculation as the only viable and all-encompassing category for understanding the gendered implications of castration.[18]

This chapter revisits the history of eunuchs not in terms of their immersion in state affairs but through the prism of their bodily experience. It foregrounds and contextualizes a dimension of their experience that has heretofore attracted scant scholarly attention—simply put, what can we *know* about the history of castration itself? This is not to deny that the political nuance of castration as a Chinese cultural practice evolved over time. The distaste for eunuchs and the antipathy for the Chinese imperium became isomorphic during the peak of Western overseas imperial and colonial expansions.[19] In the nineteenth century, China appropriated the label of the "Sick Man of Asia" from the Ottoman empire. In the twentieth century, commentators began to describe China as a "castrated civilization" (被閹割的文明, *beiyange de wenming*) and eunuchs the "third sex" (第三性, *disanxing*), both of which reinforced the "Sick Man" stereotype.[20] Rather than classifying eunuchs as third sex subjects defying the boundaries of maleness and femaleness, a more robust historical inquiry would unpack castration's multiple layers of gendered content in the process of its demise. In fact, by delineating what castration meant for different groups of Chinese men, such a revisionist effort could evaluate the practice itself, contrary to standard accounts, as a source of masculine identity and a mechanism of its social reproduction. This invariably broadens the way we interpret the meaning of castration, from the typical epistemological confines of modern biomedicine to a new conceptualization of reproduction as socially and culturally meaningful.

The emphasis on the masculinity of castration revises the diverse scholarly literature on Chinese manhood that has drawn on legal, medico-scientific, literary, family reform, homoerotic, theatrical, and diasporic examples.[21] It is perhaps worth noting that the gendered subjectivity of eunuchs has escaped the attention of the most definitive historical studies of Chinese masculinity.[22] If we borrow the queer theoretical insight of Judith Halberstam, who has narrated the first comprehensive history of female masculinity in Euro-American literature and film, we might be better equipped to entertain a more radical analytical separation of masculinity from men as fecund agents.[23] Chinese masculinity can thus be understood as neither a social extension of biological maleness nor the social meanings assigned to men per se but as a social relational index of such politically suffused cultural practices as castration. By focusing on the history of knowledge production about castration, this chapter begins to question the naturalness often assumed to be present in the immediate and productive relationship of men to manliness.

With respect to footbinding, historians have recently begun to revise its popular conception as a tool of gender oppression. In *Cinderella's Sisters* (2005), for example, historian Dorothy Ko shows that women as much as men participated in the perpetuation of this cultural practice with complex and nuanced historical agency. That footbinding was often a marker of ethnic and national boundaries, a practice of concealment and adornment, and a sign of civility and culture before the nineteenth century betrays our modern depictions of it as a form of bodily mutilation, an "unnatural" practice, and a barbaric (even perverse) custom.[24] In a similar spirit, anthropologist Angela Zito has demonstrated that twentieth-century discourses of the bound foot only reflect variations of its modernist fetishization, even thresholds of feminist theorization and intercultural displacements.[25] Taking cues from Ko and Zito, this chapter departs from outside the anticastration discourse, attempting to balance the historiographical condemnation of Chinese eunuchs. To bring to light the historicity of eunuchism and to situate castration in its proper historical and technical contexts, I pay particular attention to the discrepant paradigms of masculinity emerging from the broader confusion over the nature and consequence of castration, the ways visual milieu reciprocated the politics of these gendered meanings and its thresholds of cross-cultural transmission, and the problem of narrating the historical experience of eunuchs based on the modern nationalist bias of our sources and informants. To paraphrase Ko, there is not one castration but many.[26] A central motif in the construction of knowledge about castration (and, as the rest of this book shows, sex more generally) in the twentieth century is the indigenization of Western medico-scientific norms by Chinese writers. By reading against the grain, this chapter traces the formation of a textual and visual archive that documents the methods of Chinese castration, something that was distinctively absent before the nineteenth century and that, I suggest, directly led to the social and cultural demise of eunuchism. We begin, therefore, with the ending of an historical epoch.

Archiving Castration

Despite our best intentions, the reconstruction of an archive based on the sources available about Chinese castration is itself an inherently mediated and problematic project.[27] First, where do we end? If we assume that the metanarrative history of political change determines the broad trajectory of cultural transformation, we might conclude that the unequivocal demise of castration after the fall of the Qing empire in 1911 was a matter of course. However, even after the last Manchu emperor, Puyi (溥儀, 1906–1967), was expelled from the Forbidden City in 1924 by the warlord Feng Yuxiang (馮玉祥, 1882–1948), he was declared by the Japanese army as the Kangde Emperor of the puppet state of Manchuria in 1934. As the Kangde Emperor, Puyi was still surrounded by a dozen or so Chinese eunuchs.[28] After the Pacific War ended in 1945, these eunuchs' bodies still served as a crucial reminder of the past and their stories the lived experiences of castration, to both themselves and the global public. In 1964, for instance, the Committee for Learning and Cultural and Historical Data (中共政協文史資料研究委員會, *Zhonggong zhengxie wenshi ziliao yanjiu weiyuanhui*), a division of the National Committee of the Chinese People's Political Consultative Conference, convened a meeting in the interest of collecting valuable historical data on the late Qing dynasty. The event brought together the final cohort of surviving eunuchs who lived in various corners of Beijing. The eunuchs were interviewed so that their oral histories could be officially transcribed, published, and circulated to a worldwide audience.[29] Even the death of the last surviving Chinese eunuch, Sun Yaoting (孫耀廷, 1902–1996), might still be a misleading signpost for where the story of Chinese castration ends.[30] This is because the afterlife of eunuchism—especially the emergence of transsexuality in Sinophone communities—is indebted to the genealogical precursors such as the ones discussed in this chapter that culminated in the thresholds of its beginning. Before we examine how the body morphology of eunuchs and transsexuals operates within shifting realms of scientific truth claims and geopolitics, our story must unravel the process whereby the normative regime of eunuchism lost its aura, meaning, and cultural significance.

Apart from the puzzling question of a precise endpoint, the reconstruction of the archive relies on the type of sources that are available. Here is where the parallel between footbinding's disappearance and castration's dissolution is most striking: the abundance of textual and visual sources from the nineteenth and early twentieth centuries almost always represent the bound foot and the castrated body by *exposing* them. This mode of representation runs against the very reason of their existence in Chinese history. After all, the naturalness of footbinding and castration depended on concealing the female and male bodies because concealment links these customs to Chinese ideals of civility and culture (文, *wen*).[31] Therefore, upon reading the wealth of visual and textual documentations of the bound foot or the castrated body, the historian must avoid a telos of knowledge production that extracts a certain kind of historicity from these sources blinded by the hegemonic limits of their cultural existence. As feminist theorist Anjali Arondekar has reminded us in a different context, "even though scholars have foregrounded the analytical limits of the archive, they continue to privilege the reading practice of recovery over all others."[32] It might be more useful to read the archival remains not as the ultimate arbiter of historical recuperation but as "traces" of the past that enable alternative epistemological arrangements of the way the past and the present conjoin. In other words, we must not retell a story about eunuchs that bluntly identifies with the kind of story that the sources themselves suggest at face value. Instead of sidestepping their presumed allegorical status, we need "to read archival evidence as a recalcitrant event [that] reads the notion of the object against a fiction of access, where the object eschews and solicits interpretive seduction."[33] Where they leave us is not something to be "recovered," but something to be self-reflexively (re)configured.

Precise endpoints and the nature of the sources aside, the repository of data about Chinese castration is mediated by their availability. Three available "voices" unique to the historical period under consideration contributed to the making of this archive: Western spectators, eunuchs themselves, and members of the last imperial family.[34] Together, the textual, photographic, and oral records they left behind disclose an increasing disparity between two registers of eunuchism as a mode of historical experience: on the macro level of global narration on the one hand, and on the micro level of individual embodiment on the other. An anti-eunuch sentiment arose out of this growing disjuncture between castration as a form of public memory and castration as personal experience. This nascent sensibility that casts the practice of castration and the existence of eunuchs as indicators of national backwardness reverberated through the rest of the twentieth century. As eunuchs' gender identity was evaluated anew in the modern era through the lens of Western biomedicine, China's global image as a weak and lacking nation intensified over time. The period between the 1870s and the 1930s thus constituted a transitional phase when the castrated male body—like women's bound feet and the leper's crippled body—seemed out of sync with the modernizing Chinese body politic at large.[35]

Open Secrets

In the decades following the Opium Wars and the Taiping Rebellion, China entered an era of unequal relations with Western and, eventually, Japanese nations. The clash between the Qing and British empires, for example, was reflected in not only the semiotic legacy of sovereign thinking but also the nascent Eurocentric world order that "produced a new 'China.'"[36] According to historian James Hevia, British activities in late Qing China can be more accurately understood as a sweeping process of de-territorialization and re-territorialization, transforming China's position in the context of nineteenth-century internationalism. Through collecting, archiving, indexing, cataloging, and cross-referencing new information about Chinese people, objects, and places, agents in the British legation in Peking, the various branches of the Imperial Maritime Customs, and the dispersed chapters of the Oriental Society and the Royal Asiatic Society constructed an imperial archive that reconfigured China for the world. The orthographic achievements of the diplomat Sir Thomas F. Wade (1818–1895) and the standardization of knowledge about porcelain manufacture by Stephen W. Bushell (1844–1908) are but only two stellar examples of this broader transformation. The effort of these and other British subjects in generating new information about China, and in transporting such knowledge from the periphery of the British Empire to the repositories at the cosmopolitan metropole, ultimately imposed a new order of China onto itself. They recoded Qing sovereignty, "taught" the Chinese imperial lessons, and made them "perfectly equal" in a new age of global nation-states.[37] It was within this larger context of knowledge reconfiguration that the formation of a corpus of materials documenting the methods of Chinese castration marked a point of no return in the social and cultural demise of eunuchism.

The surgical details of Chinese castration first made it into the global public eye in 1869 when Georges Morache (1837–1906), a physician of the French legation in Peking, published a full-length study on hygiene in the city. In *Pékin et ses habitants: Étude d'hygiene*, Morache discussed eunuchs after beggars, infanticide, and prostitution in a chapter on poverty. For Morache, poverty stood as the root cause of major social unrests in China, including the devastating Taiping Rebellion.[38] Poverty also provided him the broader framework for discussing castration because it was usually the poor who could not afford a proper education to compete in the civil service examinations and who would therefore be enticed into offering themselves up to bodily mutilation for a small sum of money. "Receiving few taels (8–10 or 70–80 francs) in return," explained Morache, "led countless poor families to send their children to the infamous trades operating at two gates of the city, away from prying eyes." In fact, "adult men also came forward to escape poverty," Morache added, and they were more sought after because "they have all the attributes of manhood without offering the disadvantages."[39]

Other French, British, and American physicians followed Morache's lead in comparing eunuchs to beggars and other social outcasts and framed this comparison as a medical concern. In 1899 Jean-Jacques Matignon (1866–1928), whose work to which we return in greater detail, placed his discussion of eunuchs in the broader context of an analysis of suicide, infanticide, pederasty, and beggars in his *Superstition, crime et misère en Chine*. As late as 1910 the medical missionaries W. Hamilton Jefferys (1871–1945) and James L. Maxwell (1836–1921) classified both beggars and eunuchs as examples of "artificial deformities" and "diseases peculiar to China."[40] The idea that the nonworking poor who dwelled on the margins of urban China were "social parasites" would eventually define how native intellectuals and sociologists approached the issue of poverty in the 1920s.[41]

In depicting castration as a social problem, Morache presented possibly the earliest known account of the Chinese castration procedure in

Western language. He revealed that, due to the "indiscretion of a eunuch," he was able to "collect the most accurate indications on the operating manual, in addition to very complete information on the palace and its manners."[42] According to Morache, the patient needed to be fed, pampered, and treated well before the operation. Then, on the day of the procedure, he would be submersed in a "very hot bath" so that his "senses are numbed," after which his genitals would be wrapped in a small band made up of silk. After the operator excised the patient's genital organ with one cut of knife, the assistant would immediately apply styptic powder to the wound to stop the bleeding. The assistant continued the compression, "adding more hemostatic powder if necessary, until the hemorrhage appear[ed] to have stopped." The most serious potential side effect of the operation was the obstruction of the urethra channel. If the patient had not urinated by the end of three or four days, he was "regarded as lost and no one cares more"; otherwise, if the bandages were soaked in urine, the operator would "wash the wound with care," and "the patient is regarded as out of danger."[43]

Morache's account initiated a wave of Western fascination with Chinese castration starting in the 1870s. According to historian Melissa Dale, the textual and visual evidences left behind by these Western physicians and foreign observers "are the only known historical accounts of the emasculation procedure during the Qing dynasty."[44] In fact, it was not Morache but George Carter Stent (1833–1884) who penned what would become the best-known source on Chinese castration. Not a medical doctor, Stent arrived in China in the mid-1860s as a member of the British Legation Guard. Due to his proficiency in colloquial Chinese, he attracted the attention of the then–British minister Thomas Wade and was recruited into the Chinese Imperial Maritime Customs Service in 1869.[45] In 1877 Stent published the most elaborate description of the operation in the *Journal of the North China Branch of the Royal Asiatic Society.*

Stent first read a version of his paper, more than forty pages in length, before the Royal Asiatic Society on March 26, 1877. His opening sentence stamped the intention—and the abiding significance—of his study, namely, to bring something invisible to visibility, to demystify a vague impression: "Much has been said and written about eunuchs at various times, but very little seems to be really known concerning them." "In fact," Stent continued, "everything relating to them is described so vaguely that one is almost tempted to believe that eunuchs exist only in the Arabian Night's

Entertainments and other eastern tales, or in the imaginations of the writers, rather than actually belonging to and forming no inconsiderable portion of the human race."[46] Assigning Chinese eunuchs a textual status of reality, Stent's words epitomized the effort to expose publicly the private experience of eunuchs.

Neither opinions about the existence of eunuchs nor attacks on the tradition of castration were new to Chinese discourses. But the novelty of Stent's endeavor in making Chinese eunuchs a reality stems from its unambiguous Christian and Orientalist overtone. In his words, "eunuchs are only to be found in eastern despotic countries, the enlightening influence of Christianity preventing such unnatural proceedings being practiced in the countries of those who profess it." For Stent, the "unnatural proceedings" of castration in China reveal "at least one beneficial result of the spread of Christianity; for while we [Christian Westerners] are free from the baneful practice, it is a vile blot on less fortunate countries."[47] Similar to the discourse surrounding *tianzu* (天足, natural foot) in the anti-footbinding movement, the significance of Stent's words lies in his explicit juxtaposition of China against a more enlightened West with an overt Christian justification.[48] However, Stent's assertion that Christianity and monogamy saved the West from the "unnatural proceedings" of castration is an erroneous interpretation, especially if one considers the important role that the eunuchs played in Byzantine history or the more recent example of castrati singers in Italy.[49] Defining China as one of the "less fortunate countries," Stent's project was unmistakably Orientalist in nature. It ultimately signaled the arrival of a rhetoric according to which China "lacked" the tools of narrating and recognizing its own deficiency, for which castration, like footbinding, typified an unnatural corporeal practice that was out of both place and time. As media scholars Yosefa Loshitzky and Raya Meyuhas have observed, "eunuchs are perceived by the modern Western audience as grotesque rarities of the past that are associated with the 'otherness' of exotic cultures."[50] Eunuchism has often been regarded as "a barbaric, archaic, and uncivilized phenomenon and therefore as an anachronism."[51]

The aspect of Stent's study that has made the deepest impact is not his message about Western superiority, however, but his discussion of the castration procedure. To this day his description of how, where, and by whom Chinese eunuchs were made remains the most cited reference on this procedure since its debut in the 1870s. In fact, one would be looking in vain for a serious treatment of the subject that does not follow Stent's footsteps in one way or another. His words thus deserve quoting in full and a serious reappraisal.

The place where men or boys are made eunuchs is just outside the inner Hsi-'hua gate (內西華門) of the palace, and within the imperial city. It is a mean-looking building, and is known as the Chang-tzu, 廠子, *the shed.* Within this building reside several men recognized by government, yet drawing no pay from it—whose duty consists in emasculating those who are desirous of becoming, or are sent to become—eunuchs.

These men are called tao-tzu-chiang, 刀子匠, "*knifers,*" and depend entirely for their living on making eunuchs. They get a fixed sum—six taels—for every operation they perform on boys sent or brought to them, and for keep and attendance till the patients are properly recovered.

Grown up men desirous of becoming eunuchs, but who are too poor to pay the necessary fees, make arrangements with the "knifers" to repay them out of their salaries. But in any case the "knifers" dare not operate on them unless they (the candidates) have securities to vouch for their respectability.

The "knifers" have generally one or two apprentices to learn the profession; these are almost invariably members of their own families, so that the profession may be said to be hereditary.

When the operation is about to take place, the candidate or victim—as the case may be—is placed on a *kang* in a sitting—or rather, reclining position. One man supports him round the waist, while two others separate his legs and hold them down firmly, to prevent any movement on his part. The operating "knifer" then stands in front of the men—with his knife in his hand—and enquires if he will ever repent. If the man at the last moment demurs in the slightest, the "knifer" will not perform the operation, but if he still expresses his willingness, with one sweep of the knife he is made a eunuch.

The operation is performed in this manner:—white ligatures or bandages are bound tightly round the lower part of the belly and the upper parts of the thighs, to prevent too much haemorrage. The parts about to be operated on are then bathed three times with hot pepper-water, the intended eunuch being in the reclining position as previously described. When the parts have been sufficiently bathed, the *whole*,—both testicles and penis—are cut off as closely as possible with a small curved knife, something in the shape of a sickle. The emasculation being effected, a pewter needle or spigot is carefully thrust into the main orifice at the root of the penis; the wound is then covered with paper saturated in cold water and is carefully bound up. After the wound is dressed the patient is made to walk about the room, supported by two of the "knifers," for two or three hours, when he is allowed to lie down.

The patient is not allowed to drink anything for three days, during which time he often suffers great agony, not only from thirst, but from intense pain, and from the impossibility of relieving nature during that period.

At the end of three days the bandage is taken off, the spigot is pulled out, and the sufferer obtains relief in the copious flow of urine which spurts out like a fountain. If this takes place satisfactorily, the patient is considered out of danger and congratulated on it; but if the unfortunate wretch cannot make water he is doomed to a death of agony, for the passages have become swollen and nothing can save him.[52]

As this passage includes a wealth of details about the method of Chinese castration, it has remained the most widely referenced source on the topic. For both Western and Chinese doctors writing in the late nineteenth and early twentieth centuries, Stent's graphic descriptions offered an exception to the rule that Chinese medicine fell far behind Western medicine in the areas of anatomy and surgery. In the late 1870s R. A. Jamieson, a consulting surgeon at the Imperial Maritime Customs, wrote: "As a rule Chinese practitioners are both timid and tardy in their use of the knife. . . . The one exception to the general rule is the boldness with which the Chinese castrate men and animals." Addressing to the readers of the *Lancet*, Jamieson disclosed his "indebted[ness]" to Stent for valuable information on castration, based on which he presented what appears to be a mere paraphrase of Stent's documentation.[53] The Atlanta-based *Southern Medical Record* reprinted Jamieson's paraphrase in an article titled "How the Chinese Make Eunuchs."[54] Jamieson's account also received the attention of Wu Lien-teh (伍連德, 1879–1960), the distinguished Cambridge-trained Chinese physician who played a central role in controlling the Manchurian plague of 1910–1911.[55] At the First Pan-Pacific Surgical Meeting held in Hawaii in

August 1929, Wu co-presented a paper with Tsai Hung, the department chief of the Ministry of Health in Nanjing, titled "The Practice of Surgery and Anesthetics in Ancient China." According to Tsai and Wu, "our survey would not be complete without some reference to the operation of castration." Before citing Jamieson's description of the technique, they contextualized Chinese castration by tracing its origins to 1100 BC, when it was adopted as a mode of punishment. They reiterated Stent's claim that "the operators were known as 'knifers'" but also added that "besides men the Chinese have always boldly castrated animals."[56] An analogous passage appeared in the renowned textbook that Wu coauthored with K. Chimin Wong, *History of Chinese Medicine* (1936), under the section on early Chinese surgery.[57] In the 1970s, non-Chinese and nonmedical writers continued to draw on Stent without questioning the validity of his narrative. For example, in Taisuke Mitamura's controversial *Chinese Eunuchs* (1970), Stent's original article was the only source that the author relied on in describing how castration was carried out in late imperial China.[58] The entirety of Stent's piece also made its way into one of the most humanist studies of eunuchs, by Charles Humana, *The Keeper of The Bed: A Study of the Eunuch* (1973).[59]

Despite its richness, Stent's account raises more questions than answers. Its implicit claim of originality and validity is difficult to prove, but this difficulty has not been sufficiently acknowledged by scholars of Chinese eunuchs to date. Most critics have taken Stent's words for granted as a firsthand account of an actual castration procedure. Yet, in order for his documentation to be considered credible, it would have been necessary for Stent to be present during one of such sessions. What if Stent did not witness any of the castration surgeries? As Melissa Dale has pointed out astutely, "it remains unclear if Morache or Stent actually witnessed the emasculations described in their writings."[60] Even if Stent did pay a visit to the "knifers" for just a single case of castration, did he stay for the entire duration (at least three consecutive days or longer)? His narrative merely *implied* that he had personally observed at least two types of operation—successful and unsuccessful—to differentiate survival in the former case and potential death in the latter. He never explained in detail how he obtained such information, leaving little room for verification.

In a slightly different way, the content of Stent's words already betrayed their implicit claim of originality and validity. If the knowledge and skills required for performing castration were transmitted among knifers through hereditary apprenticeship, how was it possible for the operation to be described so openly by a Westerner in the first place? If part of the social integrity of the Chinese knifers came from maintaining a guild custom of oral instruction and personal demonstration, it seems highly improbable that a nonfamilial or nonprofessional member, let alone a foreigner, would be allowed to witness the surgical protocols in such remarkable detail. To borrow literary critic William A. Cohen's insight from a different though highly relevant context, accounts such as Morache's and Stent's "[recast] secret activities into a public story of exposure, it makes questions about truth almost impossible to answer, however it deliberately mobilizes truth-determining institutions."[61] An empirical substantiation of the existence of "the shed," where these operations were supposedly performed by the knif-ers, would elevate the validity of Stent's description. However, in their study of eunuchs in Qing and Republican China, different historians have conceded that no discussion of the knifers could be found in the Qing palace archives.[62] This uncertainty surrounding Stent's sources and infor-mants, as we will see, nested a larger confusion over the part of the body that Chinese castration actually removed. As late as 1991, two urologists from Beijing Medical University still confessed that "most people, includ-ing urologists, do not have a clear understanding of what is actually done to a man or boy to produce a eunuch."[63]

Discrepant Masculinities

We might consider the question of what Chinese castration eliminated moot. Stent's emphasis that a complete surgical castration involved the removal of *both* the testes and the penis may seem unremarkable to modern readers. His emphasis is subtle because this requirement sounds so natural to our ears. But as Gary Taylor reminds us, if the ultimate purpose of castra-tion is to impair a man's fertility, it is not necessary to destroy the penis but only the testes.[64] In fact, the earliest extant medical description of the oper-ation, by the seventh-century Byzantine Greek physician Paul of Aegina (625–690), makes it clear that only the testicles, not the penis, were targeted by the techniques of contusion and excision.[65] Similarly, modern reapprais-als of the operations performed on the European castrati singers indicate that only testicles were severed.[66] In his ambitious survey of the cultural history of the penis, David Friedman carefully incorporated a broad definition of the organ "not merely as the penile shaft and glass, but encompassing the testes, sperm, and all the other parts and products of the male genitalia."[67] This inclusive definition was fruitful for Friedman's undertaking precisely because the penile shaft had not always been the sole locus of biological masculinity since the beginning of Western civilization.

With respect to Chinese eunuchs, Stent's pronouncement that both the penis and the testes were eliminated from their bodies drew polarized reactions from those claiming to have interacted with the palace eunuchs on a personal level. Dong Guo, author of a pioneer study on the history of Chinese eunuchs, argued that Stent's account was outright erroneous.[68] According to his conversations with eunuchs, "the key [to castration] is this: when someone is made a eunuch at a relatively young age, the procedure resembles the gelding of a pig by removing or protruding the testicles. This operation is at least not fatal, and because there is no major concern over bacterial infection from the cut, the person recovers in three to five days."[69] Jia Yinghua, the biographer of the last Chinese eunuch, Sun Yaoting, seconded Dong's contention (figure 1.1). In addition to crushing, Jia described

Figure 1.1 Late Qing eunuch Sun Yaoting (left) and his biographer, Jia Yinghua (right). *Source:* Courtesy of Jia Yinghua.

two additional methods of annihilating testicular function that did not require a sophisticated magnitude of surgical execution: by strangling the testicles with a rope or by stabbing it persistently with a needle until it became dysfunctional.[70]

On the contrary, based on their physical examinations of the surviving eunuchs in Beijing in the 1930s and the 1960s, respectively, a Western doctor and two Chinese urologists defended Stent's observation, confirming that both the penis and the testes were detached from the bodies of the eunuchs they investigated. The German anatomist F. Wagenseil attempted to examine Chinese eunuchs during his first visit to Peking in 1924 but failed due to the lack of access and resources. In 1930 he accomplished this goal with the help of Paul Krieg, the director of the German Hospital in Peking, who permitted Wagenseil to use the rooms in the hospital. Formally affiliated with the Anatomical Institute of the University of Bonn and the Tungchi University in Shanghai by that point, Wagenseil conducted an extensive medical study of thirty-one Chinese eunuchs and published his findings in the German *Journal of Morphology and Anthropology* in 1933.[71] On the first page of his report Wagenseil remarked, "it is important to mention that virtually all of the investigated Chinese eunuchs underwent a total removal of their external genitalia, including penis and scrotum with its content."[72] On February 7, 1960, two faculty in the Institute of Urology at Beijing Medical University, Wu Chieh Ping and Gu Fang-Liu, reviewed and examined twenty-six eunuchs (with an average age of seventy-two years) living in a house provided by the Chinese government.[73] Wu and Gu observed, "the pubic region of an eunuch, looking from the front, resembles that of a female."[74] These clinical vignettes contradicted the information that Dong and Jia derived from their conversations with eunuchs. Although both verbal confirmation and clinical evidence were established on the basis of personal correspondence with eunuchs, their discrepancy left a historical residue of ambiguity surrounding the surgical parameters of Chinese castration. This exemplifies how "micro" accounts of eunuch corporeality do not surrender to and cannot be subsumed under "macro" narrations.

This inconsistency underscores the multiplicity of both the meaning and the practice of castration. Wu and Gu, for instance, noted the popular "erroneous use of the term 'castration.'" "Although the Greek root of the word 'eunouchos' does indicate a castrated person," they explained, "the eunuch is not only castrated. . . . 'Emasculation' should be the right term to describe the procedure . . . We think it is better to define 'emasculation' as 'removal of external genitalia in man or boy,' leaving 'castration' for removal of the testes.' "[75] This shift in conceptual preference from "castration" to "emasculation" highlights an important historical transformation in the biological definition of manhood: from a cultural regime of the scrotum to a regime of the penis. Between the sixteenth and twentieth century, the anatomical measure of manliness changed from whether a man has balls to whether a man has a big stick.[76] This fall of the scrotum and rise of the penis was accompanied by the process by which desire and libidinal pleasures replaced status and reproduction as the organizing principle for assigning social meanings to sexual acts.[77] For example, by the mid-twentieth century, the sexual drive of male sex offenders became the central focus of manipulation via chemical castration.[78] Such was the case when the progesteronal hormonal compound diethylstilbestrol was first prescribed to lower the male testosterone level in 1944.[79]

One of the cultural forces that cemented the transformation from a scrotum-centered to a penis-centered regime of masculinity in Western Europe and America was the popularization of Freudian psychoanalysis in the early twentieth century. For Sigmund Freud (1856–1939), castration anxiety was symptomatic of a psychogenic fear, or at least recognition, of "the lack of penis."[80] His most influential and controversial French disciple, Jacques Lacan (1901–1989), would subsequently prioritize the symbolic meaning of the *phallus* in lieu of the anatomical penis.[81] But the phallus is nothing more than a figuration of the physical organ, a transcendental penis, so to speak, that extends rather than subverts its anatomical register. To quote Taylor's astute insight, "castration—in humanist Europe, as in previous human societies—attacked the scrotum. In twentieth-century psychoanalysis, by contrast, castration has been redefined as an attack on the penis."[82]

Although psychoanalytic ideas gained uneven footing in early twentieth-century China, the question of which body parts were targeted by castration fascinated Chinese thinkers. In the 1930s Zhou Zuoren (周作人, 1885–1967), Lu Xun's younger brother and a leader in the New Culture movement, contributed to this debate by comparing the indigenous practice of *gongxing* (castration for punishment) with legalized sterilization in the West. Written in November 1934, Zhou's essay, "On Castration" (關於宮刑, *Guanyu gongxing*), argued that the practice of *gongxing* in ancient China differed from the "castration" operations reported in the contemporaneous West.[83] Specifically, Zhou identified his commentary as a "corrective" to a recent news article that used the word "*gongxing*" to describe the kind of sterilization operations carried out in Nazi Germany. As a culmination of the eugenics movement in Europe and North America, compulsory sterilization represented an intrusive form of negative eugenics practices.[84] On July 14, 1933, the first German sterilization law—the Law for the Prevention of Genetically Diseased Offspring—was enacted. However, forced sterilization at the time merely consisted of ligation of the fallopian tubes in women and a vasectomy for men. Zhou insisted that these swift, "scientific" operations "had nothing to do with *gongxing* in China," and "anyone hoping to assign the subjects of these operations to palace service will be extremely disappointed."[85] In a separate essay on eunuchs, composed in May 1934, Zhou interpreted the eunuch system as a distinctively Chinese phenomenon—something that "could not be found in European and Japanese nations." Ironically, in embracing the Orientalist bias of Stent, Zhou suggested that "one of the more humane aspects of [the Western and Japanese] nations" was their supposed "refusal to destroy bodily parts in order to maintain a coherent system of gender difference."[86] Evidently, the impact of Western medical perspectives did not stop within the circle of Chinese physicians, as exemplified in the writings of Wu Lien-teh, but penetrated into the thinking of cultural elites like Zhou. This marked one of the pivotal features of the cultural demise of castration in twentieth-century China: the internalization of Western perceptions by the Chinese themselves.

Before we return to late nineteenth-century Western representations of Chinese castration, three further sets of data can help us complicate a monolithic gender framing of castration and substantiate the claim that there are not one but many forms of castration. First, the ceramic figures excavated at the Han Yang Ling Mausoleum site bring us back to the era immediately before the "first epoch of the rise of eunuchs." Located on the bank of Weihe River in the northern suburb of Xi'an city, Shaanxi Province, the mausoleum is a joint tomb of Liu Qi (Jingdi), the fourth emperor of the Western Han dynasty (206 BC–AD 24), and his empress, Empress Wang. It was built in the year AD 153 and covers an area of twenty square kilometers.[87] Found within the tomb were pottery statuettes of unclothed male, female, and eunuch attendants excavated from the emperor's tomb. Chinese archeologists have discovered that the eunuch servant appears to have been castrated before puberty: as shown in figure 1.2, his genital

Figure 1.2 Ceramic figure of a eunuch servant from the Western Han dynasty.
Source: Courtesy of the Han Yang Ling Museum.

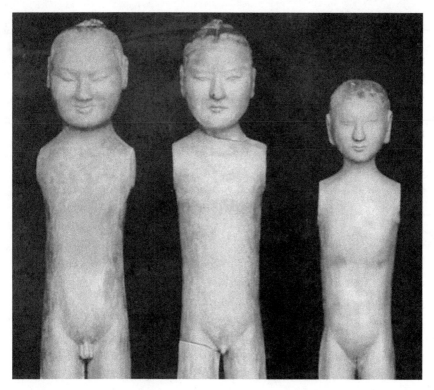

Figure 1.3 Ceramic figures of male (left), eunuch (center), and female (right) attendants from the Western Han dynasty.
Source: Courtesy of the Han Yang Ling Museum.

area reveals that only his testicles but not his penis were removed.[88] In figure 1.3, the eunuch figure in the center is distinguished from the male statuette on the left, showing both the penis and the scrotum, and the female statuette on the right, showing neither. Therefore, this archeological finding challenges any hasty characterization of Chinese castration as by way of total ablation only.[89]

Another dissenting voice came from the French Army colonel Emile Duhousset (1823–1911), who published a scholarly study of female circumcision in the "Orient" in the 1870s.[90] In 1896 Duhousset brought forward a personal anecdote that challenged the common view of Chinese castration as targeting both the penis and the testicles. In the early 1860s Duhousset had returned to France from Persia with Ernest Godard, from whom he learned a great deal about castration in Africa. In his work, he cited the following passage from Godard's book, *Egypte et Palestine: Observations médicales et scientifiques* (1867): "A colonel of artillery from China, tells me we saw two eunuchs in the palace which was taken by the French. One was killed, the other was taken alive; they were pretty well beaten. These eunuchs had small penis and no testicles; they were swollen, yellowish, and loaded with fat. They were Chinese; their voice was of a high tone. The prisoner was shown for money."[91]

In his 1896 anecdote, published as a "discussion" to an article on Chinese castration in the *Bulletins de la Société d'anthropologie de Paris*, Duhousset further qualified that the person from whom "Godard has taken the previous quote" was likely "the officer of [General de Montauban]."[92] The fact that this objection to a penis-centered understanding of Chinese castration came from scholars of not China but Africa and the Near East suggests that, from early on, Western observers were already keen to study eunuchs in a comparative and transcultural framework. The physical target and outcome of castration had long been a focal point of their scholarly investigation.

Finally, the writings of the sinologue Edmund Backhouse (1873–1944) provide a colorful though controversial window onto the disparate types of masculinity associated with eunuchs in the waning years of the Qing dynasty. Ever since Backhouse's unpublished memoir, *Décadence Mandchoue*, was discovered in the Bodleian Library of Oxford University in the 1970s, critics have repeatedly and legitimately questioned its veracity.[93] Historian John K. Fairbank characterized it as a "scabrous" script that Backhouse "wrote for the Swiss representative in Peking shortly before his death there in January 1944."[94] Robert Bickers, an expert on the history of empire and colonialism in China, described it as "drearily pornographic" and "largely constructed around imaginary interactions between himself and the blue-blooded and infamous."[95] Indeed, how can one prove—or disprove, as the case may be—that Backhouse slept with the Empress Dowager Cixi nearly two hundred times? Literary scholar Ross Forman, on the contrary, proposes an alternative reading strategy that broaches the genre, content, narrative, and functional similarity between Backhouse's manuscript and other late Victorian-era pornographic and sexological texts.[96] For our purpose, *Décadence Mandchoue* must not be treated as a bona fide historical source on life in the late Qing Forbidden City. Backhouse's words are even less trustworthy than Stent's. However, the chapter "Eunuchs Diversions" shows that Backhouse was astutely aware of the ongoing bewilderment about the nature of Chinese castration procedures and their physical effects. Backhouse's memoir bears close substantive and functional symmetries to other Western writings on the subject in the late nineteenth and early twentieth centuries. For that reason, the confessional tales of his erotic encounters with palace eunuchs bring to light a more banal, if no less interesting, range of perspectives on the relationship between eunuchism and masculinity. His stories serve as a reminder of the potential caution with which we must approach other contemporaneous accounts of castration, medical or otherwise.

Backhouse's opening story featured Li Lianying (李連英, 1848–1911), the infamous last Qing head eunuch (總管太監, *zongguan taijian*) who rose to power during the de facto rule of Cixi. The details of their sexual encounter dispelled the myth that all Qing eunuchs were expected to have their genitals removed entirely. Backhouse began by setting Li apart from a typical eunuch in terms of physical attributes: "He was . . . only nominally a 'castrato': his voice was not falsetto; his face lacked the flabbiness of a eunuch; his eyes were luminous: lust and passion radiated from them like twin candles shedding their beams in a naughty world . . . he walked like a conqueror on foreign land, taking hold of me as if already in possession."[97]

Then, in a detailed description of their intercourse, Backhouse portrayed Li's masculinity in a way not unlike himself.

> Lien reminded me of my undertaking to let him repay my sallies in kind and it was a case with him of "out-heroding Herod" (Hamlet). His tool was long, voluminous and substantial; but his scrotum had been partially enucleated and I could only feel a sort of testicle within an apology for a 'Hodensack.' It in no wise impaired his sexual activity, however, and the ejaculation of thick, healthy semen left nothing to be desired . . . after inflicting upon me very severe pain which reminded of school-days floggings . . . [he] graciously bent over the couch, while I dealt a dozen running but middle strokes on his admirable 'fesses.' He bore the punishment with an ill grace, groaning and claiming . . . "spare me: stay your hand."[98]

The reciprocity between Backhouse and Li distinguished their sexual relationship from other hierarchically oriented male homosexual relations more common at the time.[99] In this story, neither sexual positioning nor genital biology proved to be a meaningful arbiter of manliness. Li had been castrated, but, even so, he possessed a "long, voluminous and substantial" tool (a comment coming from a man whose genital organ was surgically untouched). Instead, the rather egalitarian nature of their sexual interaction places Li and Backhouse on the same axis of gender embodiment.

In another recollection that involved a threesome with Prince Gong and his eunuch Chu En-ming, Backhouse challenged the widespread rumor that eunuchs gave off an unpleasant odor. In his 1877 essay, Stent indicated that eunuchs often "wet their beds." "This habit," Stent explained, "has originated the expression—which Chinese use when wishing to convey in forcible language how their olfactory nerves have been offended when speaking of them—'He stinks like a eunuch.' Chinese also speak of them as 'stinking eunuchs,' and aver that they can smell a eunuch for half-a-*li*."[100] After making love to Chu Eng-ming, Backhouse wanted to put behind such misperception: "Persons unknowing have said in my hearing that the eunuch (in China or in Turkey) was malodorous and urine leaking; it is not so: all the 'châtrés' I encountered both in Constantinople in the nineties and in China since (and I have seen and known scores) not only were innocent of any 'Harngeruch'—as a fact, they were cleaner than the entire male—whatsoever, but were delicately and aromatically perfumed, both as regards their frontal 'void' and the erotic foyer that is their anal cavity."[101] Drawing on his personal experience, or so he claimed, Backhouse dismissed the idea that one can smell a stinky eunuch from a distance. In contrast to Stent's suggestion, he gave the impression that eunuchs were cleaner and even smelled nicer than uncastrated men.

In addition to commenting on the sexual and hygienic effects of castration, Backhouse openly documented the procedure itself by turning to the experience of Prince Qing's chief eunuch, Yin Hao-jan.

I asked [Yin] about the "ordalie" of being made a eunuch; he told me that he was aged twenty-nine at the date of the operation and still retained a measure of sexual capacity. The neighbouring region had been anaesthetized by some Chinese balsam and he felt practically no pain; a sort of tourniquet, "ou instrument pour comprimer les artères," was employed, so that haemorrhage was reduced to a minimum; the severance of the penis cost him no distress but the orchidectomy connoted a second or more of acute agony. The wound was seared with a hot iron and a styptic ointment applied. His chiefest trouble during subsequent days was in urination; as he said, he leaked like a sieve, but on the whole suffered little pain. He kindly exhibited to me his genital region: the orifice, a tiny passage not in itself unsightly nor uncomely, seemed free from inflammation and there was no oozing from the urethra which was elaborately perfumed.[102]

This passage confirmed several aspects of what Westerners already knew about Chinese castration, including the possibility of undergoing the operation in adulthood, the application of a tightened bandage before excision to prevent hemorrhage, the postoperative problem with urination, and so forth. But it also included new information that contradicted prevailing assumptions, such as the eunuch's retention of sexual capacity and maintenance of an "elaborately perfumed" orifice. These maneuvers show that Backhouse squarely placed his interventions, if not contributions, in direct dialogue with the then rampant narratives about eunuchs circulating among his Western peers. If his writings are viewed with an equal measure of skepticism as we do with other Western accounts, the value of these archival remnants comes from not their "factual" status but the way they enable us to envision discrepant paradigms of masculinity converging at the unstable conjuncture of a cultural practice in demise. At the very least, Backhouse's memoir cautions us the danger of adopting an uncritical approach to reading other surviving Western medical sources, to which we now return.

The Power of Image

In the 1890s Western physicians working in Peking began to report on Chinese eunuchs who came to them for treatment. This took place within a rapidly changing context in which Western-style hospitals, clinics, and dispensaries became more accessible in the capital and treaty-port cities.[103] Unlike the earlier reporting of Morache and Stent, these clinical vignettes drew on genuine contacts with eunuch patients. All of the eunuchs who visited foreign doctors came with urinary problems. The doctors used these rare encounters as the basis for presenting their observation to an international community of interested readers. A major outcome of their effort was the production and circulation of medical images that revealed the details of castration and imposed conformity to a penis-centered paradigm of masculinity. Because these images were also intended to capture the cultural baggage of castration perceived as specifically Chinese, the visual milieu emerged as a cross-cultural threshold for shaping the global perception of the mutual backwardness of China and the body of eunuchs.

Foreign physicians leaped on the rare and valuable opportunity to acquire deeper knowledge about Chinese castration when eunuchs actively sought them out. Eunuchs did this typically because they suffered from urological problems, such as urinary retention, urinary tract infection, and bladder stones. In 1898 Vladimir Vikentevich Korsakov, the physician at the Russian embassy in Peking, explained that "the opening of the urethra narrows in most eunuchs despite the use of the dilators," for which he blamed "the lack of proper care and the well-known Chinese lack of hygiene." Although some eunuchs were willing to consult foreign doctors in Peking, they rarely followed through a full course of treatment. Korsakov was not alone when expressing frustration with one of his patients: "One eunuch who I saw in the French hospital in Peking was only able to urinate with a very fine stream or dribbling. The opening of the urethra was not larger than a needle-tip. He had only been castrated one year before I saw him. After a slight dilation was achieved by urethral dilatation, he failed to report following 15 sessions for further treatment."[104] Although this comment provides a telling view of how foreign doctors dealt with eunuchs, Korsakov regurgitated much of Stent's version when documenting the exact steps of Chinese castration.

Other doctors began to produce a visual record of what they deemed important. Despite the plurality of how castration was practiced and how it defined masculinity, images published by late nineteenth-century Western physicians uniformly depicted Chinese castration as the eradication of the entire male sexual organ. In the early 1890s one of the foremost pillars supporting the characterization of Chinese castration as an attack on the penis can be found in the reports of the American missionary doctor Robert Coltman (1862–1931). Born in Washington, Coltman received his medical training at Jefferson Medical College in Philadelphia. He was appointed professor of anatomy at the Imperial School of Combined Learning (Tongwen Guan, 同文館, which later merged with Peking University) in 1896 and professor of surgery at the Imperial University of Peking (later Peking University) in 1898, and he served as both the personal physician to the Chinese royal family and the surgeon at the Imperial Maritime Customs and the Imperial Chinese Railways.[105] Coltman drew international acclaim for his two books, *The Chinese, Their Present and Future: Medical,*

Political, and Social (1891) and the more famous *Beleaguered in Peking: The Boxer's War Against the Foreigner* (1901), which earned him a reputation for being one of the first Westerners to establish communication with people outside China during the Boxers' siege of Peking.[106] In 1894 Coltman published a hand-drawn image of the site of castration as it appeared on the body of a eunuch patient (figure 1.4).[107] The expository text indicates that the image was produced by a Chinese assistant, a "xylographist."

Coltman included this image in an article called "Peking Eunuchs" in 1894. It was in fact a sequel to his discussion of three eunuchs in an earlier article, "Self-Made Eunuchs" (1893), published in the *Universal Medical Journal*. Together the two entries mentioned a total of six eunuchs who came to him for clinical treatment. These eunuchs came with the intent to alleviate the obliteration of urethral opening, as all of them experienced difficulty in urination due to the closing up of the orifice. Based on the six eunuchs whom he treated, Coltman observed that "the majority of the eunuchs here [in China] have penis and testicles removed entire."[108] This statement was remarkable to him because, as Coltman himself conceded, he "never for a moment supposed the mutilation extended beyond the

CICATRIX OF EUNUCH.

* In reproducing the above, our Chinese xylographist has somewhat improved on the original sketch, with regard to geometrical nicety.—(ED.)

Figure 1.4 Drawing of a eunuch's castration site (1894).
Source: Coltman 1894, 28.

testicles."[109] We will revisit Coltman in greater detail when we discuss the way eunuchs narrated their own castration experience. For now, suffice it to say that the textual descriptions and the image of castration that he presented helped (re)define Chinese castration specifically in terms of a penis-centered paradigm of masculinity.

Similar to Stent's justification of a superior West, Coltman expressed "disgust and contempt" toward his Chinese eunuch patients.[110] His final words on them were "Do such specimen of humanity deserve sympathy?"[111] If one reads the experience of Chinese eunuchs through the lens of the Coltman papers, it is tempting to subscribe to the view that the castrated male body undoubtedly needed Western biomedical assistance. We might hasten to add that the enlightenment nature of Western medicine was endorsed even by Chinese eunuchs themselves, as demonstrated by their very decision to turn to Coltman for help. However, it turned out that all of these eunuchs expressed a considerable measure of resistance to being treated by a Western doctor, even an extremely prestigious one such as Coltman, who was the personal physician of the Chinese imperial family. None of them returned to Coltman after their initial visit even if they were explicitly instructed to do so for recovery assessment. In exposing both the enthusiasm and resistance of eunuchs, the Coltman reports generated a fracturing of castration into divergent public and private meanings. The failed mutuality and reciprocation between the eunuchs and Western doctors marked the rise of a disjuncture in the experience of eunuchs—a disparity between foreigners' totalizing condemnation and their own embodied selves.

An even more powerful piece of evidence that construed Chinese castration as the removal of penis and not just the testicles came from another foreign doctor, Jean–Jacques Matignon. Matignon had been a physician to the French legation in Peking since 1894. Having established an esteemed reputation among European colonial officials, Matignon was about to be made Knight of the Legion of Honor. The unfortunate news of his probable victimization in the Peking Massacre, the Boxer Uprising, reached Europe in July 1900.[112] In 1896 Matignon published an extensive study, "The Eunuchs of the Imperial Palace in Peking," which included a photograph that exposed the naked body of a Chinese eunuch and revealed the physical site of castration (figure 1.5).[113] Based on his clinical consultation of a young eunuch with urethral stricture, Matignon described the castration procedure by repeating many of the details documented by Stent. In

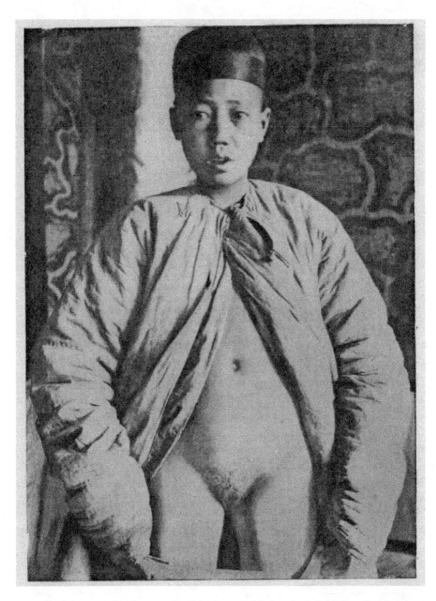

Figure 1.5 Photograph of a eunuch of the Imperial Palace in Peking (1896).
Source: Matignon 1899.

other words, Matignon added credibility to Stent's original account with the new visual evidence he provided. But he also unearthed a crucial step in the procedure that had not been discussed by others: "Preliminary precautions are taken and those details are not available, but some degree of testicle atrophy is obtained and drugs are given to reduce the degree of discomfort." To support this, Matignon explained that the eunuch treated by him "when interrogated on this particular issue replied that [the operation] was not that painful."[114] He also cast doubt on the common view that eunuchs had an intimidating facial physiognomy. Having been invited to the palace twice, he "saw a multitude of eunuchs" and "did not notice any particular abnormal facial appearance."[115] Three years later, when this piece was reprinted in his book, *Superstition, Crime et Misère en Chine*, Matignon included an additional illustration that sketched the surgical instruments used in castration (figure 1.6).[116]

In figure 1.5, the photographic proof of a Chinese eunuch's "lack of penis" makes it difficult for any viewer to deny its captured reality. The difficulty largely stems from the indirect cultural labor of the photo, which turns the beholder's gaze into the object of the eunuch's gaze. The eunuch's reciprocal gaze forces anyone looking at his exposed body to surrender to a tacit transaction of knowledge that, if neglected, would indicate a betrayal of one's own eyes. To deny that the eunuch's corporeal experience, thoroughly exploited in a docile standing posture, was marked by "the lack of penis" would mean to disqualify the very spectatorial relationship that made

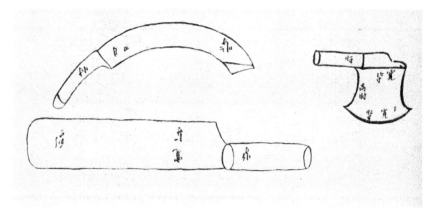

Figure 1.6 Instruments used for Chinese castration (1896).
Source: Matignon 1899, 182.

the (mis)recognition possible in the first place. On the eve of the twentieth century, Matignon's photo thus consolidated a visual layering of "truth" about Chinese castration—that it involved the elimination of the male genitalia in its entirety. This ocular evidence gave weight to Stent's earlier textual description, establishing the absence of both the penis *and* the scrotum as an indisputable reality in a castrated Chinese body from this point onward. It paved the way for twentieth-century considerations of Chinese castration to forget any castrated corporeality outside a penis–centered regime of masculinity.

The broader import of this amnesia cannot be overstated. The aforementioned cultural mechanics fundamental to its shaping were part of a global circuit of power relations, one that mediated the rise of Chinese medical photography in the late nineteenth century. According to art historian Sarah Fraser,

> Photography's role in shaping China's image from 1860 to 1900 is evident in the visual transformation of the Chinese subject of over a half-century of colonial intervention. In these shifts related to China's visual culture, the camera was an instrument of the contemporary practice to create types, classify peoples, and impose hierarchies upon the world as it was being observed. . . . By the turn of the century, the photographic lens was focused on larger statements about "the Chinese" and national character. Scenes of itinerant workers, destitute people, and military captives at the time of the Boxer Uprising reflect racial debates about the modern Chinese subject prevalent in international power relations.[117]

In analyzing the translational politics of visualizing the Chinese, Ari Larissa Heinrich has similarly pointed out that "in early medical photography in China we see the convergence of those colonial, commercial, ethnographic, and scientific ideologies that marked the indisputable entrance of the 'Chinese specimen' into global discourses of race and health."[118] Through its heterogeneous modes of circulation (e.g., archives, museums, private collections, and publications) and deployment of stylistics (e.g., the "before and after" clinical contrasting trope, portraiture, battlefield documentary, and erotic thematization), photographic images of the ill decontextualized and recontextualized Chinese identity by "representing supposedly specifically Chinese pathologies to a global medical community."[119] In the formative years of China's nation formation, the increasing popularity of clinical photography gave representational claims of Chinese pathology a new set of cultural valence and ideological relevance. The diseased ontology of the photographic specimen came to be absorbed by the very medium of its cultural production and naturalized as emblematic of the inherently pathological quality of the Chinese political identity. Over the course of the nineteenth century, China was granted entrance into the global system of nation-states on the condition of being racially stereotyped as "the Sick Man of Asia" with growing intensity.[120]

The evolving relationship of the camera to its object of representation relied on, among other things, the circulation of certain medical beliefs about Chinese identity, which bolstered the "Sick Man" stereotype: in the nineteenth century, China was blamed for being the original home of the bubonic plague, cholera, small pox, and, eventually, leprosy.[121] Through its photographic presence as medical specimens, the castrated male body joined the bound feet, the leper's crippled body, and other exotic corporeal "types" as exemplars of the material figuration of diseased embodiment peculiar to China. In this sense, Matignon's photograph could be viewed as a "confession of the flesh" whereby the penis-absent eunuch body displayed and circulated through it helped solidify an ideological portrayal of China as intrinsically deficient, problematic, and in need of Western (biomedical) assistance. Going back to figure 1.5, the image that we are staring at asks not what is wrong with the body of this eunuch per se, but *what is wrong with China*. Or, to borrow Jacques Derrida's terms, the ghost of the penis affirms the spectral presence of the Eurocentrically commodified body; the hauntology of this absent presence revalues the global ontology and epistemology of *being* Chinese and *knowing* what Chinese is.[122] In fact, we can take a step further and draw on Judith Halberstam's remark concerning the queer art of failure: this photograph, then, leads the viewer "to the site of dispersal and then leave[s] us there, alone, to contemplate all that has been lost and what remains to be seen."[123] If history tends to be written from the perspective of the victors, perhaps a story about failure and the defeated can help subvert our reading practice when confronted with an evidence of colonial violence. Figure 1.5 captures that violence that constantly reminds us what is lost and who the failure represents.

The legacy of this colonial violence can be assessed by comparing two sets of photographs. Figure 1.7 consists of images taken from an article on castration published in an English medical journal in 1933, while figure 1.8

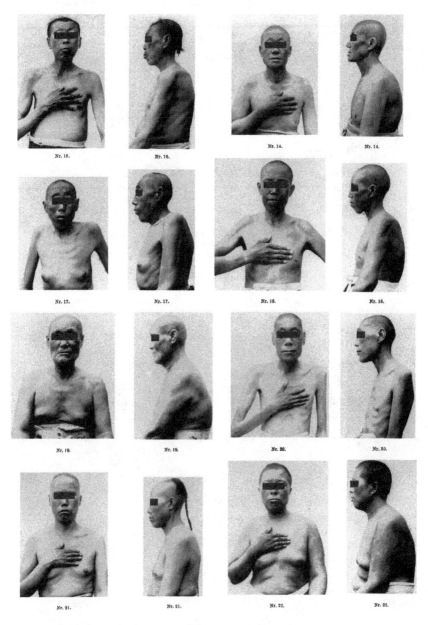

Figure 1.7 Western medical images of Chinese eunuchs (1933).
Source: Wagenseil 1933, reprinted in Wilson and Roehrborn 1999, 4329.

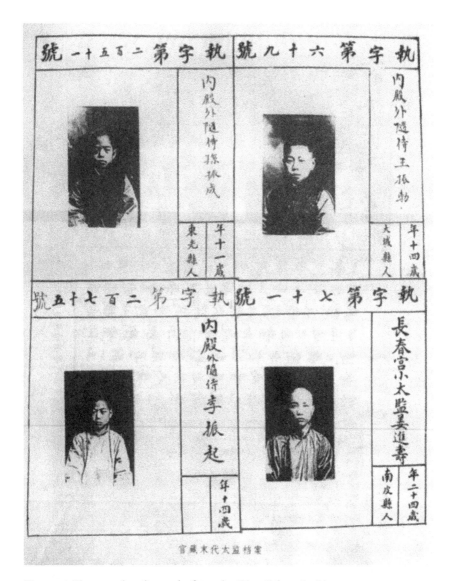

號一十五百二第字執　號九十六第字執

內殿外隨侍孫振成　內殿外隨侍王振勃

東光縣人　年十一歲　大城縣人　年十四歲

號五十七百二第字執　號一十七第字執

內殿外隨侍李振起　長春宮小太監姜進壽

年十四歲　南皮縣人　年二十四歲

官藏末代太監檔案

Figure 1.8 Photographs of eunuchs from the Qing Palace Archive.
Source: Jia 1993. Courtesy of Jia Yinghua.

presents photographs of eunuchs filed in the Qing palace archive.[124]
In comparing these two sets of images, we witness a distinct contrast
in the operation of their epistemological claims. Although both objec-
tify the eunuch's body, the former carefully structures the viewer's position in the subjective terms of clinical and, one must not forget, colonial

gaze. As the object of this particular kind of gaze, the naked bodies constitute the pathological material ground on which the ramification of spectatorship was made possible and intelligible. These unclothed bodies are intended to be compared, deciphered, and scrutinized in every minute detail, and such an attempt on the part of the viewer is comforted, or at least made less guilt-laden, by the complete erasure of any indication of the eunuchs' identity. The only available information from which any sort of personal identification can be inferred is their "specimen number." Carefully presented, their anonymity thereby makes the power imbalance of this entire visual stimulation all the less threatening to the viewer. The images of eunuchs in the Qing palace archive, on the other hand, defy the foreigner's clinical and colonial spectatorship. The fully clothed body and the extensive information (name, age, job title, place of birth, etc.) accompanying each photograph depict these young eunuchs in the normative terms of the native population, not an ostensibly mysterious, deficient object waiting to be investigated and treated according to the normative metrics of Western biomedicine.

Agency and Reproduction

When castration was a widely accepted practice, intentions to expose its targeted bodily site were never committed to visual imaging, perhaps because primary instructions for the practice were transmitted orally and demonstrated corporeally. Starting in the late nineteenth century, however, detailed textual and visual documentations of the castration procedure became available and thus signaled the creation of new knowledge about eunuchs' bodies and new venues of its circulation. This unprecedented repository of information unveiled the secrecy surrounding the operation, transforming a private matter into something public. But in bringing a corporeal practice as private as castration into the public domain, Westerners' textual description and visualization had elevated, rather than diminished, the tension between the private and the public awareness of Chinese eunuchism. The foreigners' standardization of the castrated body simultaneously made its personal relevance all the more imperceptible, silencing any corporeal embodiment of eunuchism that did not match their globalizing narrative. The increasing irrelevance of certain forms of corporeal experience thus went hand in hand with the collapse of eunuchism as a contested subject of experience. Their effort, in other words, constituted the first major step in making a practice as incendiary as castration one of the most *un*controversial issues in and out of China.

From this standpoint, what remains utterly inadequate in the existing literature on Chinese eunuchs is the singular meaning critics have assigned to and extrapolated from the act of castration—the permanent elimination of procreative function. Here it might be useful to borrow the insight of historian Nancy Rose Hunt from a different context (colonial Congo) to help us appreciate the significance of castration in Chinese history: namely, to "broaden our focus from reproduction narrowly defined in demographic and medical terms as fecundity and the birth of children, to social and cultural reproduction."[125] Insofar as our perception of the consequences of castration remains inside the framework of biomedical reproduction, the other half of the historical story completely escapes our attention: the castration of male bodies also reproduces eunuchs socially and culturally in imperial China.

Similar to the role of the civil service examination and footbinding in Chinese society, castration represents a mechanism of systemic gendering whereby certain social, cultural, and political norms were reproduced.[126] To the extent that scholars have neglected the social and cultural reproductive aspect of castration, they have served, however implicitly, as passive agents of Western biomedicine to reinforce its epistemic authority. Since the mid-nineteenth century, the rise of Western anatomical and scientific knowledge has led many to privilege the biological consequence of castration as the only indicator of its sociocultural function and reality.[127] To overcome this limited reading, it is important to acknowledge the constructed nature of Chinese castration as a bifurcated mode of historical experience. On the level of individual experience, castration denotes a ritualized episode where the death of a man's biological fertility intersects with the birth of his new life as a eunuch. The necessity to register with new official names upon entering palace service clearly illustrates this. On the level of public collective memory, castration symbolizes a boundary of cultural difference that deceptively claims to unify the multiplicity of how eunuchs were made while compressing the significance of its history into a global perception of China's natural "lack." By reducing the importance of castration to the realm of biology on both the macro and micro levels of historical experience, critics have inevitably fallen short in addressing the agency of eunuchs in their social and cultural reproduction.

Whether we consider the scrotum to be the seat of male virility or the penis the locus of male pleasure, the intended effect of castration on eunuchs is the deprivation of their power to reproduce biologically. And that is it. They were not impotent in any other sense. Historian Jennifer Jay, for instance, has shown that Chinese eunuchs retained an overtly "male" gender in aspiring to a traditional Confucian lifestyle: "From both the historical sources and the anecdotal reminiscences of Qing eunuchs, it seems that with very few exceptions, the Chinese eunuchs were without gender confusion at the time of castration, and after the operation they experienced physiological changes but no apparent shift in their gender identity and male-oriented role in society."[128] Many eunuchs got married, adopted children, or had kids before offering themselves to the imperial court, suggesting that their masculine social role remained intact as they continued to embrace Confucian family norms.[129] Quite simply put, undergoing castration did not indicate, to them and their surrounding community, a complete erasure of their masculine identity. Moreover, in contrast to other female servants (宫女, gongnü) inside the palace, eunuchs enjoyed a greater prospect for attaining socioeconomic opportunity and political power.

Eunuchs could not reproduce biologically, but the practice of castration made their social and cultural reproduction possible. Even if it was not physically feasible for them to give birth to future generations of their own kind, eunuchs frequently occupied a leading role in overseeing the nuts and bolts of castration, the most important procedure that defined their identity. The presence of eunuchs in monitoring castration operations came to the attention of Westerners as early as 1908, when the French physician Richard Millant published his full-length study on eunuchism, *Les eunuques à travers les âges*. Millant's description of the procedure in fact borrowed heavily from Stent and Matignon, including the necessary step of eradicating both the penis and the testicles together. Millant even went so far as to reproduce a hand-drawn picture of Matignon's photograph, giving further visual credence to the penis-centered interpretation of castration (figure 1.9).[130] However, Millant mentioned in passing one crucial aspect of the operation that was absent in other Western accounts. In the post-operation stage, according to Millant, "Repair of the wound, which affects a triangular form from top to bottom, takes on average three months to complete. The surgery is then examined by an old eunuch, responsible for ensuring that the mutilation was complete."[131] For the first time in a Western language source,

Figure 1.9 A hand-drawn replicate of
Matignon's eunuch photo (1908).
Source: Millant 1908, 234.

eunuchs had been given a decisive role within the overall scheme of their social reproduction.

A Shanghai lithograph from the late nineteenth century also sheds light on the involvement of eunuchs in castration (figure 1.10).[132] According to the author of this *Dianshizhai* (點石齋) lithograph, when instances of self-castration occurred on the streets of late imperial Peking, eunuchs were the authorities to whom people often turned for assistance. The title of the lithograph is "How He Lost His Significance One Morning," and the textual description of the incident reads as follows:

> There once was a man named Tang who lived outside the Shunzhimen Gate in Peking. Though in his early twenties, Tang had already acquired the evil habit of gambling and on one recent occasion had lost all of his money. He had no place to flee to, nor any way to repay his debts.
>
> On the ninth of last month, Tang proceeded to the Changyu Pawnshop with the intention of obtaining two strings of cash by pawning a pair of short pants. The pawnbroker on duty told Tang that his pants weren't worth that much, and that he would have to add something more substantial if he hoped to obtain the desired amount. To this Tang replied, "But all I've got to my name are my balls (卵袋)!"
>
> "That would be just fine!" replied the pawnbroker with a laugh.

Figure 1.10 Late Qing Shanghai lithograph "How He Lost His Significance One Morning."
Source: Cohn 1987, 36–37.

Tang walked away in a huff. When he got home, he sharpened his knife—which had a blade sharp enough to fell a kingdom—and returned to the Changyu Pawnshop. When he got there he removed all of his clothing and stood there as naked as when he was proceeded to turn himself into a sawed-off shotgun with a single energetic slash of his knife, losing enough blood in the process to float a pestle.

Tang passed out immediately, whereupon the pawnbroker, frightened out of his wits, rushed off to a local official's residence to find a eunuch who could come to Tang's rescue. On the way, he stopped at North City police headquarters to report the incident. Within minutes, the police had dispatched a runner to arrest the pawnbroker, and subjected him to a thorough questioning. Only through the intervention of an intermediary was he able to extract himself from a potentially burdensome lawsuit.

In the meantime, Tang had been carried home on a wooden plank, but he had lost so much blood that his life hung in a delicate balance. The proceedings described above cost the pawnshop some four hundred taels of silver.[133]

Late nineteenth-century lithographs have long been considered by historians as a vivid source for interpreting the tangled social and cultural history of late Qing China. Published between 1884 and 1898 as a supplement to *Shenbao* (申報), *Dianshizhai* pictorials were the first of its kind to center on current affairs. Although most of them acquainted their readers with melodramatic stories, leaving a strong impression of sensationalism and shock, a careful reading suggests that these stories often embraced a consistent ambition of finding social order in chaos.[134] The lithograph shown in figure 1.10 can be viewed within this context. First, Tang's choice of the word "*luandai,*" which is translated here as "balls" but literally means an "egg bag" (i.e., the scrotum), goes a long way to show that people had not always considered the entire male genital organ the bodily target of castration. This lithograph might in fact be the only visual representation of Chinese castration before Matignon's photo. In the mid-nineteenth century, when the English word "testes" was translated into Chinese for the first time, the medical missionary Benjamin Hobson (1816–1873) left no room for ambiguity when he remarked that the "outer kidney" (外腎, *waishen*)—his terminology for the male gonad—was the organ responsible for "the generation of semen," and for "the change in voice and facial features alongside the elimination of reproductive power when castrated (閹之割之, *yanzhi gezhi*)."[135] This is not surprising since some Daoist and Chinese medical texts since the twelfth century had already interpreted the female breasts and the male testicles as the primordial markers of sex difference.[136] The implication of both the lithograph and Hobson's translation is clear: before the rise of the penis, the anatomical target of castration was the scrotum.

Moreover, the lithograph implies that in matters involving castration, eunuchs were the everyday experts from whom people sought advice. That knifers were not consulted in this incident is reasonable because they would have been located too deep inside Peking at the time to be a source of assistance, if they existed at all, according to Stent.[137] But for a health issue as serious as loss of blood, and potentially death, it is interesting to note that no physician is either mentioned in the expository text or present in the lithographic staging of this male-dominated event. Curiously enough, the individual who had the best view of what Tang actually detached from his body is the child located at the center of the drawing, and most certainly not the eldest man on the left who seems to focus more on Tang's upper body. If the lithograph is an informed representation of common attitudes toward castration in the 1880s and 1890s to any degree, even if limited, one can infer from it that the persistence of castration as a cultural practice depended, at least partially, on the role of eunuchs as a key mediator of its historical continuation.

The most striking example of the agency of eunuchs in perpetuating the culture of castration is a custom that Republican-era commentators called "born from within the inner quarters" (門裏出身, *menli chushen*) or "passage through the chambers" (過房, *guofang*). This referred to a specific process in the official designation of palace eunuchs, according to which individuals were first appointed by established eunuchs as their apprentices and underwent castration only upon the approval of their mentors. If the individuals were castrated before puberty, they were known as *naidi'er* (奶地兒). Otherwise, they would be identified as *banlu chujia* (半路出家), which was the same label commonly assigned to people who became a monk or nun late in life. In fact, in an article titled "Why We Look Down on Monks and Eunuchs" (1941), one writer compared monks to eunuchs by construing the former as betraying male reproductive power psychologically and the latter physically. According to this article, Chinese society bemoaned both groups of men as they shared similar defiance to carrying on the family bloodline, a central tenet of the Confucian expectation from a filial son (不孝有三 無後為大, *buxiao yousan wuhou weida*, "there are three unfilial things in life, not having an offspring is the most serious offence").[138] The idea of "born from within the inner quarters" suggests that eunuchs held agency in selecting, nominating, and authenticating future generations of their own kind. In this regard, the social reproduction of castration depended as much on the autonomy of eunuchs in protecting and defining their membership as on the biological elimination of their fecundity. At the same time, the idea of "born from within the inner quarters" clearly distinguished itself from a phenomenon that acquired increasing popularity in the late Qing period known as voluntary castration (投充, *touchong*, or 投幕, *toumu*).[139]

Evidence of the Self

At the turn of the twentieth century, the mutual perception of China's "lack" and the castrated body's deficiency rested on a form of ontological omission circulating within the global production of knowledge. As I have been suggesting, one way to redress the critical absence of eunuch's agency is by broadening our conceptualization of reproduction to incorporate a social and cultural definition and, by extension, turning to the actual participation of eunuchs in castration operations. Western doctors' interaction with eunuchs soon triggered a similar interest in the self-narration of eunuchs among the Chinese themselves. In the twentieth century, eunuchs' recollection of their castration experience assumed the center of public spotlight for the first time. Ironically, the oral testimonies they left behind—eventually transcribed, published, and made available to a worldwide audience—ended up reproducing many of the substantive and formal elements of the earlier Western narratives. Nonetheless, this exceptional set of materials also brought to sharper focus a form of castration that had not been discussed in depth by previous writers: the phenomenon of voluntary castration.

In the historical record, one of the most powerful ways through which eunuchs exerted a significant measure of agency in their social and cultural production was by narrating their own experience. This began arguably as early as when Stent was collecting materials for his study. Stent mentioned twice about the existence of native "informants."[140] But Chinese eunuchs did not narrate their experience through the voice of outsiders alone. Besides the textual and visual repository created by foreigners such as Stent, Coltman, and others, additional historical information about the operation came from the personal testimonies of eunuchs who worked in the palace in the late Qing period and lived in the various corners of Beijing after the fall of the Qing. Ren Futian (任福田) and Chi Huanqing (池焕卿) were among the oldest surviving eunuchs in the twentieth century. Based on their recollection, the two most well-known places that offered professional services in castration prior to the 1890s were Biwu (Bi "the Fifth") and Xiaodao Liu ("pocket knife" Liu). Bi operated an establishment on Nanchang Street, whereas Liu's was located inside the Di'an Gate in the imperial city. "Each season," Ren and Chi explained, "they supplied the *Neiwufu* [Imperial Household Department] forty eunuchs.

Together the two families were responsible for all the formal procedures pertinent to castration."[141] The presence of these professional castrators was also mentioned in scattered Chinese commentaries in the 1940s.[142]

According to Ren and Chi, "registration" with Bi or Liu was the first step required of parents who wished to turn their boys into palace eunuchs. In turn, the boys would be "examined—for his appearance, conversational skills, intelligence, and genital organ (done with his pants on)—and admitted only if considered appropriate." Although Bi and Liu "had many years of experience and the necessary utilities," Ren and Chi insisted that "the overall experience remained painful for the subject of operation, since they possessed neither pain relievers nor adequate medicinals that would help stop the bleeding. Antiseptic preparation was done simply by heating up the surgical knives with fire."[143] Committed to print almost a century apart, Ren and Chi's discussion of Bi and Liu seem to provide solid evidence for the existence of the knifers described by Stent.

However, their words verify Stent's account only by augmenting, rather than subsiding, the distance he first established for the historical experience of castration between a personal realm of embodiment and a public domain of collective memory. Evident from this example, eunuchs themselves participated in the archival rendering of the knifers as primary operators of Chinese castration. This historiographic validation adds another layer of complexity to the historian's task, to quote medievalist Gabrielle Spiegel, of "solicit[ing] those fragmented inner narratives to emerge from their silences."[144] For any eunuch whose castration experience deviated from this global narrative would require additional explanation and narrative space for inclusion. One of the most popular alternative routes to serving in the palace, for instance, was voluntary castration, a category that included self-castration (自宫, zigong). Whereas voluntary castration was a general term used to describe any method of castration that resulted in the self-referral of eunuchs (including procedures performed by fathers), self-castration referred more specifically to an individual's castration of his own body.

More prevalent in the Ming dynasty, self-castration became illegal under early Qing law. The lessons from Ming eunuchs' political corruption were difficult to ignore, so up to the second half of the eighteenth century, Qing emperors made it illegal for civilians to castrate themselves, which simultaneously curbed the number of available eunuchs. However, the legal codes that imposed death penalty for voluntary castration were not strictly enforced throughout the first century or so of Qing rule. In June 1785 the Qianlong Emperor took a step further in loosening the codes to allow the Imperial Household Department to accept individuals who offered themselves to serve as eunuchs after voluntary castration.[145] Actually, this only reflected the strictness of the regulations imposed on eunuchs by early Qing rulers, which facilitated the decline in the formal supply of eunuchs and the growing number of eunuchs who fled. Although Qianlong promoted a policy of eunuch illiteracy, the case reports of eunuch misdeeds housed in the Qing archive "reveal a more lenient and compromising system."[146] According to historian Norman Kutcher's observation, "Over time, the system that had been set up to control eunuchs deteriorated."[147]

Famous late Qing eunuchs whom the court admitted as a result of voluntary castration included Ma Deqing (馬德清), Zhang Lande (張蘭德, more popularly known as Xiaode Zhang 小德張), and the last Chinese eunuch, who died in 1996, Sun Yaoting.[148] To be sure, persons born with ambiguous or dysfunctional genitalia were often categorized by physicians as "natural eunuchs" (天閹, *tianyan*) and, as a typical solution, recommended to the imperial palace as young eunuchs. However, in many cases of voluntary castration, the father was the person who performed the operation. Such was the experience of Ma Deqing, one of the last surviving Chinese eunuchs, who recalled:

When I was nine, roughly in 1906, one day my father succeeded in persuading me to lie on the bed and castrated me with his own hands. That was a really agonizing and scary experience. I can't even recall the exact number of times I passed out. I've never been willing to discuss the incident with anyone, not because I'm shy, but because it was way too painful. . . .

Think about it: in those years, no anesthesia, needles, or hemorrhage-preventing medicines were readily available. . . . Consider the kind of pain inflicted on a restless kid by holding him down on the bed and cutting his *yaoming de qiguan* (要命的器官, "organ for life") from his body! Every single vein was connected to my heart, and, with the kind of pain involved, I almost puked my heart out. Ever since, my reproductive organ and I became two separate entities.

After the surgery, it was necessary to insert a rod at the end of the surgical opening. Otherwise, if the wound seals up, it becomes impossible to urinate and will require a second surgical intervention. . . .

Seriously, [in those years,] the meds applied to facilitate the healing of the wound were merely cotton pads soaked with white grease, sesame oil, and pepper-powder. Changing and reapplying the meds was always a painful experience.

I remember I was on the dust *kang* all the time, and my father only allowed me to lie on my back. Sometimes I wished to readjust myself when my back started to sore, but how could I? Even a mild stir would bring up extraordinary pain from the cut.[149]

When asked about his reason for becoming a eunuch, Sun Yaoting recalled a similar experience. Born on September 29, 1902, in Jinghai County, Hebei, Sun readily admitted that his decision to become a eunuch was greatly influenced by a role model, Zhang Lande. In 1909, when Zhang returned home from Beijing for a short visit, the endless conversations praising his eminent status in the palace carried over a sustained period of time in the village and made a huge impression on Sun. According to Sun, Zhang was also a victim of voluntary castration, except that he became a eunuch by way of volitional self-castration rather than being castrated by his father. "One day," Sun explained, "the twelve-year-old Zhang castrated himself with a sickle in one hand and a rope in another. After the bleeding stopped, Zhang offered himself to the palace."[150]

Hoping to emulate Zhang, Sun decided to become a eunuch too. After a lengthy deliberation, his father agreed to perform the procedure with great reluctance.

Dad made me a huge bowl of noodle soup with two eggs, telling me that I need to be well fed. After I ate, he asked me to take a walk outside the house until he cleaned up the kitchen and was ready. I did not want to go somewhere far, so I just waited near the front entrance. It did not take long before he summoned me back. After I came back inside, I saw a door plank in the middle of the floor. It was covered in a used cloth with my pillow on the one end and some herbal powder on the other. Covered in tears, my dad asked me to take off my pants while holding a thick rope in his hands.

"Son . . . dad is awfully sorry! I need to . . . tie you up . . . in order to make the process less disquieting!"

When I saw his posture, the speed of my heartbeat rose, but I still maintained a smile on my face: "Dad, I won't run!"

Dad wiped the tears off his face and shook his head: "Good son . . . dad . . . knows that you . . . won't run away. Dad is just afraid that . . . you . . . can't bear the pain."

"Dad! Tie me up! Go ahead. I am not afraid!"

Dad seemed mad with his hands trembling. He tied me up firmly on the wooden plank and blindfolded me with a black cloth. To be honest, by this point, I was seriously afraid. My heart was beating like a drum roll; I was sweating all over. Every single move that dad made caused me to twitch. Then dad said, "good son! Scream as loud as you want. . . ." Suddenly, I felt my dad's hand circling twice around my growing area, immediately after which he inflicted the utmost pain. Then I could see nothing but darkness. I was unconscious . . .

Following the instructions provided by other people, dad performed this surgery on me. He then inserted a small rod into the urethra meatus in order to prevent it from closing up, which could lead to urination problems and would require a second operation.[151]

Similar in function to Ren and Chi's account, these recollections of childhood castration experience confirm aspects of the procedure openly described by late nineteenth-century Western writers. For instance, both Ma and Stent mentioned the *kang* on which castration was operated, and both Sun and Morache documented the custom of offering the castrated subject a good meal before the surgery. The personal testimonies of Chinese eunuchs further corroborated the necessity to place a rod inside the main orifice after the operation in order to ensure successful urination. From the lithograph to these life narratives, then, eunuchs actively monitored the details of what it took to become and live as a eunuch, historically and historiographically—that is, in both historical real time and as vanguards of their own body history.

One notable difference, though, was that whereas knifers or professional castrators took the center stage in previous documentations of the operation, their role was replaced by the young boy's father in these later reminiscences of childhood castration. Unlike the narratives that centered on professional castrators, eunuchs' personal recollections captured a more humane image of castration when carried out between a father and a son. This further underlines one of the most significant parallels between footbinding and castration in Chinese history: the cultural survival of both practices entailed a homosocial environment in the occasion and demonstration of their corporeality. Footbinding was a custom conducted by women and on women; castration was a practice performed by men and on men. But whether mothers bound the feet of their daughters, fathers castrated their sons, or male knifers operated on boys, neither footbinding nor castration should be understood as a timeless, spaceless practice with a universal raison d'être. In recalling the actual measures of castration and the degrading ways in which they were treated by the imperial family, the stories Chinese eunuchs told of themselves ultimately joined the public repository first developed by foreign "outsiders," constituting the second major step in defining their own bodies as templates for national histories.[152] Yet it is also when we depart from a strictly biomedical understanding of reproduction and redefine it instead as a social and cultural process that can we gain a better insight into the different, multiple forms of castration as experience and not just fact.

A preeminent example of how Western perspectives became deeply embedded with the self-narrations of Chinese eunuchs can be found in the Coltman reports. Recall that Robert Coltman, a personal physician to the Chinese imperial family, treated six Chinese eunuchs in Peking and, based on his clinical encounters, provided an image of the castration site from one of his patients (figure 1.4). Coltman revealed a transformation in his feelings toward eunuchs from "sympathy" to "disgust and contempt."[153] In the two medical articles that he published in the 1890s, Coltman conceded that this transformation may be explained by his realization that a surprisingly high number of Chinese eunuchs, at least during the late Qing period, actually castrated themselves. In all of the cases he reported, the patients did not merely undergo voluntary castration, but they became eunuchs through the more specific measure of self-castration. In 1894 Coltman wrote: "I am now fully convinced, that many of the eunuchs employed in and about the palace, have made themselves so, for the purpose of obtaining employment."[154] In light of the personal recollections of eunuchs as discussed earlier, one might assume that self-castration was rather rare, and most voluntary castrations were executed by their father. Coltman presented a wealth of contrary evidence that revealed the prevalence of self-castration in an era when eunuchism gradually came to an end.

One of Coltman's patients, over fifty years of age and who went to him "for the obliteration of the urinary meatus," was once with the Tongzhi

Emperor and, after the death of the emperor, took service with the seventh prince. This eunuch "stated that at twenty-two years of age, he being married and the father of a year old girl baby, resolved to seek employment in the palace. He secured a very sharp *ts'ai-tao-tzu* [chopper], and with one clean cut removed his external organs of generation entire."[155] Coltman also operated on a thirty-two-year-old eunuch "who emasculated himself eighteen months ago." "This man," according to Coltman, "is a large framed sturdy fellow who could earn a good living in any employment requiring strength, but he deliberately emasculated himself for the purpose of getting an easy position in the Imperial employ."[156] Interestingly, some eunuchs castrated themselves to irritate their fathers. One of his patients, sixteen years old, "had an elder brother who had been made a eunuch at an early age, and was in service at the imperial palace. Knowing that his father depended on him (his second son) for descendants to worship at his grave, this lad, after a quarrel with his father, on the 23d of March, took a butcher-knife and cut off his penis close to the symphysis pubis."[157] Another eunuch, aged twenty-two, cut off his penis and testicles "with a razor," explaining that "he was the only son of this father, and having had a quarrel with him, he had, to spite him, thus deprived him of all hopes of descendants at one blow—the dearest hope known to an elderly China-man."[158] To stress the relatively high incidences of self-castration in the 1870s and 1880s, Coltman concluded that "many able bodied men voluntarily submit to the operation by others, and not a few perform it upon themselves."[159]

These examples confirm a number of the crucial insights that we have developed thus far concerning the historical demise of Chinese eunuchism: the foregrounding of the penis in the biomedical (re)definition of masculinity with respect to Chinese castration procedures; the separation of the eunuchs' masculine subjectivity in the social sphere (as husbands, fathers, and sons) from the gendering effect of the castration operation itself in the realm of embodiment; the crucial role of foreigners—especially Western doctors—in relating the castration experience of Chinese eunuchs to a global community of observers; and the self-narration of eunuchs, though often conveyed through the writings of foreign observers in the early phases, as a cornerstone in the shaping of twentieth-century common understandings of their own experiential past.

Abolishing the Past

Adding to the public discourse on the corporeal experience of castration sustained by Western commentators and eunuchs themselves, members of the last royal family completed the process of turning the eunuch body into homogenous anchors of anticastration sentimentality. Strictly speaking, there was no anticastration movement comparable to the anti-footbinding movement that acquired a national urgency in the final years of the nineteenth century. The eunuch system was simply terminated when the last emperor, Puyi, decided to do so. Puyi's ad hoc explanation for his decision, supplemented by the detailed recollection of his cousin Pujia (溥佳, 1908–1949), thus brought an end to the social and cultural production of Chinese eunuchs. Once the eunuch system was abolished, the cultural existence of castration also came to a halt in China.[160] With Puyi's and his relatives' autobiographical words printed and circulated globally, Chinese eunuchism became a truly historical experience.

According to Puyi, his motivation for expelling palace eunuchs originated from a fire incident inside the Forbidden City during the summer of 1923. By 1923 more than a decade had passed since Sun Yat-sen (孫逸仙, 1866–1925) inaugurated a new republic. Puyi and the imperial family were nonetheless protected by the Articles of Favorable Treatment of the Emperor of Great Qing after His Abdication (清帝退位優待條件, *Qingdi tuiwei youdai tiaojian*), an agreement reached between his mother, Empress Dowager Longyu (1868–1913); Yuan Shikai (袁世凱, 1859–1916), who was then the general of the Beiyang Army; and the provisional Republican government in Nanjing. The articles guaranteed Puyi and his family the right to continued residence in the Forbidden City and ownership of Qing treasures as well as a $4 million per year stipend and protection of all Manchu ancestral temples. Under these conditions, Puyi retained his imperial title and was treated by the Republican government with the protocol and privileges attached to a foreign monarch. Hence, the overthrow of the Qing dynasty did not end the institutionalization of eunuchs immediately. The corporeal experience of Chinese eunuchs existed almost a quarter into the twentieth century as the demand for them survived with the royal family in Peking.

The fire swept across and destroyed the entire surrounding area of Jianfu Palace (建福宮) at the northwest corner of the Forbidden City, including

Jingyixuan (靜怡軒), Huiyaolou (慧曜樓), Jiyunlou (吉雲樓), Bilinguan (碧琳館), Miaolianhuashi (妙蓮花室), Yanshouge (延壽閣), Jicuiting (積翠亭), Guangshenglou (廣生樓), Nihuilou (凝輝樓), and Xiangyunting (香雲亭).[161] The timing of the event coincided with Puyi's effort to catalog his official assets. Indeed, Jianfu Palace stored most of his valuables, including the wealthy repertoire of antiques, paintings, pottery, and ceramics collected by the Qianlong Emperor. One day when he came upon (and was astonished) by a small portion of Qianlong's collection, he asked himself: "How much imperial treasure do I actually possess? How much of it am I aware of, and how much of it has slipped through my fingers? What should I do with the entire imperial collection? How do I prevent them from being stolen?"[162] Ever since the founding of the Republic, Puyi and members of his extended family had confronted repeated reporting of theft. The frequency of palace robbery rose rapidly by the early 1920s, which galvanized the increasing value of the Qing collection of artistry and material goods in the global market. In hoping to control the situation, Puyi decided to tabulate and document his inventory at Jianfu Palace. "On the evening of June 27, 1923," Puyi recalled, "the same day when the project was just under way, a fire took off, and everything was gone, accounted for or not."[163]

Puyi formally abolished the palace eunuch system on July 16, 1923, only twenty days after the fire. Still relying on their service and loyalty at the time of the fire, Puyi held eunuchs responsible for the episode. In the words of his cousin, Pujia, who had been taking English classes with him since 1919, "the fire undoubtedly had a direct bearing on [this decision]."[164] Pujia recollected that after what happened to Jianfu Palace, many eunuchs were interrogated, and through these interrogations Puyi learned about their previous stealing and selling of his possessions. "According to the fire department," Pujia added, "the firefighters smelled gasoline upon arriving at the palace. When Puyi heard about this, he became even more suspicious of the eunuchs whom he accused for having started the fire in order to cover up what they had stolen from the Jianfu Palace."[165] Initially, Puyi's uncle Zaitao (載濤, 1887–1970) suggested that the eunuchs who guarded Jianfu Palace should be sentenced to death. The court advisor, Shaoying (太保紹英, 1861–1925), considered this punishment too cruel and advised instead for the eunuchs to be legally trialed according to the newly established Republican criminal justice system. However, Puyi's other uncle, Zaixun (載洵, 1885–1949), pointed out that the new legal system relied heavily on the kind of evidence that they lacked and, as such, this would only further damage the reputation of the Qing clan, so he strongly opposed the idea of subjecting the eunuchs to legal investigation. Caught in this frustration, Puyi issued a statement, which he asked Shaoying to put into immediate effect. In the statement, Puyi identified in the Hundred Days' Reform the recommendation to abolish the eunuch institution but also observed, "due to political volatility at the time, this policy reform was soon dropped." In order to avoid similar chaos directed by eunuchs in the future, he ordered "all of the eunuchs to be evacuated from the palace at once."[166] Initially met with great resistance from his father, wife, uncles, and other relatives, Puyi eventually won them over by asking: "If the palace is on fire again, who's willing to take the responsibility?"[167]

Interestingly, Puyi himself revealed a different set of reasons for terminating imperial eunuch employment. Although he realized how rampant theft was inside the palace, he was more concerned about his life than his possessions. Not long after the Jianfu Palace incident, another fire was indeed started right outside his own bedchamber, Yangxindian (養心殿). This incident escalated Puyi's anxiety, leading him to speculate that "there were people who were trying to not only cover up palace robbery but also murder me."[168] Given how badly he treated eunuchs, Puyi's real motivation, therefore, came from his growing suspicion that eunuchs tried to take his life for revenge.

Moreover, Pujia suggested that Puyi's decision to end the eunuch system was likely shaped by the influence of their English teacher, Reginald Johnston. Johnston's distaste for eunuchs, for example, resulted in the swift eviction of bystanding eunuchs in their initial tutorial sessions.[169] In 1923 Johnston informed Puyi about the activities of eunuchs who smuggled palace treasures and auctioned them in antique shops.[170] As an "outsider" and a non-Chinese, Johnston was able to remind Puyi frankly of the rampant corruption among his palace eunuchs. In addition, Puyi constantly expressed his admiration and respect for Johnston: "over time, Johnston has become simply the best in my mind."[171] It is therefore reasonable to attribute Puyi's motivation for disbanding the palace eunuchs, at least partially, to the way he was moved by his English teacher's attitudes toward Chinese culture.

But whether it was Puyi's own paranoia, frustration with palace theft, or intentional self-refashioning and self-Westernization under Johnston's influence, the historical reason for dissolving the eunuch system was minimally concerned with how eunuchs felt or how they were treated. The displacement of eunuchism in China proceeded on one final precondition:

the imposition of historiographic hegemony by members of the imperial family, especially the last emperor, Puyi. With respect to castration, the growing distance between a public domain of collective memory and a private realm of individual embodiment was so firmly in place by the 1920s that even when we are confronted with the concrete reasons and motivations for discontinuing eunuchism, a phenomenon with thousands of years of history in China, the reasons and motivations had no bearing on the actual embodied lives of castrated men. Eunuchism and castration are perceived as backward, oppressive, shameful, and traditional not because they impose violence onto men's bodies, not because they punish men corporeally, not because they hurt men's psychological well-being, and not because they demonstrate inflicted cruelty of the flesh: these are not the most immediate reasons why eunuchism and castration elicit negative attitudes in Chinese nationalist sentimentality. Eunuchism and castration come across as undesirable because they unveil a host of social values—lagging behindness, oppression, shame, tradition, and even disregard for human rights—that give Chinese civilization a negative history in the global consciousness. When one enters this global platform to reflect on China's past, one is tempted to neglect the personal voices of those castrated servants who lived next to the epicenter of that very history.

After Eunuchs

Nearly a decade and half after the termination of the eunuch system, detailed descriptions of the castration procedure appeared in vernacular Chinese. In a newspaper article titled "The Eunuchs of the Qing Dynasty" (1936), author Liu Zhenqing (劉振卿) provided a captivating account:

> Legend has it that castration targets one specific pressure point (穴道, *xuedao*), which would lead to fatality if missed even by a very small distance. . . . Twenty-five years ago, when bandits robbed people, they would cut off people's fingers or flesh or, as the severest form of threat, perform castration. If a eunuch was present to attend the injury, the castrated person could avoid bleeding to death. Regardless of the veracity of the pressure point theory, the process of creating a eunuch does not involve a mysterious method. Nonetheless, the post-operation "fluid conveyance" (接水, *jieshui*) procedure, which involves

a careful application of special medicines, is a vital step whose details have been hidden from the purview of outsiders and yet publicly known only by word of mouth. The two key medications are the dark *jiulong huishengsan* (九龍回生散) powder, which has styptic and pain-killing functions, and the red *zhupo shengjisan* (珠珀生肌散) powder, which inhibits the growth of bodily flesh. These two major ingredients have been prepared by [the chief eunuch] Li Lianying himself. . . . Not very much is known about castration procedures other than the above mentioned two unique medicines. Although they are invaluable, one must not forget the "fluid conveyance" step, which involves the insertion of a wooden stick into the urethra opening after genital cutting. In conjunction with the cautious disbursement of special medicines, the stick can be removed only when the flesh in the surrounding area has contracted. Otherwise, the urinary tract may be blocked.[172]

Printed at a historical juncture when the aura and cultural significance of castration had been displaced, these words signaled the culmination of decades of global interest in Chinese eunuchism. Aspects of this textual description brought to light new information, such as the pressure point theory and the names of the special medicines used in castration operations. But taken together they reinforced much of what had been said by foreign observers since the late nineteenth century. The absolute necessity of inserting a rod inside the orifice immediately after the operation, despite its acquired new name "fluid conveyance," further corroborated the longstanding understanding that castration was a practice by and large performed in the private realm: concealing details of its praxis anchored its very cultural endurance. As the social reproduction of eunuchs came to an end, this mechanism of concealment was quickly turned inside out first by Western spectators and eventually by Chinese natives themselves.

From the Self-Strengthening movement to the May Fourth era, whereas the anti-footbinding movement was built on the rhetorical power of newly invented categories such as *tianzu* and *fangzu* (放足, letting foot out), the demise of eunuchism depended on the collapse of the saliency of existing categories such as *tianyan* and *zigong*. The inappositeness of these categories thus carved out a space in Chinese culture for new ontological concepts to be associated with the practice of sex alteration, such as transsexuality. Viewed from this genealogical perspective, both eunuchism and transsexuality are categories of experience whose historicity is contingent rather than foundational or uncontestable.

The critical reflections on the meaning and value of evidence throughout this chapter are attempts to demonstrate, quoting feminist historian Joan Scott, "the possibility of examining those assumptions and practices that excluded considerations of difference in the first place." They ask "questions about the constructed nature of experience, about how subjects are constituted as different in the first place, about how one's vision is constructed—about language (or discourse) and history."[173] Insofar as the available sources claimed for themselves some representativeness of the social reality of castration in late Qing and Republican China, they metamorphosed from being a form of evidence operating on the level of individual embodiment to a source type functioning on the level of global historical narration. They likely stand for most of what we know about the administration and operation of Chinese castration in the past.[174] As such, on the register of historiography, this textual and visual archive not only exposes the castrated body of the eunuchs in the public sphere but also conceals its intimate existence in the personal historical realm.

I have adopted a specific reading strategy in approaching the archive assembled in this chapter by "underscore[ing] the grids of intelligibility within which claims of both presence and absence have been asserted and questioned."[175] To queer the historical experience of eunuchs and our knowledge of them as such, this reading strategy pursues "an exploration of cultural texts as repositories of feelings and emotions, which are encoded not only in the content of the texts themselves but in the practices that surrounded their production and reception."[176] Queer theorist Ann Cvetkovich's notion of "an archive of feelings" is pertinent to our analysis because the traumatic experience of castration "serves as a point of entry into a vast archive of feelings, the many forms of love, rage, intimacy, grief, shame, and more that are part of the vibrancy of queer cultures."[177] Scrutinizing an archive of feelings surrounding castration highlights the unstable positionality of eunuchs in relation to Chinese history. This method of archival problematization brings us closer to, rather than blinding ourselves from, the core issues of proof, evidence, and argumentation that define the historian's task.

The sedimentary effect of the sources laid out in this chapter belongs to the global episteme of historical narration and is occasioned outside the pulses of men's embodied lives. Just like *tianzu* or *fangzu* are " 'gigantic' categories formulated from a vintage point outside the concerns and rhythms of the women's embodied lives," the dissonance between the public records of castration and the private experiences of eunuchs in the past becomes constitutive of an ongoing sentiment that considers Chinese castration backward, traditional, shameful, and oppressive.[178] As a truly historical specimen, the castrated male body has become completely out of sync with the Chinese body politic at large. When transgender sex workers commanded public fascination in postcolonial Singapore and the news of the "first" Chinese transsexual eventually broke in Cold War Taiwan, the glamour of these new queer subjects saturated the lingering culture of a "castrated civilization." The birth of a corpus style is predicated upon another's death. A fuller assessment of this transition must examine how the grounds for truth claims about the body had changed over time. The next chapter shifts our attention from the historicity of a corpus style to the epistemology of a scientific concept.

CHAPTER II

Vital Visions

Sexing Nature

The contested meanings of deficiency and sovereignty embroiled anxious late Qing reformers, many of whom were quick to embrace industrial technocracy at the very moment when it was inextricably entwined with the violence of foreign imperialism.[1] In this age of great political upheaval, the demise of eunuchism buckled an evolving gradation of civilizational measure, according to which China came to embody the metaphoric projection of a "castrated civilization" and eunuchs a "third sex." In the last chapter, I argue that the historical production of knowledge about castration is best understood by reading against the evidential grain. This is to avoid a teleological assumption that China had always remained an unchanging "castrated civilization" and that eunuchs had always possessed a transhistorical "third sex" identity. This flattened perspective, in other words, would fix the two historic metaphors onto a sufficient basis for the universalizing claims predicating the mutually generative histories of modern China and the body corporeal. It forces into the past the discursive contingencies upon which Westernization and modernization came to resemble one another in "most of the world."[2] In this chapter, I refocus the connection between the two metaphors by evaluating their transformations under the impact of Western biomedical knowledge.

By delving into a moment in Chinese history that scholars have variously characterized as "colonial modernity," I aim to tabulate the grounds of new knowledge on which China evolved from a "castrated civilization" into a modern nation.[3] The previous chapter foregrounds the transformation of "China" by making eunuchs' sexual identity an ancillary, rather than a primary, object of historical inquiry. This and the subsequent two chapters follow their collateral changes in a reverse manner: namely, by centering on conceptual issues surrounding the subject of sex over issues relating to China's changing political sovereignty. If eunuchs had become, rather than having always already been, a cultural relic in this political unfolding, the grounds of knowledge production must have also shifted accordingly so that new kinds of claims, especially scientific ones, could be made about the human body, China's "geobody," and their codeterminacy.[4] Central to the evolving epistemological grounds of truth about the body, I suggest, lies the emergence and transformation of the concept of sex.

This chapter explores the incremental inception of this "epistemic nexus": the culmination of new layers of visual evidence that made it possible for sex to become an object of observation. A nascent visual literacy developed in this period that leveraged the remapping of gender boundaries and the consolidation of sex as a biological and biomedical concept. As discussed in the introduction, between the 1860s and the 1890s the urban center of Chinese culture and society shifted from the heartland to the shore.[5] Along coastal China, missionary doctors dedicated themselves to translating Western-style medicine, including asylum practices and modern anatomical knowledge.[6] Their work stamped the first sustained effort in redefining Chinese understandings of sexual difference in terms of Western reproductive anatomy. Focusing specifically on the first Western-style anatomy text introduced to China, Benjamin Hobson's *A New Treatise on Anatomy* (全體新論, *Quanti xinlun*; 1851), I begin with the mid-nineteenth century as a crucial turning point for the modern visual representations of sex. By comparing Western-style with Chinese-style anatomical studies, I suggest that the visual realm occupied a central role in the reconceptualization of sex and provided a point of commensurate and universal reference for the modern definition of the body. The ways in which the medical visualization of sex had been reoriented, in other words, grounded the formation of a Chinese body politic on the verge of national modernity.

The gradual spread of the Western biomedical epistemology of sex from elite medical circles to vernacular popular culture reached a crescendo in the 1920s. In the years surrounding the New Culture movement (1915–1919), Chinese biologists learned from their Euro-American colleagues to promote a popular understanding of sex dimorphism. Their writings strengthened the visual evidence of anatomical drawings that first appeared in the work of late Qing missionaries. Refining the older drawings with allegedly more "accurate" translations and more diffused apparatuses of observation, they construed the bodily morphology and function of the two sexes as opposite, complementary, and fundamentally different. Republican-era life scientists also provided the first topographic drawings that divided all life forms into *ci* (雌, female) and *xiong* (雄, male) types. They bestowed upon the concept of *xing* (性, sex) an added stratum of visual persuasion by establishing congruity between what they called "primary," "secondary," and "tertiary" sexual characteristics. Like Western biologists, they extended these connections to all organisms across the human/nonhuman divide, explaining hermaphroditism with genetic theories of sex determination.

When the "life" of the Chinese nation rose to an unprecedented degree of urgency and uncertainty, scientists offered more intricate ways of depicting sex visually. By the 1940s three techniques of visualization operated conterminously in transforming sex into a scientific concept, the essence of life, and a fundamental object that can be seen and easily identified: what I call the anatomical aesthetic of medical representation, the morphological sensibility of the natural history tradition, and the subcellular gaze of experimental genetics.[7] I argue that these three techniques supplied the pivotal pillars for building a new form of visual proficiency that reorganized *gender* in naturalist terms, establishing the first and foremost conditions under which *sex* became an object of empirical knowledge.

Anatomical Aesthetic

In the early modern period, medical understandings of the body both shaped and reflected the status, role, and experience of women in society. Conceptualizations of sex in Europe took a decisive turn during the eighteenth century, when the "one-sex" model (which viewed women and men as two versions of a single-sexed body) was eventually taken over by the "two-sex" model (which treated men and women as incommensurable opposites).[8] Similarly, the androgynous "Yellow Emperor's body" gave way to the female gestational body that distinguished itself when *fuke* developed in

Song dynasty China.[9] Whether we consider the emergence of the two-sex model in Enlightenment Europe or the rise of gynecology in Song China as a paradigmatic turning point in the history of medicine, the significance of these shifts were not confined to the narrow realm of medical ideas and practices. Rather, these junctures anchored broader conceptual transformations in the relationship between the body proper and the body social.

In *fuke* medicine, Chinese physicians emphasized the importance of blood in the diagnosis of female-specific ailments. This was best captured by the omnipresent medical cliché, "in women, the blood is the ruling aspect." According to Charlotte Furth, between the late sixteenth century and the nineteenth century, a "positive model of female generativity" depicted female bodily experience not around symbols of purity and pollution but vitality and loss. Medical texts often associated blood (血, *xie*) with the female body and described *qi* (氣) primarily as a male essence. Construed in these dialogical terms with roots in the yin and yang cosmology, women's medical problems were perceived from late Ming onward as characteristic of bodily depletion and loss. From menstruation and pregnancy to gestation and childbirth, Qing doctors frequently described women as having a physically (and to some degree, emotionally) weak body due to their serious manifestations of blood depletion and stagnation. This shift from pollution to depletion significantly narrowed the scope of gynecological focus on female blood and its reproductive aspect, making it increasingly important to identify the root cause of female blood disorder. Women, unlike men, were often dubbed as the "sickly sex": Chinese doctors considered female bodies more prone to sickness in general, not only in the reproductive realm but also due other physical processes such as menstruation. Nonetheless, Furth contends that "female gender in the medical imagination implied sources of symbolic power," since it was represented by a range of images from that "of the 'prenatal' cosmic vitality of earth, to the constructive energy of the growing and reproducing body, to the dangerous efficacy of reproductive substances able to cure or kill."[10] By being the "sickly sex," women served their proper role in the web of Chinese social relations—as powerful agents of reproduction.[11]

Focusing on childbirth, historian Yi-Li Wu has deepened our understanding of the way late imperial Chinese doctors envisioned the female reproductive body. Although blood was overwhelmingly the central focus of discussion when doctors referred to the female body, Wu shows that "the womb" occupied an equally significant, if slightly different, role in

Chinese medicine. "Unlike blood, whose protean nature made it an obvious focus for investigation and therapeutic manipulation, the womb seems to have been largely taken for granted as a relatively stable object whose range of functions and pathological states were more narrowly defined. But fully understanding the intellectual architecture of *fuke* requires us to acknowledge what Chinese writers took for granted: that blood health and womb health were both essential for successful childbearing."[12]

Tracing the earliest medical discussion of the womb to the appearance of the term "*bao*" (胞) in *The Yellow Emperor's Classic of Internal Medicine*, Wu notes that doctors also debated on the womb's actual shape. Beginning in the seventeenth century, some doctors even believed that both men and women had wombs. This de-articulation of symptoms and diseases as being specifically female-linked supported the larger trend in the Qing to retreat to an "androgynous" understanding of the medical body.[13] According to Wu, "Even as elite medical doctrine subsumed the female womb into a rhetoric of bodily universality, the treatment of female diseases still assumed that women's bodies had special morphological features and functions. The dynamic functions of *qi* and Blood in women, in other words, were inevitably patterned by the physical layout of the female body, and the womb was a key node in the system of hydraulic flows that enabled female fertility."[14] Although Chinese physicians did not share the European obsession with anatomical dissection, late imperial doctors developed a sophisticated lexicon around the condition of the womb to underscore the importance of bodily morphology in their understanding of female reproductive health.

Whether blood or the womb had been the intellectual focus of *fuke*, Chinese doctors never attributed the cause of sexual difference to an isolated organ and delivered visual representations of distinct male and female bodies on that basis. The introduction of Western, dissection-based anatomies to China, in this regard, denotes an important turning point in the modern Chinese understandings of the body. As many have noted, Benjamin Hobson's *A New Treatise on Anatomy*, the earliest of these anatomies, was a landmark contribution to the systematic development of Western anatomical knowledge in the second half of the nineteenth century.[15] Hobson was a British surgeon who served as a medical missionary in China under the auspices of the London Missionary Society from 1839 to 1859. Hobson's text was important because it brought into existence terminologies that crossed Chinese and Western medicine (although many of them were later revised or dropped altogether); it synthesized and distilled a wide range of Western anatomical texts for the Chinese audience (since it was not a translation of one specific text); and it posited an universal corporeal referent (the emerging discourse of race notwithstanding) as the plausible and necessary ground for the cross-cultural translations of bodily meanings.[16] In explaining his authorial motives, Hobson commented on the notable absence of refined anatomical knowledge and practice among Chinese doctors: "The human anatomy of internal viscera, bones and muscles, and blood and pulsation is identical in China and the West. Yet, a sophisticated knowledge of it and the mastery of the application of that knowledge are present only in Western countries. There is no comparable phenomenon in China. Is it not a pity?"[17]

Above all, the transmission of Western-style anatomy to China, as exemplified by the publication of Hobson's text, produced a radical transformation in "the philosophical priorities and ways of seeing or imagining the body" from the principle of relative function to that of scientific observation so that "concepts of surface, depth, and scale took on a newly finite flavor."[18] This, as Ari Larissa Heinrich has shown, was accomplished through the introduction of a new mode of representation, or what he identifies as the "anatomical aesthetic," grounded in dissection-based realism.[19] Indeed, Hobson's *New Treatise* was among the first of a steady stream of illustrated texts in Chinese that in the following decades would find their way into the curricula of medical school classes, the academies affiliated with the arsenals established as part of the Qing Self-Strengthening movement, and even the hands of practitioners of traditional Chinese medicine.[20] By the 1910s and 1920s, leading writers like Lu Xun would incorporate the new dissection-based anatomical aesthetic into their own production of literary realism and visions of national modernity.[21]

In *A New Treatise*, Hobson used the term "outer kidney" (外腎, *waishen*) to refer to testicles and "yang essence" (陽精, *yangjing*) to translate "semen." He described the anatomy of the outer kidney as follows:

> Outer kidney, more popularly known as *luanzi* [卵子], is the producer of *jing* [精, "essence"] and the conduit of reproduction; its removal [or "castration," 閹之割之] will transform the vocal pitch and facial appearance of a man and eliminate his reproductive power entirely.
>
> The scrotum has two layers—inner and outer. There is a middle region, which separates the two *luanzi* into two halves—or, two sacs.

In each sac, there is a [double-sided] membrane region: one side of the membrane connects to the inner layer of the scrotum, and the other side operates as the protective layer of *luanzi*.

The membrane is often filled with water to maintain moisture. If there is too much water inside the membrane region, the scrotum will appear swollen and luminal. This disorder is called [scrotal] hernia [水疝, *shuishan*].

The physical appearance of *luanzi* looks like the flatter version of a bird's egg. Its length is about an inch, and its width is about eight *fen* [分, 1 *fen* = 0.33 cm; roughly 2.64 cm]. A testicle weighs about four to five *qian* [錢, 1 *qian* = 3.75 g; so roughly 15g].[22]

By locating the seat of masculinity in the male gonads this way, Hobson established the conceptual grounds for the physiology of their secretion:

Jing [精, "essence"] is produced from blood, and it appears in the form of a liquid. When one examines it under the microscope, one will discover that it contains many vital entities [活物, *huowu*] that look like tadpoles. These tadpole-like entities [sperm] have long tails and swim really fast, but their life lasts only a day. These are true for the *jing* of all kinds of beasts and animals, with the exception that the physical appearance of the vital entities varies.

For teenage boys before the age of twenty, blood does not produce semen. After twenty, blood enters the outer kidney. It moves from testicular arteries [微絲管, *weisiguan*] into seminiferous tubules [眾精管, *zhongjingguan*], where the sperm cells are produced. Sperm cells then move from the tubules to [the] epididymis [卵蒂, *luandi*], and from there to below the bladder through [the] vas deferens [精總管, *jingzongguan*]; they are stored inside the seminal vesicles [精囊, *jingnang*, actually at the epididymis].

After elucidating the nature of the male reproductive organ, including its production of the seminal fluid, Hobson unearthed its role in sexual intercourse:

During sexual intercourse, semen is released from the seminal vesicles [through the ejaculatory ducts]. Semen is difficult to harbor, and yet easy to lose [or dispense?]. Adolescents usually lack the maturation of blood and *qi* and various body parts. So if they allow themselves to indulge in sexual intercourse, the consequences of such behavior range from signs of physical weakness to the possibility of death.

As a practice of remaining lustless, *yangsheng* [養生, the cultivation of health] comes from reducing the level of desire. If one masturbates to accomplish this, it will be more detrimental to physical health, including the possibility of becoming blind or deaf.[23]

With such a detailed explanation of the male reproductive organ, Hobson provided multiple illustrations of its surrounding area in the body to give the reader both a cross-sectional perspective and a more complete impression (figure 2.1). He also included illustrations of *yin* (陰), his term for the womb, which contained other crucial parts of the female reproductive

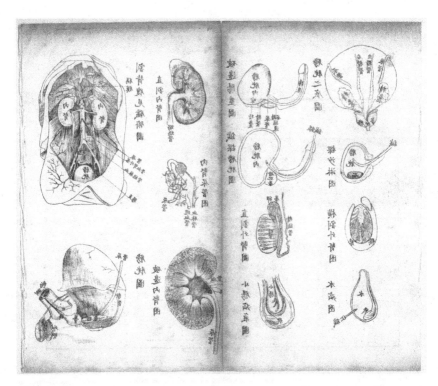

Figure 2.1 Hobson's illustrations of male reproductive anatomy (1851).
Source: Hobson 1851. Courtesy of the National Library of Australia, Bib ID 1869894.

organ, such as *zigong* (子宮, uterus) and *yindao* (陰道, vagina) (figure 2.2).[24] In his *Outlines of Western Medicine* (西醫略論, *Xiyi luelun*), Hobson guided the reader with visual demonstrations of the surgical treatment of scrotal hernia (figure 2.3).[25] Out of the range of terms that Hobson used to introduce human reproductive anatomy to the Chinese, his words for semen (*jing*), scrotum (*shennang*), uterus (*zigong*), and vagina (*yindao*), among others, remain today as the standard Chinese translations of the corresponding English terms. Most of these terms were already in use in China, but the main difference between Hobson's use and earlier Chinese use was that he imposed a one-to-one correspondence between the terms and their anatomical referents. This was accomplished semantically by restricting the meaning of terms that were previously less anatomically specific (e.g., *jing*) and selecting one term as the only proper term for identifying an anatomical part (e.g., *zigong* becomes *the* term for uterus).[26]

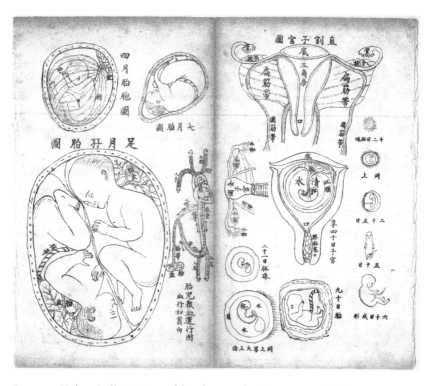

Figure 2.2 Hobson's illustrations of female reproductive anatomy (1851).
Source: Hobson 1851. Courtesy of the National Library of Australia, Bib ID 1869894.

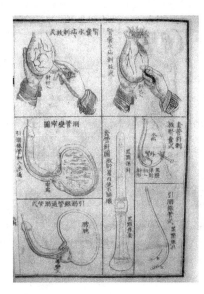

Figure 2.3 Hobson's illustrations of the surgical treatment of scrotal hernia (1857). *Source:* Hobson (1857) 1858. Courtesy of the East Asian Library and the Gest Collection, Princeton University.

These illustrations were the first of its kind in Chinese to visually depict the male and female reproductive organs in consonance with the anatomical aesthetic of Western medical representation. Unlike the discussions of gender-specific ailments in earlier gynecological texts, these images established for the reader a certain way of knowing about the body in which concrete anatomical terms could be *seen* in the physical-visual sphere. The profound influence these anatomical images had on their viewers went beyond the mere effects of representation. Even for doctors of professional standing, as medical historian Shigehisa Kuriyama has reminded us, "Dissection is never a straightforward uncovering of truths plain for all to see. It entails a special manner of seeing and requires an educated eye. The dissector must *learn* to discern order, through repeated practice, guided by teachers and texts. Without training and long experience, Galen insists, one sees nothing at all. . . . The anatomist aspires to see beyond the immediate, unpleasant material stuff of the body and behold the end (*telos*) for which each part is fashioned."[27] To grasp the distinctions between Western and Chinese-style anatomical representations, in other words, requires different ways of looking, and this is only possible with the carefully trained "educated eye."[28]

Indeed, Hobson's anatomical images train the eye to perceive the body in a way radically different from before. Whereas Chinese physicians were accustomed to imagining the organs of the body in relative terms within an elaborate system of conceptual correspondence (figures 2.4–2.5), Western anatomy introduced a new concrete sense of depth and closure to the dissected body (figure 2.6). In compiling a compendium like *New Treatise*, Hobson and other Western missionaries essentially developed a new mode of conceptual and visual engagement, one that relied not only on a critical distance between the viewer and the image of representation but also on an exact sense of the physical locations of what is being represented. Contrary to the anatomical representations found in earlier Chinese medical texts, which assumed no precision in the distance between the viewer and the visual object, Western-style anatomical images turned that distance into the very mechanism of its persuasion.[29] Here the introduction of a new mode of representation began to stake an epistemological claim that had been emphatically absent. The anatomical aesthetic of Western medicine, put differently, undermined the link between the represented and the real in Chinese medicine by claiming the physical distance between the viewer and the visual object as the ultimate source of its scientific authority.

To be sure, we are not merely facing a simple difference in the forms and conventions of representation between Western versus non-Western anatomy. It is worth pointing out that, unlike the anatomical illustrations

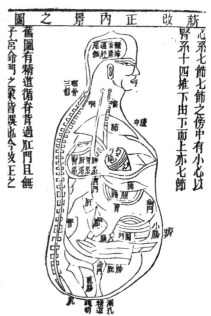

Figure 2.4 Diagram of the internal organs in *Huangdi Neijing* (side view). *Source:* Veith 2002, 41.

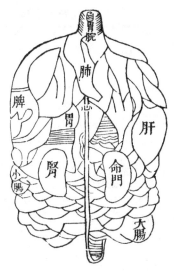

Figure 2.5 Diagram of the internal organs in *Huangdi Neijing* (front view).
Source: Veith 2002, 38.

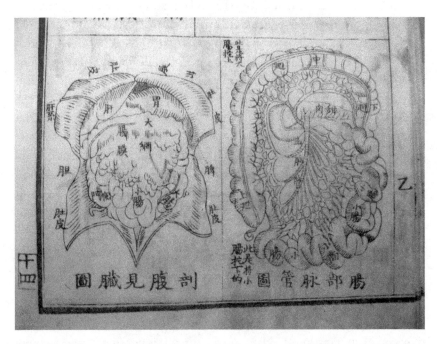

Figure 2.6 Hobson's diagram of organs visible from anatomical dissection (1851).
Source: Hobson (1851) 1857. Courtesy of the East Asian Library and the Gest Collection, Princeton University.

that appeared in Andrea Vesalius's *De humani corporis fabrica libri septem* (1543) or Bernhard Albinus' *Tabulae sceleti et musculorum corporis hominis* (1747), most of the anatomical drawings in Hobson's *New Treatise* embodied a more nuanced kind of natural realism. That is to say, the kind of anatomical illustrations underpinning the works of Vesalius and Albinus were "ideal"—absolute perfect but imagined—composites because some of their artistic fabrications were intended to imitate "the best patterns of nature." Hobson's anatomical depictions, in contrast, resembled the drawings found in William Hunter's *Anatomy of the Human Gravid Uterus* (1774): these were "corrected idealizations" that, although aiming to map the internal details of a perfect body, still reflected the effects of nature in slightly more pronounced ways, as demonstrated by the none-too-subtle violence wrought upon the cadaver.[30] Decisively absent in these later anatomical illustrations, for instance, include the standing posture of a skeleton or the unfragmented body in whole. These critical eliminations of certain artistic techniques render the body as a direct specimen of the clinical gaze *as it is seen* rather than with enriched and elaborate imaginations.

Moreover, the absence of Western-style anatomy in China before the nineteenth century does not mean Chinese doctors lacked faith in visual knowledge. A handful of surviving records prove that a small number of dissections were performed in the ancient and medieval periods.[31] It is also more useful to stress what Chinese physicians actually *saw* in a body instead of what they did not see: meridian tracts rather than nerves or muscles, the palpation of *mo* instead of the circulation of blood, and the color of the living rather than the cadaver of the dead. The first of each of these pairs of interest and focus involves a way of conceptualizing the somatic body different from Greek anatomy. Furthermore, the existence of male and female reproductive organs in texts such as *Ishimpo* (醫心方) since the tenth century and the *Manchu Anatomy* (康熙硃批臟腑圖, *Kangxi zhupi zangfutu*) since the eighteenth century, though with limited circulation imposed by the imperial household, implies the possibility that some gynecological experts in imperial China were familiar with these drawings before they read Hobson.[32]

Nonetheless, the biological basis of visible sexual difference became an object of sustained medical scrutiny only after Western reproductive anatomy had been introduced in late Qing China. It is not that Chinese physicians fell short in differentiating the development of maleness from that of femaleness. In fact, if we turn to the passage in *The Yellow Emperor's Classic of Internal Medicine* that attempts to achieve the closest to what we are trying to do here, we find evidence for a systematic way of explaining sexual differentiation.

Huang Di asked, "When people grow old then they cannot give birth to children. Is it because they have exhausted their strength in depravity or is it because of natural fate?"

Ch'I Po answered, "When a girl is seven years of age [*sui*], [her kidney *qi* (腎氣, *shenqi*)] become[s] abundant, she begins to change her teeth and her hair grows longer. When she reaches her fourteenth year she begins to menstruate and is able to become pregnant and the movement in the great thoroughfare pulse (太衝脉, *taichongmai*) is strong. Menstruation comes at regular times, thus the girl is able to give birth to a child.

"When a girl reaches the age of twenty-one years [her kidney *qi* is stabilized], the last tooth has come out, and she is fully grown. When the woman reaches the age of twenty-eight, her muscles and bones are strong, her hair has reached its full length and her body is flourishing and fertile.

"When the woman reaches the age of thirty five, the pulse indicating [the region of] the 'Sunlight' (陽明, *yangming*) deteriorates, her face begins to wrinkle and her hair begins to fall. When she reaches the age of forty-two, the pulse of the three [regions of] Yang deteriorates in the upper part (of the body), her entire face is wrinkled and her hair begins to turn white.

"When she reaches the age of forty-nine she can no longer become pregnant and the circulation of the great thoroughfare pulse is decreased. Her menstruation is exhausted, and the gates of menstruation are no longer open; her body deteriorates and she is no longer able to bear children.

"When a boy is eight years old [his Kidney *qi* is replete]; his hair grows longer and he begins to change his teeth. [At sixteen his Kidney *qi* is] abundant and he begins to secrete semen. He has an abundance of semen which he seeks to dispel; [he can begin unite yin and yang and so beget young].

"At the age of twenty-four [his Kidney *qi* is stabilized]; his muscles and bones are firm and strong, the last tooth has grown, and he has reached his full height. At thirty-two his muscles and bones are flourishing, his flesh is healthy and he is able-bodied and fertile.

"At the age of forty [his Kidney *qi* begins to wane]; he begins to lose his hair and his teeth begin to decay. At forty-eight [the yang energy of the head begins to deplete, the face becomes sallow, the hair grays, and the teeth deteriorate.] At fifty-six [his liver *qi*] deteriorates, his muscles can no longer function properly. [At sixty-four the *tian kui* dries up and the *jing* is drained, resulting in kidney exhaustion, fatigue, and weakness.]"[33]

According to Charlotte Furth's reading, this passage describes two parallel trajectories for the development of the same, homologous Yellow Emperor's body. In contrast to Thomas Laqueur's early European "one-sex model," what Furth calls the Yellow Emperor's body is "more truly androgynous" because it "has no morphological sex, but only gender."[34] Furth's interpretation is compelling especially in the context of her broader argument about the development of *fuke*. But from reading the above passage alone, one does not need the sophisticated language of gender theory to recognize pregnancy and menstruation as crucial analogs of female bodily process and semen/essence secretion the homologous signifiers of the male body. The mere presence of hair, teeth, bone, and other fleshy body parts does not constitute the concrete ground upon which gender difference can be inferred, although the timing of their development does. In the medicine of systematic correspondence, it seems, the body is truly androgynous insofar as our conception of sex adheres to a modern Western anatomical axiom.

We can at least conclude from the above quotation that the developments of male and female bodies share one thing in common: the kidney being the most important of the five viscera responsible for the regulation of vitality and growth. In fact, this "master system" is the only viscous that is mentioned in the passage in respect of "the generative powers of both sexes."[35] Likewise, the expository text of a *Neijing* (內經) illustration (figure 2.7) suggests that one of the kidneys is understood to govern the flow of the life forces that contribute to the bodily manifestations of sexual difference. Below is the full passage from *The Classic of Difficult Issues* (難經, *Nanjing*) from which part of that expository text has been excerpted: "The thirty-sixth difficult issue: Each of the depots is a single [entity], except for the kidneys which represent a twin [entity]. Why is that so? It is like this. The two kidneys are not both kidneys. The one on the left is the kidney; the one on the right is the gate of life [命門, *mingmen*]. The gate of life is the place where the spirit-essence [精, *jing*] lodges; it is the place to which the original influences are tied. Hence, in males it stores the essence; in females it holds the womb [繫胞, *xibao*]. Hence, one knows that there is only one kidney."[36]

According to his commentary, Hua Shou seems to fully endorse the passage: "There are two kidneys. The one on the left is the kidney; the one on the right is the gate of life. In males, the essence is stored here. The essence [transmitted] from the five depots and six palaces is received and stored here. In females, the womb is tied here. It receives the essence [from the males] and transforms it. The womb is the location where the embryo is conceived."[37]

Interestingly, this *Nanjing* passage appeared almost verbatim in Hobson's *A New Treatise* under the section "Inner Kidney" (內腎, *neishen*). Hobson opened the section with this passage to introduce the subtle distinction between the inner, "real" kidney and the outer kidney, a term he reserved for translating "testes."[38] Figure 2.5 presents a *Neijing* illustration of the relative locations of the five viscera in the body. Right below the center of the diagram, we see a "kidney" (*shen*) on the left and a "gate of life" (*mingmen*) on the right. If figure 2.7 leaves room for conflating the two, figure 2.5 allows for no ambiguity. Still, it is difficult to single out any hint of sexual

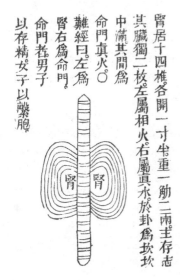

Figure 2.7 Diagram of the kidneys in *Huangdi Neijing*.
Source: Veith 2002, 26.

difference (or differentiation for that matter) simply by looking at either of these anatomical illustrations. This is because, insofar as they are intended to support medical claims about gender difference, these *Neijing* images operated within a structure of knowledge for which the *visual* depictions of male and female sexual anatomy fell outside its primary "epistemological function."[39] Even in the neo-Confucian elaborations of Zhang Jiebin (張介賓, 1563–1640), a late Ming proponent of the "supplementing through warm" intellectual tradition (溫補派, *wenbu pai*), the depiction of the "gate of life" derives from the morphology and position of a uterus that is not sex-specific because "men's essence and women's Blood are both stored here" (figure 2.7).[40]

In contrast, Hobson's images posit the relation of truth to sexual nature in terms of physical, visually discernable differences between male and female anatomy. So what his anatomical illustrations enabled was an epistemic shift in the conceptualization of sexual differentiation away from relative theoretical terms and toward concrete visual depiction. Images of Western reproductive anatomy trained Chinese people to connect what could be said about sex to what the eye could see of the physical body, eroding the vested privilege of organ development or functional relativity. With the dissection-based realism, it was now possible to infer anatomical meanings from a static body in its present state rather than having to bring together multiple layers of meaning behind the bodily flow and transformation of vital energy. No longer discussed in such vague, invisible, and even highly esoteric terms, sex emerged as a universal, visible entity from the production of Western medical images. These illustrations reoriented the burden of proof away from the system of theoretical correspondence and into the realm of anatomical appreciation and its attendant techniques of visual comprehension.[41] The availability of the more "scientifically objective" images of Western anatomy, which depended on the implicated precision in their distance from the viewer, transmuted into a more "objective" image of Western anatomical science itself. Whereas earlier bodily processes such as menstruation and depletion arbitrated Chinese understandings of sexual difference, the impossibility of mapping such conceptualizations onto the realm of visual depiction is precisely what the new anatomical illustrations sought to overcome. In carrying the weight of science, these images began to define the part of the body that is knowable. As instruments of a new form of visual literacy, they created—and not merely represented—truth.

Morphological Sensibility

Toward the end of the nineteenth century, China's rapid and unexpected defeat by Japan completely repositioned the two countries' international standing. The signing of the Treaty of Shimonoseki in 1895 represented a watershed event in the cultural imagination of China's power and weakness on both domestic and international fronts. According to Benjamin Elman and Ruth Rogaski, the turn of the twentieth century reversed the positioning of China and Japan in which one "acquired" scientific knowledge and conceptions of health and diseases from the other.[42] As the key to maintaining social order, classical learning and natural studies slowly gave way to Western scientific, medical, and technological expertise, which Chinese educated individuals began to master through Japan, as opposed to the longstanding convention in which the Japanese appropriated learned knowledge from the Chinese.[43] To be sure, after their interactions with the Jesuits in the "investigation of things and extension of knowledge" during the seventeenth and eighteenth centuries, Chinese literati were exposed to Western science by coming into contact with Protestants in the nineteenth century.[44] Yet, to bring Japan into the larger East Asian picture, after waves of Self-Strengthening efforts, many Chinese officials and reformists alike concluded from their global humiliation that Western science and technology held the distinct key to effective modernization. This soon became imbricated with the bourgeoning discourse of nationalism.

The survival of the Chinese nation emerged as one of the foremost preoccupations of government officials, local elites, and educated thinkers after the fall of the Qing. Although immensely shaped by the imported discourse of Social Darwinism and the adjacent discussion of "species," this preoccupation nonetheless raised a separate but fundamentally related question: the question of life itself.[45] At this point, however, the status of popular religion and natural science was so volatile that it is difficult to discern in retrospect whether one or the other was regarded as the ultimate authority for answering questions about life. A top-down state-driven campaign against religious medical practices would come into full force only during the Nanjing decade (1928–1937).[46] In the immediate post-Qing period, the relations between science and religion were perhaps not consistently antagonistic, but a variation of this certainly surfaced in the famous 1923 "science versus metaphysics" debate in the New Culture movement.[47]

The debate reflected a growing tendency among many Chinese intellectuals to approach Western science with a more refined sense of appreciation and commitment. The emerging urban-based, broadly educated class of entrepreneurs and managers, too, grabbed onto the language of "survival of the fittest" and applied the principles of free market competition to international relations.[48] It was within this broader political and cultural context that Western biology gained epistemological grounding over classical neo-Confucian cosmology for the empirical understanding of life.[49] Chinese thinkers' gravitation toward the natural sciences led to an exponential growth of translations of foreign biology texts and pictures, which included not only diagrams of human anatomy but also various depictions of the animal kingdom and the natural world. A new technique of visualization emerged from this wave of popular biology books: the morphological sensibility of the natural history tradition. This system of visualization assigned the scientific meanings of sex to all forms of life. By sexualizing the human/nonhuman connection, this normalizing technique expanded and recalibrated the kind of visual objectivity that was still evolving from the anatomical aesthetic of Western medicine.

In the 1920s popular life-science writers first and foremost categorized all higher-level organisms into two distinct types: *ci* (雌, female) and *xiong* (雄, male). Chai Fuyuan (柴福沅), author of the popular booklet *ABC of Sexology* (1928), identified "the two unique organs in higher-level species" as one being "*xiongxing* [雄性], which produces sperm" and "another *cixing* [雌性], which produces egg."[50] Similarly, in the succinct words of Feng Fei (馮飛), author of *Treatise on Womanhood* (1920), "*ci* organisms are organisms that generate egg; *xiong* organisms are those that generate sperm."[51] The *ci/xiong* distinction therefore portrayed higher-level life forms in a dualistic framework. According to Wang Jueming (汪厥明), translator of a Japanese textbook called *The Principle of Sex* (1926), "There is no *xing* distinction among lower-level unicellular organisms. . . . The reproductive cells of the more evolved species are called sperm and ovum, and they mark the difference between *ci* and *xiong* vital beings." Wang continued, "The morphological distinction between *ci* and *xiong* is present in all animals, but in varying degrees. We even have terms that reflect this notable difference. For example, the male chicken is called a 'rooster' and the female a 'hen'; the male deer is called a 'buck' and the female a 'doe'; the male cattle is called a 'bull' and the female a 'cow,' etc. . . . 'Sex-dimorphism' is a term that denotes this difference in biological morphology."[52] To sharpen his point, Wang synthesized the main points of Patrick Geddes and J. Arthur Thompson's book, *The Evolution of Sex* (1889), and, based on that information, delineated a list of differences between female (雌體, *citi*) and male (雄體, *xiongti*) organisms in binary opposites (table 2.1).[53]

In ways that had not previously held sway, the human/nonhuman divide came to anchor the entire Chinese biological discourse of *ci* and *xiong*. This divide defined what was so decisively different between these two terms and others such as *nü* (女)/*nan* (男) or *yin* (陰)/*yang* (陽), both of which appeared more regularly in popular discourses. Indeed, the epistemic function of *ci* and *xiong* amassed an unprecedented scope of cultural authority in China only after Western biology gained a robust metaphysical foothold. For instance, one finds in Charlotte Furth's and Judith Zeitlin's work that the notion of *yin* and *yang* prevailed in much of the literary, legal, and medical discussions of hermaphroditism in the late imperial period.[54] As *ci* and *xiong* became the most widely used pair of biological concepts for

TABLE 2.1

Sex Differences in Terms of Binary Opposites

雄體 (Male)	雌體 (Female)
精子生產著 (sperm producer)	卵子生產者 (egg/ovum producer)
生殖之消費較小 (lower "output" in reproduction)	生殖之消費較大 (higher "output" in reproduction)
新陳代謝激烈 (higher metabolism)	新陳代謝不激烈 (lower metabolism)
較為易化的 (affinity for difference)	比較為同化的 (affinity for similarity)
間有壽命較短者 (lower life expectancy)	間有壽命較長者 (higher life expectancy)
間有身體小者 (smaller body size)	間有身體較大者 (larger body size)
色彩多壯麗 (more colorful)	色彩多質素而不鮮明 (less colorful)
能力之激發者 (more able to stimulate)	較有耐忍力 (more able to withstand)
性急而為試驗的 (more impatient and experimental)	較為固執的保守的 (more stubborn and conservative)
變異性較大 (more mutable)	變異性小 (more stable)
求滿足性慾之意志甚強 (stronger sexual desire)	務求作家族 (more focused on the family)
較為好鬪 (more ambitious)	堅固家族 (more domestic)

Source: Reproduced with my own translation from Wang 1926, 118–19.

conveying sexual difference in the early twentieth century, they gradually replaced *yin* and *yang* as the definitive rubric for understanding the relationship between sex and life in the natural world. In fact, the congruency between these two pairs—*ci/xiong* and *yin/yang*—precisely relied on their similarity in denoting sex as a *form* of life.

Based on the biologizing discourse of sex, writers began to define men (男性, *nanxing*) and women (女性, *nüxing*) as human equivalents of *ci* and *xiong*. For Feng Fei, "humans represent the most complex biological organisms. *Xiong* and *ci* humans are called man (*nan*) and woman (*nü*) respectively, and they constitute the most telling example of morphological dimorphism. As such, man and woman are sheer manifestations of the material aspect of the biological world."[55] In an essay written in 1927, "The Evolution of *Xing*," Zhou Jianren (周建人, 1888–1984), the youngest brother of Lu Xun, similarly remarked that "in the evolution of sex, after the first step of making a distinction between an egg and a sperm, the second step is thereby to differentiate *ci* from *xiong* on the individual organismal [個體, *geti*] level—the individual organism that produces sperm is identified as *xiong* and the organism that produces eggs is identified as *ci*."[56] Connecting *ci/xiong* to *nü/nan* and stressing the mutual exclusivity of sex, Zhou defined humans as "animals that are either *ci* or *xiong* and not both [雌雄異體的動物, *cixiong yiti de dongwu*]: those who generate sperms are called *nan*; those who generate eggs are called *nü*."[57] According to these formulations, the biological basis of sex dimorphism defined *nü* as the human counterpart of *ci* and *nan* that of *xiong*. Neither the words "*nan*" and "*nü*" nor the concept of *shengzhi* (生殖, reproduction) were new to Chinese discourses.[58] But the novelty of *xing* in the twentieth century stems from its conceptual operation around which all three terms coalesced to mean sex.

In the 1920s and 1930s, biology books made available visual illustrations of the *ci* and *xiong* morphologies of animals and plants (figures 2.8–2.11). These images were not "typical" (Goethe's archetype), "ideal" (absolute perfect but imagined composite), or "corrected ideal" (Hobson's anatomical drawings) but were "characteristic" because they located the "typical" in an individual and made the organism depicted to stand for a whole kind of class.[59] By pairing the organisms, these drawings situated the qualitative difference between *ci* and *xiong* animals on the physical-morphological register. Many of these illustrations may have been direct appropriations from foreign scientific publications. Nevertheless, one notable feature distinguishes them from the pictures produced by the European naturalists

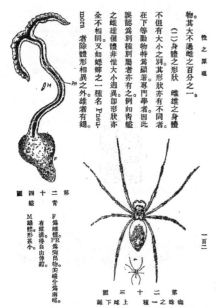

物。

（二）身體之形狀。雌雄之身體。不但有大小之別。其形狀亦有不同者。在下等動物特為顯著。初專門學者因此誤認為別種別屬者亦有之。例如青蠅之雌雄個體非惟大小迥異。即形狀亦全不相同。有一種名Pneumora者。除體形相異之外。一種雄者有翅

其大不過雌之百分之一。

第二十三圖　青蠅　F為雌蠅。PR為突出物。尖端分為兩端。有縱溝得自由伸縮。M雄體形甚小。

第二十三圖　蜘蛛之一種　上雌下雄

一百二

Figure 2.8 *Ci* and *xiong* morphology of Echuria (spoon worm) and spider (1926). *Source:* Wang 1926, 102.

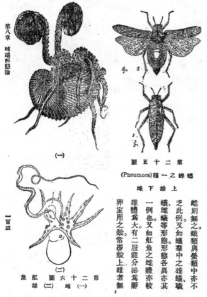

第八章　雌雄形態論

（一）

第二十五圖　蝗蟲之一種（Pneumora）　上雄下雌

（二）　紅魚圖第二十六圖　雌（一）雄（二）

雌則無之。蛾類與螢類中亦不乏此例。又如螽葦中之雄蟻職蟻。雄蟻蟻等形態各異其一例也。又如紅魚之雌體亦較雄體為大。有二肢能分泌為孵卵室用之殼。常覆殼上雄者無。

一百三

Figure 2.9 *Ci* and *xiong* morphology of argonaut and grasshopper (1926). *Source:* Wang 1926, 103.

第 三 十 三 圖 蠑螈之一種

雄 M 雌 F

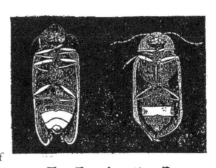

Figure 2.10 *Ci* and *xiong* morphology of
salamander and *yusha* fish (1926).
Source: Wang 1926, 110.

第 三 十 四 圖 玉 沙 魚

上 雄 下 雌

Figure 2.11 *Ci* and *xiong* morphology of
Lampyridae (firefly) (1926).
Source: Wang 1926, 111.

第 三 十 五 圖

(左)雄 與(右)雌 之 螢

in Qing China. Whereas the nineteenth-century naturalists did not systematically label their drawings with *ci* or *xiong*, these twentieth-century images invited its viewer to "compare and contrast" the *ci* and *xiong* versions of all animal types. This "compare and contrast" effect fundamentally depended on, and in turn crystallized, the new scientific concept of sex.[60]

The mapping of a new concept onto the visual representation of nature was an important step in bolstering the objective image of Western bioscience. This connection between the visual and the objective imported new rules for the production of truth about the natural world. Whereas the earlier anatomical drawings allowed people to see sexual difference in the physical human body, the new illustrations reinforced their embedded visual objectivity by broadening the conceptual appositeness of sexual difference. Sex, these new images declared, was an essential aspect of life, so it could be identified not only in humans but across the entire biological kingdom. The visual illustrations did not merely correspond to words or sentences that made up new claims of truth and falsehood, although that, too, was a definitive element of their validity. These pictures "became more than helpful tools; they were the words of nature itself."[61] Viewers of these images learned to accumulate "a form of cultural capital," to borrow communication scholar Scott Curtis's remark from a different context, as "scientific observation functions very much like the cultivation of distinctive aesthetic taste."[62]

The kind of morphological sensibility groomed in these images persisted well into the period after the War of Resistance (1937–1945). The best example comes from none other than the work of one of the leading authorities in reproductive biology in twentieth-century China, the embryologist Zhu Xi (朱洗, 1899–1962), best known for his study of the parthenogenesis of frogs. Throughout the late 1930s and the 1940s, Zhu authored and revised a total of six monographs in a book series called Modern Biology (現代生物學叢書, *Xiandai shengwuxue congshu*; published by Wenhua Shenhuo Chuban She [Cultural Life Publishing House]). The series introduced various subjects in Western biology to the Chinese lay public.[63]

According to historian Laurence Schneider, the history of Chinese genetics and evolutionary biology can be seen as an example of "how modern science was transferred to China, how it was established there and *diffused throughout culture* and institutions."[64] Indeed, numerous Republican-era Chinese magazine articles, periodicals, books, and pamphlets published under the banner of "biology" were not always written by individuals belonging to formal establishments of natural scientific research, such as at Peking, Qinghua, Yanjing, or Nanjing Universities. Zhou Jianren, for instance, was one of the most reputable popular life-science journalists at the time. Unlike his two elder brothers, Zhou remained loyal to his interest in science (rather than literature for example) and earned his bachelor's degree from the Agricultural School of Tokyo Imperial University. He frequently published articles and opinion pieces in popular periodicals such as the *Eastern Miscellany* (東方雜誌, *Dongfang zazhi*) and the *Ladies Journal* (婦女雜誌, *Funü zazhi*). His writings that defended Lamarckism during the Republican period and Lysenkoism after the rise of the Chinese Communist Party attracted a much wider readership than the technical writings of professional geneticists.[65] In tackling the riddle of life, Zhou belonged to a global community of science translators who challenged the "universal language" of science outside the modern West.[66]

In this respect, Zhu Xi and other notable Chinese geneticists such as Tan Jiazhen (談傢楨, 1909–2008) and Chen Zhen (陳楨, 1894–1957), a student of E. B. Wilson, represented a group of more formally trained but esoteric biological scientists.[67] Born in Linhai, Zhejiang Province, Zhu went to France with several friends in May 1920 as participants in the anarchist Li Shizeng's (李石曾, 1881–1973) "work-study programme." According to his autobiographical account, they received no assistance from the Sino-French Education Association upon arriving in France, so they had to live in tents on the lawn in front of the Chinese Federation. Eventually, they were allowed to sleep on the floor of the building on a temporary basis. During the first five to six years of his life in France, Zhu's experience was quite typical of Chinese young adults who decided to join the work-study program and travel overseas: frequent job changes, difficult physical labor, poor living conditions, unending negative encounters with Westerners, and an increasingly entrenched sense of disappointment and despair. Nevertheless, Zhu eventually attended the University of Montpellier and studied embryology under J. E. Bataillon from 1925 to 1932. After earning his doctorate in biology from Montpellier, Zhu returned to China and began his academic career as a professional biologist. From 1932 on, he was associated with the National Zhongshan University, the Beijing Academy of Sciences, and various private research organizations. Zhu became a member of the Experimental Biology Institute of the Chinese Academy of Sciences in 1950 and was appointed as its director three years later.[68]

Before joining the institute, Zhu authored and expanded his six monographs for the Modern Biology book series, the aim of which was to introduce a wide spectrum of Western biological ideas to the Chinese nonexpert public. In his general preface to this six-volume project, Zhu declared:

My intention in publishing this series of monographs is to offer my knowledge in biology to the lay reader in a systematic way, hoping that it will encourage a better understanding of human life. The topics of our investigation include the origins of human beings, the evolution of their ancestors, and the development of human thinking, behavior, and moral consciousness. Simply put, we need to analyze ourselves, study ourselves, and understand ourselves; after cultivating this sort of understanding we need to improve ourselves, allowing humans to be part of science and to march forward in a more reliable destiny.[69]

This statement shows the conviction Zhu shared with many other modernizing thinkers (not only scientists) in the Republican period that it was important to acquire a general knowledge about life through a scientific way of thinking rooted in Western biological, especially evolutionary, ideas.[70] He therefore opened his series with a volume called *Humans from Eggs and Eggs from Humans* (蛋生人與人生蛋, *Danshengren yu renshengdan*), which described various aspects of the developmental phases of life, including detailed accounts of male and female reproductive anatomy (as well as an interesting chapter on teratology).[71] In this first volume, Zhu distinguished humans from animals, plants, and other living species in ways that would become even more vivid throughout his subsequent writings—by holding up sex as an integral dimension of life.

In *Changes in Biological Femaleness and Maleness* (雌雄之變, *Cixiong zhibian*; 1945), the fourth volume of the series, Zhu began his scientific investigation of sex with an opening chapter called "The Conceptualization of *Ci* and *Xiong*" (雌雄的概觀, *Cixiong de gaiguan*).[72] He argued that the distinctions between *ci* and *xiong* (or *nü* and *nan* when he referred to humans) bespoke the most important calibers of differentiation in living species—animals, plants, and humans.[73] Similar to the visual illustrations of *ci* and *xiong* species circulating in other popular biology books, his hand-drawn images of different organisms served one simple purpose: to enable the viewer to grasp from a critical distance the nature of sexual difference across a spectrum of life forms (figure 2.12). Above all, these images must be understood as the product of visualization rather than mere representation.[74] They teach the eye to recognize specific patterns and assist the reader to discriminate *ci* from *xiong* through the knowledge of seeing.

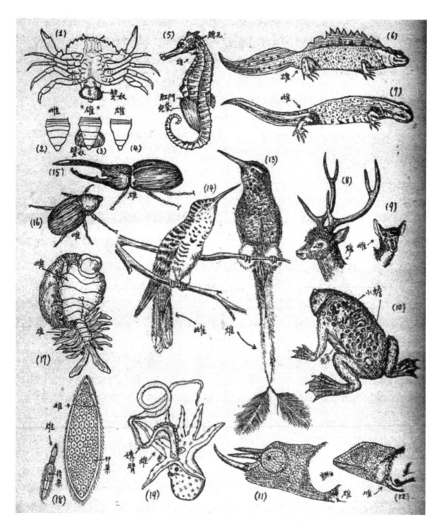

Figure 2.12 "The Morphological Differences Between *Ci* and *Xiong* Animals" (1945).
Source: Zhu (1945) 1948, 33.

Whereas pictures 6, 8, 11, 13, and 15 all refer to the *xiong* versions of a particular species, pictures 7, 9, 12, 14, and 16 indicate their *ci* counterparts. The very marking of such characters as "*ci*" and "*xiong*" on the diagrams indexes these words' epistemo-*logicality*, according to which their semantic logic, usage, coherence, and possibility became visually equated with the dualistic physical structures of life.

Similarly, although the visual illustrations in other biology books (figures 2.8–2.11) were presented as if they truthfully described reality, they were in fact establishing new boundaries within which claims of truth and falsehood about sex could be made. As ethnomethodologist Michael Lynch has noted, visual representation in science is really about "the production of *scientific* reality."[75] Since these images did not simply represent nature, they had deep implications for the negotiation of truth claims. In the case of *xing*, they claimed an autonomy of sorts by showing that morphology nested the seat and nature of sex. Figures 2.8–2.12 were not just passive aids for learning, but they prescribed for the reader the conceptual boundaries of life and the forms in which it took shape, such as through the binary manifestations of *ci* and *xiong*. In other words, the "scientific" reality of sex, or sexual difference, depended on the way the morphological sensibility of these images reinforced the anatomical aesthetic of the earlier medical representations. Although the word "*xing*" did not yet mean "sex" before the twentieth century, the visual mappings of biological form made it possible for the earlier anatomical drawings and the later naturalist illustrations to be merged into the concept of sex by the 1920s. By claiming a scientific status for the images they produced, both techniques of visualization ultimately secured an objective image for the sciences themselves.[76]

In the new discursive context of the *ci/xiong* distinction, natural science writers incorporated the anatomical aesthetic of medical representation. They often included in their books detailed descriptions and drawings of human reproductive anatomy. Throughout the 1920s and 1930s, examples could be found in Chai Fuyuan's *Sexology ABC*, Zhu Jianxia's (朱劍霞) edited translation of *The Physiology and Psychology of Sex* (1928), Li Baoliang's (李寶梁) *Sexual Knowledge* (1937), and Chen Yucang's (陳雨蒼) *Research on the Human Body* (1937), among other professional and periodical publications (figures 2.13–2.16). The circulation of these images continued a long tradition of the cross-cultural translation and dissemination of Western anatomical knowledge, an endeavor dating back to as early as the seventeenth century. As discussed in the last section, in the nineteenth century, medical missionaries including Benjamin Hobson (1816–1873), John G. Kerr (1824–1891), John Dudgeon (1837–1901), and John Fryer (1839–1928) extended and revised this intellectual trajectory.[77] Interest in anatomical knowledge came of age in the early Republican period in part as a result of the passing of the Anatomy Law in November 1913, and by

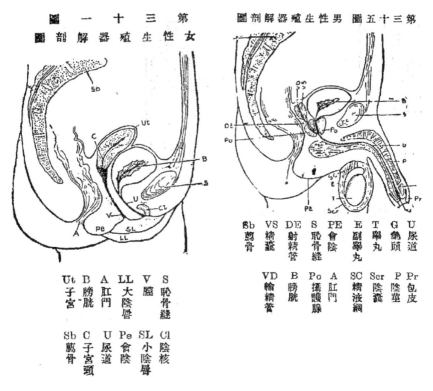

女性生殖器解剖圖　第三十一圖　　第三十五圖　男性生殖器解剖圖

Ut 子宮	B 膀胱	A 肛門	LL 大陰唇	V 膣	S 恥骨縫
Sb 薦骨	C 子宮頸	U 尿道	Pe 會陰	SL 小陰唇	Cl 陰核

Sb 薦骨	VS 精蕋	DE 射精管	S 恥骨縫	PE 會陰	E 副睪丸	T 睪丸	G 龜頭	U 尿道
VD 輸精管	B 膀胱	Po 攝護腺	A 肛門	SC 精液綢	Scr 陰囊	P 陰莖	Pr 包皮	

Figures 2.13a and 2.13b Zhu's anatomical diagrams of male and female reproductive organs (1928).
Source: Zhu 1928, 42, 48.

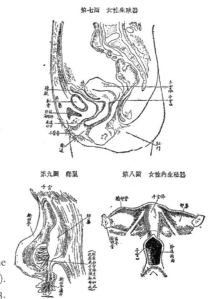

Figure 2.14 Li's anatomical diagrams of the female reproductive system (1937).
Source: Li Baoliang 1937, 33.

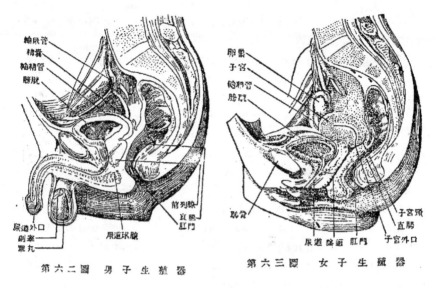

Figures 2.15a and 2.15b Chen's anatomical diagrams of male and female reproductive organs (1937).
Source: Chen 1937, 166, 171.

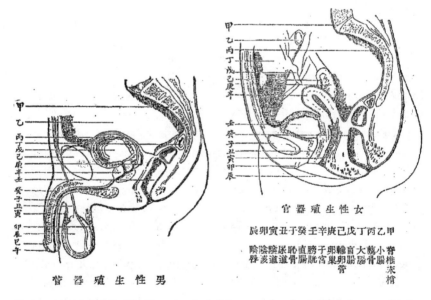

Figures 2.16a and 2.16b Chai's anatomical diagrams of male and female reproductive organs (1928).
Source: Chai (1928) 1932, 19, 37.

1920 regular, though limited, dissections were performed in treaty-port medical schools.[78]

The new illustrations of the Republican period not only updated many of the previous "errors" and "mistranslations" but also consolidated a systematic language of maleness and femaleness in the universal terms of bioscience. For example, one finds great resemblance between the diagram of the female reproductive system on the lower right hand side of figure 2.14 and the one next to the center of figure 2.2 from Hobson's treatise.[79] Whereas late Qing missionary anatomical drawings were circulated mainly among the medical elites, especially those who were less resistant to Western biomedicine, Republican-era anatomical illustrations were printed in popular publications, found a ready audience, and thereby reached a critical mass.[80] Above all, the popularization of the anatomical aesthetic demonstrates that "what is accepted as true knowledge ultimately depends not exclusively on truth claims negotiated among experts but required public mediation."[81] The sheer quantity and accessibility of this new cohort of illustrations not only reinforced the visual authority of late Qing anatomical drawings, but they also demonstrate that, at least by the 1920s, the epistemic jurisdiction of the Western biosciences was advocated even by native writers.

From Gender to Sex

In the novel language of bioscience, physical structure and morphology reflected, rather than predetermined, human gender difference. More often, popular writers claimed that the secret of masculinity and femininity resided in the gametes and gonads, which formed a crucial part of what biologists called "primary sexual characteristics."[82] In this spirit, Gao Xian (高銛), the translator of another Japanese biology textbook, *Sex and Reproduction* (1935), posited a broader definition of *ci* and *xiong* that incorporated the role of anatomical parts: "Organisms in the animal kingdom that either generate spermatozoa or have a testis that produces it are called *xiong* (male); those that generate ovum or have a functional ovarium that can produce it are called *ci* (female)."[83] For Gao, the essence of *ci*-ness and *xiong*-ness entirely depended on the "actual presence of a testis or an ovary," which constituted "*zhuyao tezheng* [主要特徵] (principal sexual character)," and those bodily features that "bear some immediate relevance to the principal sexual character are called *fudai tezheng* [附帶特徵] (accessory sexual character). Together, they are called *diyici xingtezhen* [第一次性特徵] (primary sexual character)."[84] The anatomical drawings (figures 2.13–2.16) therefore provided unequivocal visual evidence for the natural existence of primary sexual characteristics. These images educated the eye to see what normally could not be captured beyond the exterior features of the human body. In other words, the anatomical aesthetic of these medical representations allowed the viewer's gaze to penetrate the external integument of the body, a fundamental attribute of the morphological sensibility of the natural history tradition. Here we begin to see how the two techniques of visualization worked in a mutually productive fashion.

If the seat of masculinity and femininity could be inspected in these anatomical drawings, morphological appreciation was still important for distinguishing these features from what biologists called "secondary sexual characteristics." Chinese scientists often credited the British surgeon and anatomist John Hunter (1728–1793) as the originator of this idea. According to Gao's definition, "coined by John Hunter, *dierci xingtezhen* [第二次性特徵] (secondary sexual character) typically refers to variations in body size, morphology, color of physical appearance, sound production, odor and its intensity, illumination and its intensity, parts of the body that illuminate, etc. among normal animals."[85] Wang Jueming, in translating *The Principle of Sex*, summed up the definition of "secondary sexual characters" rather cogently: "a concept invented by Hunter and adopted by Darwin" that referred to "sexual characters that bear no direct relationship to biological reproduction."[86] Labeled "secondary," physical features such as the antlers of the male deer helped to build a perception of the difference between *ci* and *xiong* animals as an immutable distinction of nature (illustrations 8 and 9 in figure 2.12). The morphological sensibility found in figures 2.8–2.12 precipitated a form of visual assertion precisely by contrasting the physical appearance of male and female species. Like the anatomical aesthetic of medical representation, it put the viewer in a position where it was still possible to determine the sex of the object represented on the page; the crucial difference, though, was that this process of "sex determination" was made possible not by looking beneath the layers of the skin but by looking precisely at those physical features that are externally visible.

Compressed into the congruence between *ci/xiong* and *nü/nan*, the visuality of sex extended "secondary sexual characteristics" to human forms. In *Sexual Knowledge* (1937), Li Baoliang asserted that "the best examples of secondary sexual characteristics in humans include women's smaller physique, paler pigmentation, softer skin, richer body fat, and less well-defined muscles in comparison to men." Other features of human "secondary sexual characteristics," according to Li, included men's hairier bodies, lower-pitched voice, and narrower pelvis (figure 2.17).[87] Li reasoned that women had wider pelvises due to their procreative functions, not unlike how their biological capability to breastfeed led them to easily develop larger breasts. "Therefore," wrote Li, "many scientists have observed wider female pelvises among people of races on the higher end of the evolutionary scale. This is because the brain size of the babies of better races tends to be larger, and larger pelvises would allow a better fetus to develop inside a woman's womb."[88] The extension of secondary sexual features to the human body developed a visual framing of sex dimorphism in *all* living beings as a fundamental product of nature.

Ultimately, though, it was the naturalization of the very connection between primary and secondary sexual characteristics that cemented the visual culprit of *xing*. By naming all those sex features not directly involved in reproduction "secondary," observers described human gender difference as the natural outgrowth of "primary" characteristics, characteristics that also determined sexual difference in animals. This biologizing discourse posited the fundamentals of maleness and femaleness beneath the surface of observable bodily features. It was with this in mind that Shen Chichun (沈霽春), author of *The Life of Sex* (1935), argued that although "women have large breasts; men have beards," these "do not constitute quintessential gender difference [根本上的夫婦之別, *genbenshang de fufu zhibie*]," because "quintessential gender difference refers to anatomical difference [解剖上的區別, *jieposhang de qubie*]." Shen explained that "if an ovum

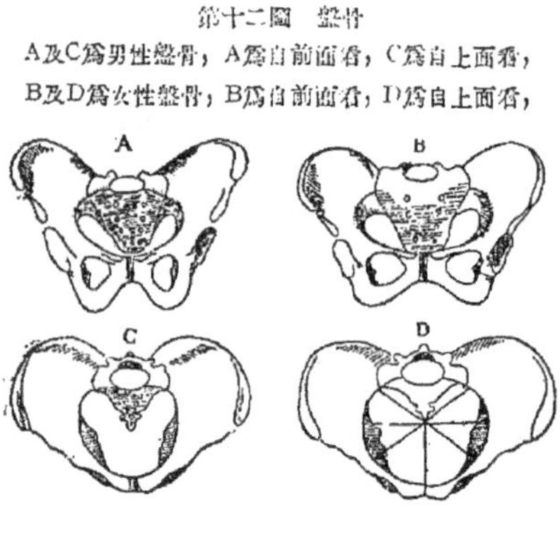

第十二圖　盤骨
A及C為男性盤骨，A為自前面看，C為自上面看，
B及D為女性盤骨，B為自前面看，D為自上面看，

Figure 2.17 "The Pelvises" of men and women (1937).
Source: Li Baoliang 1937, 41.

[卵粒, *luanli*], a yolk [卵黃, *luanhuang*], and an oviduct [輸卵管, *shuluang-guan*] are found inside an animal after dissection, then the organism is *ci*. If testes and vas deferens are found, the animal who has sacrificed his life [for the dissection] should be *xiong*."[89] As Yi-Li Wu has pointed out, late imperial Chinese physicians regularly considered breasts and beards as the primordial physical markers of gender difference.[90] Shen's assertion was thus first and foremost a riposte to this longstanding Chinese belief. In his implicit gluing of human gender difference to the *ci*/*xiong* distinction, Shen instructed his reader to consider a layer of truth, in relation to nature, beneath the visible horizon. These anatomical and morphological drawings enabled Chinese readers to comprehend sex beyond the physical markers of breast and beard and to locate the biological roots of manhood and womanhood in gonadal biology.

Apart from primary and secondary sexual characteristics, cultural commentators often spoke of "tertiary" ones, too. They univocally attributed this concept to the British sexologist Havelock Ellis (1859–1939).[91] Pursuant to Wang Jueming's translation of *The Principle of Sex*, the notion of "tertiary sexual characters" was "invented by Ellis to highlight unique features of male and female bodies. Tertiary sexual characters are not as obvious as secondary ones, but examples abound."[92] The differences in skull size, body height, level of physical activity, blood cell count, and cerebral regions in the brain between men and women were some of the examples he enumerated. Sexual difference in these somatic traits, according to Wang, may come across as less significant to a zoologist than, say, a sociologist or an anthropologist. As such, "even if they cannot all be grouped under secondary sexual characters, it is still useful to include them under the broad category of tertiary sexual characters. Although this concept has been endorsed variously by Papillault, Haeckel, P. Weber, and Kurella, each scientist outlines a different set of criteria for associating it with certain sex-specific features."[93] Tertiary sexual characteristics thus welded an important scientific vocabulary (and the cognate set of visual proof) to naturalize those sexual/gender differences that bear no immediately obvious relationship to chromosomal or gonadal sex.

Simply put, what the cultural discourse of bioscience mediated in the early Republican period was the transformation of previous bodily "gender" into the modern notion of "sex." Some historians have used the blanket term "scientism" to explain the optimism that many Chinese intellectuals expressed toward Western scientific principles and practices in the early twentieth century, but they rarely, if ever, specify the underlying mechanisms of knowledge production by which the cultural authority of that optimism came about.[94] More recently, historian Sean Hsiang-lin Lei has provided an illuminating account of how Chinese medicine became increasingly legitimated through the application of modern scientific methods—a process he termed "scientization" (科學化, kexuehua)—in the 1930s and 1940s. The effort of the Nationalist state in forcing practitioners of Chinese medicine to cope collectively with the concept of science for the first time resulted in several innovations, most notable of which involved the incorporation of the germ theories of disease into Chinese medicine and the laboratory and clinical research on traditional drugs.[95] Gaining credence in the same historical context, what I have been calling "techniques of visualization" shift to the visual authority of scientific claims. Cumulatively, these techniques explain the epistemological procedures that facilitated the envisioning of scientific optimism—primarily through the power of image. As we have seen, in their effort to challenge neo-Confucian prescriptive claims about gender hierarchy, urban elites drew on natural scientific knowledge to recast gender distinction in terms of biologically determined structures. More specifically, they relied on the anatomical aesthetic of medical representation and the morphology sensibility of natural history to establish an intrinsic nature of sex that could be identified visually and universally.

Relying on the concrete physical structures of sex, these scientific elites also reconceptualized functional processes of the body. Earlier cultural vestiges of femaleness, such as menstruation, were now reframed from the viewpoint of modern physiology. As discussed earlier, the increasing association of women with blood depletion reflected the rise of a "positive model of female generativity" in late imperial Chinese medicine. From the seventeenth century on, this model construed female health around symbols of vitality and loss. Chinese physicians considered women to be the "sickly sex" that had a physically (and to some degree emotionally) weak body more prone to sickness due to their constant association with blood loss, such as through childbirth and menstruation.[96] In the 1920s and 1930s blood discharge continued to be perceived as *the* emblematic biological symptom of femaleness. In 1935 Su Yizhen (蘇儀貞) opened her *Hygiene Manual for Women* with the statement that "menstruation is the most unique physiological difference between men and women."[97] In their book on *Women's Hygiene* (1930), Guo Renyi (郭人驥) and Li Renling (酈人麟) also structured women's life cycle around definitive turning points of menstruation: it is decisively absent before the onset of puberty; its first occurrence marks the girl's entry into young adulthood; and its permanent cessation marks the beginning of menopause, the final stage of the female life cycle.[98] Hence, bioscience universalized femininity by recoding traditional physical markers of blood and menstruation in modern anatomic-physiological terms. As a result, womanhood came to establish itself as the epistemic equivalent of manhood. The introduction of Western biology turned earlier gender signifiers into "natural" sex differences.

The transformation of women's gender into female sex corroborates feminist theorist Tani Barlow's assertion that, strictly speaking, "woman" (女性, *nüxing*) did not exist in China as a universal category before the twentieth century. The closest term available was *funü* (婦女), which referred to various female subject positions within the discursive network of family, marriage, and kinship. Women were virtuous wives, mothers, daughters, and so on, but they were never identified as a distinct group of individuals outside familial relations. It is interesting to note, for instance, that Chinese gynecology was called *fuke* (婦科) and not *nüke* (女科), implying that female bodies of generation and reproduction remained its primary clinical target. Therefore, one of the most reputable legacies of May Fourth feminism was the creation of a generic category of womanhood filtered from its earlier grounding in kin relationality. As Barlow explains it, "Feminist texts accorded a foundational status to physiology and, in the name of nineteenth-century Victorian gender theory, they grounded sexual identity in sexual physiology. Probably the most alarming of all of progressive Chinese feminism's arguments substituted sexual desire and sexual selection for reproductive service to the jia [family] and made them the foundation of human identity."[99] The shift from the clinical target of *fuke* to a collective and universal category of sex thus reflected "the passage of women from objects of another's discourse to women as subjects of their own."[100] An interesting parallel can be found in the context of late colonial India, where the rhetorical reinvention of the woman question, quoting historian Mrinalini Sinha, "offered a new construction of women not as simply saturated by their identification with kin, community, and the nation but, pointedly, as a universal and homogenous gender category."[101]

As modernizing elites began to explain gender roles and relations within a Western biomedical lexicon, the images and language of anatomy buttressed a popular vision of sex dimorphism. This turned *xing* into a dichotomous concept of humanity that manifested itself most tellingly in the physical (sexual) differences between men and women. Consider, for example, the prominent feminist writer Zhang Xichen's (章錫琛, 1889–1969) remark in 1924: "In the past ten years, there is something most powerful that is developing most rapidly—that is, a shapeless reform in consciousness. This reform is what is called women's awakening as 'human beings.' . . . Women who had some contact with new thought all have the consciousness that 'a woman is a human being, too.' The books which have been regarded women's bibles, such as *Nüjie*, *Neixun*, *Nülun*, and *Nüfan*, have all been trampled under the feet of new women."[102]

Zhang's words signaled the formation of an autonomous female subjectivity from the shadow of *funü* in the new intellectual climate of May Fourth feminism, as womanhood came to be understood no longer in strict congruence with family and kin relations but as the biological representation of half of the human population whose social status ought to be equal to men. This egalitarian view was clearly expressed in 1904 by another pioneer in the women's movement, Chen Xiefen (陳擷芬, 1883–1923), the daughter of the editor of the radical Shanghai journal *Subao*: "The inhabitants of China number about four hundred million all together. Men and women each constitute half of this."[103] To borrow Barlow's insight again, Chinese women "became nüxing only when they became the other of Man in the colonial modernist Victorian binary. Woman was foundational only insofar as she constituted a negation of man, his other."[104]

Since the late nineteenth century, Liang Qichao had emphasized the prospect of independent women wage earners contributing to the nation's economy. But in the years surrounding the New Culture movement, the new discourse of *nüxing* (meaning biologically sexed woman) proliferated and mainly drew on the ideas of the Western life sciences—an epistemic move away from metaphysics—that would not only define women and female subjectivity in terms of their biology and sexuality but also completely overturn the authority and prestige of neo-Confucian learning.[105] According to Leon Rocha, "before it was possible to have a discourse of woman based on her sexual, biological, actual differences (that is, *nüxing*), sex had to first become human nature through the creation of the neologism *xing*."[106] In this context, sex/gender, like ethnicity, class, and age, became an important marker of the self.[107]

The political, social, and cultural factors that motivated Chinese intellectuals, journalists, social reformers, university professors, doctors, and other cultural elites to replace Confucian philosophy with human biology are undoubtedly significant. Their efforts, for instance, cannot be understood as independent of the 1898 reform movement, which had already challenged the imperial institutions and orthodox ideologies in significant ways; the abolishment of the civil service examination system in 1905, which gave women greater access to education; the fall of the imperial polity in 1911; the anti-footbinding and feminist movements; the rise of the printing press; the birth of vernacular Chinese literature; the establishment of modern universities; the consolidation of an intellectual class; and, of course, the resulting famous science versus metaphysics debate in 1923, just to mention a few poignant examples. But "scientism" has often been introduced as a catchall term for rationalizing these interlocutors' interest in and commitment to the universal value of Western science. On the contrary, my analysis suggests that the universal value that guided these Chinese thinkers was mutually generative of their modernizing optic: the visual objectivity of sex emerged from and critically anchored the earnest production of images of human anatomy, *ci* and *xiong* animals, and, as we will see, genes and chromosomes. By making it possible for people to relate what they called *xing*/sex to a global vision in concrete terms, biomedical science simultaneously affirmed its status as the ultimate arbiter of truth about life and nature in ways that fell outside the tenor of Confucian philosophy or classical Chinese medicine.[108] Rather than taking for granted the rhetorical authority of anatomical sex in May Fourth feminist discourse, what I call techniques of visualization helps to explain *how* and *why* Western biological notions of sex came to constitute the new epistemological ground for authorizing claims about gender and the body. The visuality of *xing* further animated new understandings about the subjectivity and mutability of sex, to which we return in the subsequent chapters.

Man and Machine

A parallel historical transformation can be identified in the scientific reconceptualization of manhood. If blood and menstruation were reframed as the most visible cultural indicators of femaleness, the chief "natural" markers of maleness remained sperm and spermatorrhea. In 1925 the Commercial Press published a book, *The Sexual Hygiene and Morals of Adolescents*, with this inaugurating sentence: "Being the most essential ingredient of health, the internal secretion of the testicles is also known as the 'inner energy.'"[109] On the next page the authors reinforced the importance of semen conservation. They cited an example of a male student who could not perform his duties responsibly after having had a "lewd" dream the previous night, which led to "the loss of his essential internal secretions." "Based on this example," the authors concluded, "a young man's physical health is closely related to the natural product of his body."[110]

The new vocabulary of bioscience allowed the authors to explain the health implications of semen physiology in a way that would sound nearly incomprehensible to premodern ears: "Teenagers' secretion will be absorbed by blood, sent to the heart and through the arteries to the muscle fibers; through such a journey, muscles grow and strengthen. When the secreted substance is sent to the brain, it enables the brain to have thoughts, hopes, and expectations and gives the mind evidences of rationality, critical judgment, deep ambitions, strong determination, and rich volition."[111] This interpretation would not have made sense in an earlier period because its logic of reasoning relied on a style of visual imagination—for example, the anatomical representation of muscles and muscularity—that did not exist before the nineteenth century.[112] The authors stressed the importance of attaining accurate knowledge about sperm: "Research on the physiological function of sperm is the most important thing, because sperm is the most essential thing in life—it's the thing that makes someone a father—and the nature of its size makes it almost invisible unless with the help of the microscope."[113] These statements forged a neat coherence between sperm (a Western anatomical concept) and traditional notions of male essence.

Consider another example, the discussion of spermatorrhea by the self-proclaimed sexologist Chai Fuyuan in the late 1920s. According to Chai's definition, "spermatorrhea refers to the discharge of semen while the mind is in an unconscious state. It is a bodily condition unique to men, who often begin to experience it in adolescence." Chai explained that teenage boys typically experienced two types of spermatorrhea: with or without dreams. In the former case, the person was said to have "inappropriate thoughts," and this made the condition more of a "physiological" type. These subconscious "physiological" ejaculations were the natural response of the central nervous system to the bodily overflow of semen accumulation. In contrast, the "pathological" type, according to Chai, was often associated with masturbation, sexual indulgence, leprosy, diabetes, testicular infections, bladder stones, enlarged prostate, and tuberculosis, among other diseased bodily states.[114] Chai maintained that physiological spermatorrhea was neither "beneficial" nor "harmful": it was purely "the result of abstinence." On the other hand, pathological spermatorrhea was an entirely different story, not the least "because semen is the most important essence of the male body." To broach the continuing relevance of this view stemming from traditional Daoist conceptions of sexual health and longevity, Chai quoted from modern Western doctors the claim that "one drop of semen equals to forty drops of blood."[115] He cautioned his reader on the long-term detrimental effects of pathological spermatorrhea, which included dizziness, visual disturbance, auditory disorders, hand-trembling, notable drop in body weight, pale face, and lack of appetite. Hence, Chai implied a causal relation between spermatorrhea and those physical symptoms not related to sex.

The corollary example of masturbation allowed Chai to accentuate the categorical link between spermatorrhea and men. In his words, "although both [spermatorrhea and masturbation] involve ejaculation, they are completely different." Whereas masturbation referred to "a conscious experience practiced by both men and women," spermatorrhea was "an unconscious experience unique to men."[116] The analogy of masturbation thus made spermatorrhea a sex-specific biological process. This parallels how contemporary discussions of menstruation revealed the persistent labeling of certain bodily experience associated with blood loss as female-specific. Chai's discussion visualized *xing* by stressing sexual difference as the biological guarantor of an adequate understanding of semen leakage.

The examples of menstruation and spermatorrhea suggest that the new discourse of Western biology defined sex dimorphism in terms of not only physical structure but also biological function. In the context of this functionalist biology, the introduction of Western-style reproductive anatomy also gave rise to a metaphoric framework that compared the human body to a machine. Already in the late Qing, the reformer Tan Sitong (譚嗣同, 1865–1898) described male and female bodies "like a machine" that "functioned independently from any external reality; a collection of intricately assembled parts, it was imagined to be self-contained."[117] Based on this view of the body, Liang Qichao called attention to the potential contribution that women's bodies could make to the economic mode of production. And as Dorothy Ko has noted, the body-as-machine played a central role in the invention and dissemination of the discourse of *tianzu* (natural foot), based on which the early twentieth-century anti-footbinding rhetoric flourished.[118]

In the mid-1930s, the best-known author who promoted this mechanical metaphor was Chen Yucang (1889–1947). After receiving his medical education in Japan and Germany, Chen became the director of the provincial hospital of Hubei Province, the director of the Medical College of Tongji University, and a secretary to the Legislative Yuan. In his *Life and Physiology* (生活與生理, *Shenghuo yu shengli*) and *Research on the Human Body* (人體的研究, *Renti de yanjiu*), both published in 1937, Chen included visual illustrations of the human body as a mechanical entity composed of smaller parts performing distinct duties all crucial for the efficient operation of the entire unit, of the digestive system as a large factory of metabolism breaking down food material into micromolecules of nutrients and wastes, and of the heart as the epicenter of human energy. In his explanation of the sensory system, the physiology of visual perception relied on the mechanical similarities between the eye and a visual recorder, whereas the auditory process depended on the resemblance of the ear to a telephone handset. Other authors, such as Hu Boken (胡伯墾), described the human body as a macro apparatus that comprised various smaller machineries: "Although we often compare the human body to a machine, this 'single machine' metaphor is not entirely adequate. A better way to imagine the body is the interactive working of multiple machines."[119]

With the rise of this new body-as-machine metaphor, biological sexual difference both served as the basis for and mirrored social gender norms. The most striking example is the popular depiction of sperm as aggressive and eggs as passive agents. Again, in *ABC of Sexology*, Chai Fuyuan viewed the social differences between men and women as preordained by nature:

> The main difference in men and women's temperament is best articulated in the active-passive distinction. This has real connections to the nature of the sex cells. The movement of male sperm represents activity and mobility. The nature of women's ovum is completely opposite and has the characteristic of being static and latent. Therefore, men are active and women are passive, just like the nature of sperm and ovum. This sex difference is also reflected in the tradition of men proposing marriage to a woman. Even after marriage, men are often the initiator in sexual intercourse. The wife would refrain from initiating an intercourse even if she becomes sexually aroused.

Instead, she would always come up with a plan to make the husband initiate. Men always end up being the active party.[120]

In *The Life of Sex* (1934), Shen Chichun similarly stated that "sperm is super tiny. Its size is hard for me to describe. I can only tell you this, my friend: if you collect hundreds and thousands of them in an envelope and mail them out to your relatives and friends in the country, all you need is a four-cent stamp, and they will reach their destination without a problem! Its miniscule physique and proficient movement are exactly what gives it its uniqueness in life. Sperm's only job and purpose in life is the constant search of a mate—an ovum."[121] For Hu Boken, "the ovum is more quiet and inactive. Its movement relies on the tiny flagella on the wall of the oviduct." In contrast, "sperm is exceptionally active. Each sperm (there are many, each being very tiny) has a long tail. When it swings, it enables sperm cells to swim as fast as they can in semen, like how fishes swim in water."[122]

The similarity between these descriptions and the discussions of the active sperm and the passive egg penned by Western biologists is striking.[123] But their broader historical import does not simply lie on the level of metaphors or stereotypes. The techniques of visualization evident in these imageries point to something more significant: what these Republican-era authors translated was not just the science of sex or the gender stereotypes embedded in them but an entire system of scientific authority that established the empirical status of sex as an object of observation. By the time that the descriptions about the dominance of the sperm or the passiveness of the egg had emerged, these scientific discourses already took for granted the objectivity of natural observation, which, according to historians of science Lorraine Daston and Elizabeth Lunbeck, "is the most pervasive and fundamental practice of all the modern sciences" that "educates the senses, calibrates judgement, picks out objects of scientific inquiry, and forges 'thought collectives.'"[124] These depictions assumed, for their author and reader alike, that it was intuitive to imagine visually the sperm and the egg, their indirect and explicit relationship to anatomical sex, their implication for the morphological appearance of maleness and femaleness, their naturalizing effects on the role and function of men and women, and their connection to human sexual expression. Typifying the anatomical aesthetic of medicine and the morphological sensibility of natural history, these texts and drawings cannot be interpreted on the level of translated meanings or representations alone. They attest to a whole new way of looking at the human body based on a different ontological calculus. These anatomical and morphological logics of visual imagination—or what I have been calling techniques of visualization—made it no longer possible to discuss gender without sex.

The Subcellular Gaze

After 1928 the Nationalist government provided China greater unity and stability. Following the leads of the Rockefeller Foundation's China Medical Board and the China Foundation for the Promotion of Education and Culture, the two main institutions responsible for the development of a large infrastructure of scientific and medical research in the early twentieth century, the government advocated the strengthening and expansion of all areas of science research and education.[125] The first generation of Chinese geneticists, who were mostly educated at Cornell or Columbia, began to play a prominent role in the field of biological research. They made significant contributions, for example, to the neo-Darwinian synthesis of the 1930s and 1940s.[126] Although social commentators took a serious interest in evolutionary theory, experimental biologists for the most part focused their attention on establishing accurate understandings of genetic science (although both groups shared a distinct interest in the larger problem of heredity). Their professional interest in genetics, echoing their American colleagues, took them in a slightly different direction from the Social Darwinist reformers and nationalists of their time. As Laurence Schneider has remarked, an early cohort of matured geneticists "contributed to the ongoing differentiation of science from scientism in China, particularly by wresting the discourse of heredity and evolution in China from the monopoly of Social Darwinism, utopian socialism, and other social philosophies."[127]

The idea that sex was determined by chromosomal factors played a pivotal role in shaping early twentieth-century debates in genetic science.[128] Between the 1920s and 1940s, the specific topic of sex determination and the related discussions of Mendelian genetics and Thomas Hunt Morgan's theory of heredity could be found in plain-language journals like *Eastern Miscellany* and *Science* (科學, *Kexue*) as well as more in-depth textbooks such as Chen Zhen's *General Biology* (普通生物學, *Putong shengwuxue*).[129] Chinese biologists expressed strong enthusiasm toward understanding the genetic basis of sex determination. Their interest sustained a sophisticated cross-cultural conversation on hermaphroditism, which they tried to explain with the theories of intersexuality and gynandromorphism first articulated by American geneticists. Using Zhu Xi's discussion of hermaphroditism as a window into these debates, this section delves into the conceptual importance of sexual ambiguity in Republican-era biology. Zhu's work offered complex theories of natural hermaphroditism and, with that, introduced a third technique of visualizing sex: the subcellular gaze of experimental genetics. This technique added a subtle layer of visual evidence to the naturalizing discourse of sex by projecting its ontological definition on a level beneath the cell. It did not, however, replace the anatomical aesthetic of medicine or the morphological sensibility of natural history found in the earlier examples. Instead, it continued to rely on these two mutually reinforcing techniques of visualization in order to consolidate a full-scale rendering of sex as an object of empirical observation.

The topic of hermaphroditism revealed an underlying paradox in the biological theories of sex: life scientists differentiated *ci/xiong* from *nü/nan* even as they were said to be semantically and conceptually related. Again, in promoting the biological basis of sex dimorphism, scientists defined *nü* as the human equivalent of *ci* and *nan* that of *xiong*. This definition situated *ci/xiong* and *nü/nan* on two different semantic planes, a nonhuman grid and an anthropocentric intelligibility. When they applied Western genetic theories to both animal and human case studies, Chinese life scientists organized the visual objectivity of sex around the figuration of hermaphroditism, an effort that made it possible for the epistemological rendering of sex as a *form* of life. Hermaphroditism typically referred to a natural condition that combines the biological features of both sexes.[130] One could better understand how the visual appreciation of sex came about by probing the conceptual importance of hermaphroditic species— or at least how scientists understood them—because scientific definitions of the boundaries of sex were the most salient and at the highest stake in the process of discerning the sexually ambiguous status of this biological category. Scientists who studied hermaphrodite organisms entertained certain underlying assumptions about sex—or at least understandings of the "physical circumstances [that] might delimit the space in which life forms manifest."[131] In short, the significance of sexual ambiguity and the logical coherence, lexical possibility, and syntactic relevance of *ci* and *xiong* bring to the fore issues of visualization at the very threshold of what makes sex a hermeneutic object of scientific scrutiny.

The topic of hermaphroditism also opened up for scientists the possibility of visualizing sex in developmental terms. In the 1930s and 1940s Zhu Xi understood hermaphrodites to be life entities with distinct visual configurations. The two terms that Zhu used most frequently to describe the biological condition of hermaphroditism were "*cixiong tongti*" (雌雄同體) and "*liangxing tongti*" (兩性同體). The former literally means "*ci* and *xiong* in the same body," and the latter literally means "two sexes in the same body." Zhu's intricate explanation of hermaphroditism featured both the temporality and spatiality of sex, representing sex as not only a form but also a function of life. In places where Zhu began to use such terminologies as *nan* and *nü* for human hermaphroditism, the human/nonhuman divide would appear all the more crucial to the comprehensibility and epistemic functionality of *ci* and *xiong*.

In *Changes in Biological Femaleness and Maleness*, Zhu proposed that in order to understand natural hermaphroditism, it is necessary to distinguish the theory of intersexuality from the theory of gynandromorphism. Two diagrams included in his eleventh chapter, "An Analysis of the Two Sexes in Invertebrate Animals," are most representative of his effort to clarify this distinction (figures 2.18–2.19).

In introducing the theory of intersexuality, Zhu wrote:

> From 1921 to 1922, after [Calvin] Bridges, one of the foremost American Morganists, examined the reproductive results of fruit flies, apart from pure *ci* and pure *xiong* types, he observed a third kind of organism that appears to have a type of body that is neither *ci* nor entirely *xiong*. At first he was very surprised, but after careful research, he realized that they are abnormal creatures with a *ci-xiong* dual-sexed body [雌雄兩性混生的怪物, *liangxing hunsheng de guaiwu*], a condition that could be called hermaphroditism [雌雄同體, *cixiong tongti*]. However, this author specifically names them "Intersexes" [中間性個體, *zhongjianxing geti*], in order to distinguish them from the regular hermaphrodites. Although these abnormal creatures [怪物, *guaiwu*] have the features of both sexes, they can never reproduce.[132]

The point Zhu went on to make with respect to the theory of intersexuality was that through a deep chromosomal analysis of intersexed fruit flies,

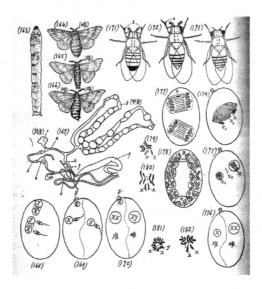

Figure 2.18 "The Gynandromorphism of Silkworms and Fruit Flies" (1945).
Source: Zhu (1945) 1948, 221.

Figure 2.19 "The Intersexuality of Tussock Moths and Butterflies and the Gynandromorphism of Bees" (1945).
Source: Zhu (1945) 1948, 222.

Bridges realized that the mere presence or absence of a Y chromosome could no longer be taken as the sole determinant factor of sex. Zhu referred to the diagrams labeled 181 and 182 in figure 2.18 as showing that "the chromosomal numbers inside these creatures' cellular nuclei are entirely different from normal *ci* or *xiong* individuals! They have instead *three pairs of autosomes and one pair of X chromosomes*."[133] As such, Zhu noted that Bridges began to incorporate the number of autosomes into his formula of sex determination. As Bridges recalculated the ratio of the number of X chromosomes to the number of autosomes in fruit flies, he further developed the concepts of "Superfemales" (過雌體, *guociti*) and "Supermales" (過雄體, *guoxiongti*) to denote those organisms that contained a ratio of X chromosomes to autosomes higher or lower than the ratio for normally sexed organisms respectively.[134] The subcellular boundaries between what counts as male and what counts as female, according to this theory of intersexuality, were unsettled by the category of hermaphroditism. The relevant agents in this technique of visualization no longer came from anatomical configurations or morphological bodies. The seat of maleness and femaleness, or what biologists called "primary sexual characteristics," now rested on the scale of chromosomes.

In introducing the theory of gynandromorphism, Zhu began with Morgan and his students: "Among the fruit flies they investigated, Morgan and his students unexpectedly discovered cases in which the features of *ci* and *xiong* collapsed in a single body [雌雄形性合璧, *cixiong xingxing hebi*]; one side of this body not only displays *xiong* secondary characters but also contains testes, while the other side of the body not only displays *ci* secondary characters but also contains ovaries."[135] To portray this "mosaic" understanding of hermaphroditism, Zhu directed the reader's attention to the diagram labeled 172 in figure 2.18. According to Zhu, picture 171 refers to the normal body of *xiong* fruit flies with white eyes; picture 173 refers to the normal body of *ci* fruit flies with red eyes; and picture 172 shows the body of a "gynandromorph" fruit fly with both *xiong* white eyes on the left and *ci* red eyes on the right. The theory of gynandromorphism, it seems, still relied on the morphological technique of visualization. It differed from the chromosomal explanation of intersexed organisms.

To explain the difference between "gynandromorphs" and "intersexes" more fully, Zhu cited the works of Richard Goldschmidt (1878–1958) and argued that one distinct feature of "intersexes" was that they could be further separated into "*xiong* intersexes" and "*ci* intersexes," whereas gynandromorphs could not. In figure 2.19, for example, diagram 183 was supposed to represent a normal *ci*/female tussock moth, with "a large abdomen, light-colored wings, and short antennas." Diagrams 184 through 187 were representations of "*ci*/female intersexed" moths. Similarly, diagram 188 was a normal *xiong*/male moth with "a small abdomen, dark-colored wings, and long antennas," while 189 through 192 represented "*xiong*/male intersexed" moths.[136] So the theory of intersexuality was now explained via the morphological technique of visualization. These images therefore suggest that it was possible to visualize intersexuality—and, by extension, sex—through different techniques on the level of either morphological bodies or chromosomal agents. The introduction of a new technique of visualization did not displace the earlier ones. As the example of intersexuality makes clear, the empirical status of sex was consolidated through the very interaction of different visual modalities.

Zhu continued his discussion of intersexuality by pointing out and clarifying its quantitative nature. Based on Goldschmidt's theory of intersexuality, Zhu explained that a "lower degree of intersexuality" simply referred to a very limited "change in sex" (變性, *bianxing*) due to a later (in the temporal sense) opportunity for inducing this developmental change in the sexual appearance of an individual organism.[137] Therefore, the extent to which this notion of a "lower degree of intersexuality" differed from a "higher degree of intersexuality" (高度的中間性, *gaodu de zhongjianxing*) only depended on the *timing* of the possibility for modifying the sexual characteristics of an organism along its developmental pathway.[138] According to Zhu, diagram 184 in figure 2.19 would represent a *ci* moth with a "lower degree of intersexuality," while 187 would be a moth with a "higher degree of intersexuality," and both 185 and 186 were simply ones that lay somewhere in between ("a medium degree of intersexuality"). This quantitative notion of intersexuality was not restricted in its applicability to *ci* intersexes; *xiong* intersexed organisms could also display different degrees of intersexuality. It follows that diagram 189 would represent a *xiong* moth with a "lower degree of intersexuality," while 192 would be one with a "higher degree of intersexuality," and both 190 and 191 resembled those that display a "medium degree of intersexuality."

In order to bring home the fundamental difference between gynandromorphism and intersexuality, Zhu explained that the origins of gynandromorphic bodies derive from the moment of conception. Due to the irregular chromosomal interactions at the time . . . some cells are *ci* types that contain a *ci*-like chromosomal make-up in the nucleus, thus displaying *ci* features. Other cells are *xiong* types that contain a *xiong*-like chromosomal make-up in the nucleus, thus displaying *xiong* features. . . .

As for the origins of intersexuality, all of the cells of an intersexed individual are either *ci* or *xiong*. . . . Intersexed bodies are the result of sex-change at some point along the developmental pathway [中途變性, *zhongtu bianxing*]; it is purely a function of the time of sex-change, which could be early or late, that the degree of transformation (high or low) corresponds to. . . . What is important here is that intersexuality is a symptom of change with respect to a developmental pathway; this can be identified as *a change in temporality* [時間上的變化, *shijianshang de bianhua*]. On the other hand, gynandromorphism is something inherent to the individual organism; this can be identified as *a change in spatiality* [空間上的變化, *kongjianshang de bianhua*].[139]

Through the example of hermaphroditism, sex was now conceived not only as a *form* of life but also as a complex *function* of life—a function of its time and space.

Zhu further clarified what he meant by "degrees of intersexuality":

To sum up, humans are like other animals: the origins of sex determination reside within the hereditary materials. Midway sex changes in humans take place in ways similar to how they occur in animals—there is nothing unique about this. Moreover, similar to the sex-transformation cases in animals, *nü-bian-nan* ("female-to-male changes") in humans tend to occur more frequently than *nan-bian-nü* ("male-to-female changes"). From this observation, we can conclude that the basis of *nüxing* ("female essentials" or "femaleness") is more mutable, similar to the cases in amphibians and other types of animals. In the past, what people meant by female human pseudo-hermaphrodites [女性的假兩性同體者, *nüxing de jialiangxing tongtizhe*] can be more accurately understood as individuals with a lower degree of intersexuality; what people meant by true human hermaphrodite [地道的男女同體者, *didao de nannü tongtizhe*] can be more accurately understood as individuals with a medium degree of intersexuality (having testis, ovaries, and the corresponding spermatic duct and oviduct simultaneously in the reproductive organ); what people meant by male human pseudo-hermaphrodites [男性的假兩性同體者, *nanxing de jialiangxing tongtizhe*] can therefore be more accurately understood as individuals with a higher degree of intersexuality that completely transformed from a female to a male.[140]

Therefore, the most significant aspect of Zhu's understanding of human hermaphrodites was that they should be best understood as intersexuals rather than gynandromorphs. His usage of technical phrases such as "a lower degree of intersexuality," "a medium degree of intersexuality," and "a higher degree of intersexuality" to describe such conditions as "female human pseudo-hermaphrodites," "true human hermaphrodites," and "male human pseudo-hermaphrodites," respectively, makes it evident that he viewed all forms of human hermaphroditism as heterogeneous manifestations of human intersexuality. He thus ended the paragraph with a clause clarifying his definition of "male human pseudo-hermaphrodites" as individuals "that completely transformed from a female to a male."

What is most striking about Zhu Xi's illustrations (figures 2.18–2.20), especially when viewed in conjunction with one another, is the way they capture the three techniques of visualization at once. Again, the subcellular gaze of experimental genetics is conveyed most powerfully in figure 2.18 (diagrams 168 to 170 and 174 to 182). In the context of explaining the mutability of life forms with the theory of intersexuality, Zhu used the images to restructure the reader's visual imagination of sex: sex was no longer visualized and conceptualized in terms of the morphological appearance of an organism (e.g., figure 2.19) or the internal anatomical configurations of the body (e.g., figure 2.20 except diagram 153). The subcellular gaze of experimental genetics locates the seat of maleness and femaleness on a level beneath the cell: on chromosomes, genes, chromatins, microtubules, and the like. Meanwhile, the significance of this subcellular epistemic grid was closely intertwined with, rather than independent of, the far-reaching effects of the other two techniques of visualization. In Zhu Xi's discussion, the anatomical, morphological, and subcellular visual depictions overlapped and worked off one another to render hermaphroditism an important nodal junction trafficking biological reasoning. Male and female chromosomal sex, the anatomical organs of human hermaphrodites, and the morphological appearance of moths, flies, and

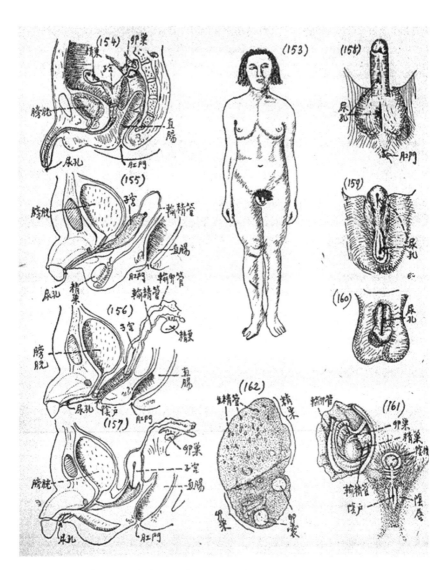

Figure 2.20 "The Morphology of Sex-Transformations in Human Reproductive Organs" (1945).
Source: Zhu (1945) 1948, 196.

the intersexed lady were all brought together in a visual matrix of scientific observation. By representing hermaphroditism this way, the three mutually reinforcing techniques of visualization reflected "a relentless [effort] to replace individual volition and discretion in depiction by the variable routines of mechanical reproduction."[141] Indeed, Chinese doctors would carry this mechanical effort to deduce sex along three different axes of scientific perception into the second half of the twentieth century (figures 2.21–2.23).[142]

From Scientific Images to the Image of Science

My objective in this chapter has been to reorient our attention from words to concepts, from modes of representation to techniques of visualization, and, above all, from an abstract notion of scientism to an account of how scientific authority took shape through the rhetorical power of images. Scholars who study the gendering of modern Chinese history through the lens of "colonial modernity" have unearthed the importance of the new discourse of Western sexual science in the early Republican period. They have shown that in the context of China's colonial modernity, new subject positions emerged corresponding to the nascent sexualized subjectivities of individualism, intellectualism, and liberalism of the 1920s.[143] In a similar spirit, the Modern Girl Around the World project highlights one of the many horizontal global ramifications of colonial modernity in the realm of cultural politics between the era of high imperialism and the period of decolonization.[144] In the Chinese context, to quote Barlow, "nüxing [woman] coalesced as a category when, as part of the project of social class formation, Chinese moderns disavowed the old literary language of power. . . . The career of nüxing firmly established a foundational womanhood beyond kin categories. It did so on the ground of European humanism and scientific sex theory."[145] Nonetheless, recognizing the metaphysical importance of this new sexual science is not the same as understanding the underlying processes of its conceptual deployment and transformation.

This chapter suggests that what colonial modernity scholars of China have not fully accounted for is not the new science of sex and sexuality per se but the visual premises of its epistemological authorization. The anatomical aesthetic, morphological sensibility, and chromosomal gaze discussed in the foregoing analysis are strategies of observation insinuated by the effects of a "visual turn" in sex.[146] These techniques of visualization form the basis of a system of shared beliefs that gives these scientific images and, by extension, science a universal power of persuasion—and principle of faith. In other words, whether all of the writers, editors, or translators of the sources surveyed in this chapter actually practiced medicine, natural

TRUE HERMAPHRODITISM

LIU BEN-LIH AND LIU KAI

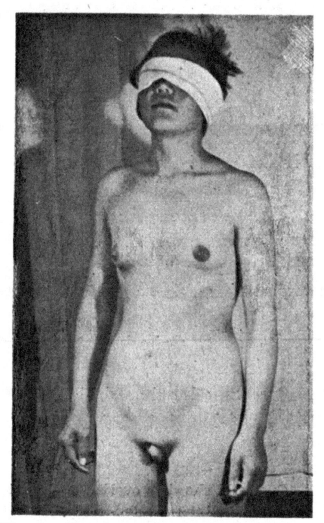

Fig. 1.   Showing patient's physical stature.

Figure 2.21 Liu and Liu's clinical photograph of human hermaphroditism I: Morphological visualization of sex (1953).
Source: Liu and Liu 1953, 153.

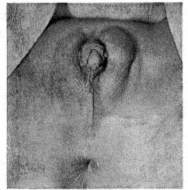

Fig. 2. Appearance of external genitalia. Notice prominence of the left labium major.

Figure 2.22 Liu and Liu's clinical photograph of human hermaphroditism II: Anatomical visualization of sex (1953). *Source:* Liu and Liu 1953, 153.

THE CHINESE MEDICAL JOURNAL

TRUE HERMAPHRODITISM
LIU BEN-LIH AND LIU KAI

Fig. 3. Ovotestis showing ovarian and testicular tissues intermingled freely.

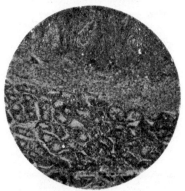

Fig. 4. Ovotestis showing thick fibrous septum in between ovarian and testicular tissues.

Figure 2.23 Liu and Liu's clinical photograph of human hermaphroditism III: Subcellular visualization of sex (1953). *Source:* Liu and Liu 1953, 154.

history, or experimental genetics is perhaps less important (although many of them, like Benjamin Hobson and Zhu Xi, did) than the fact that they had made available to the Chinese public the authority of scientific objectivity rooted in the medium and labor of perception. In effect, the neutrality of the various scientific visualizations of sex came to stand in for the alleged objectivity of the sciences themselves. Emerging from the shadow of a "castrated civilization," the modern Chinese nation learned to embrace the universalism of scientific objectivity. As many historians have pointed out, this objectivity played off on a central preoccupation with the survival of the Chinese "race" or Han "ethnicity" in the world of nation-states.[147] But no longer castrated, China also found sex in the work of nature.

Lest any reader still finds the coemergence of sex and visual objectivity unconvincing, it might be useful to revisit Matignon's photograph of a eunuch (figure 1.5) and compare it to the photo of a Chinese intersexed patient from the 1950s (figure 2.21). A comparison of the two reveals the profound nature of transformation in the cultural representations of "China" in the first half of the twentieth century. In the 1894 photo, we witness a castrated body that is supposed to resemble the "lacking," diseased Chinese body politic: the Sick Man of Asia stares back and begs for Western (biomedical) assistance. In the 1953 photo, we rest assured that Western biomedicine has finally gained footing in China, the nation itself has finally "stood up" (under Mao), and, rather than "lacking," the problem with the specimen is its excessive body parts (having both male and female genital organs). Given this anatomical configuration and the blindfolding of the eyes, the 1953 photo certainly raises the question of who is looking. Yet, more importantly, if the 1894 photo represents China's "lack," what was gained after half a century?

CHAPTER III

Deciphering Desire

Carnal Transformations

A well-known example of the rich cross-cultural interactions between Qing China (1644–1911) and Tokugawa Japan (1603–1867) is the translation of the erotic novel *The Carnal Prayer Mat* (肉蒲團, *Rouputuan*) into Japanese in 1705. The name of the author, Li Yu (李漁, 1611–1680), did not appear on the cover of the book, but most critics attribute this erotic comedy to him.[1] Written in 1657, only thirteen years after the northern Manchus took over Beijing, the novel is replete with graphic descriptions of the sexual pursuits of the protagonist, Wei Yangsheng (未央生). As the front page of the Japanese translation indicates (figure 3.1), the book was considered by many in the early modern period as "the most promiscuous story in the world" (天下第一風流小說, *tianxia diyi fengliu xiaoshuo*). The most complete surviving duplication of the original copy is archived at Tokyo University in Japan. Given its explicit content, the book cannot be sold to minors in Taiwan and is still banned in the People's Republic of China. A quick foray into the text itself provides an important historical preface to unpacking the empirical impulse of sex research in Republican China.

The Carnal Prayer Mat can be situated in the genre of literary pornography similar to the way in which other erotic novels have been perceived in and out of China's past. The late Ming *The Plum in the Golden Vase* (金瓶

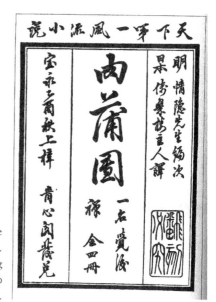

Figure 3.1 Front cover of the Japanese translation of *The Carnal Prayer Mat* (1705). *Source:* http://upload.wikimedia.org /wikipedia/commons/2/20 /Rouputuan1705.jpg.

梅, *Jin Ping Mei*), for instance, which appeared only a few decades before *The Carnal Prayer Mat*, is perhaps the best example of this kind of literature. What these seventeenth-century erotic novels capture, some observers have argued, is the hedonistic and amoral urban behaviors associated with the growing consumer culture in the waning decades of the Ming.[2] Feminist historians and other literary scholars, too, point to the loosening of gender boundaries and sexual mores of the time, as reflected in the blossoming of women's cultural creativity and alternative arrangements of love and intimacy, especially in the South.[3] But the most striking thing about these novels is the considerable degree of popular interest they continue to attract in contemporary Chinese culture. The plots of *The Carnal Prayer Mat* and *The Plum in the Golden Vase* have been adapted time and again in the production of new computer games and films, including, most recently, *3-D Sex and Zen: Extreme Ecstasy*, a three-dimensional cinematic adaptation of *The Carnal Prayer Mat* released in 2011.[4]

If one focuses on the book itself, certain episodes of *The Carnal Prayer Mat* appear surprisingly queer. Granted, as many critics have pointed out, the story brings a sense of closure to Wei Yangsheng's erotic adventure, reinstating a normative effect of Confucian discipline through eventual punishment. Having mistreated all the women with whom he had sexual relationships, including his wife, Wei eventually castrates himself and becomes a Buddhist monk to atone for his sins. However, as Angela Zito has suggested, it might be more compelling to foreground Li Yu's narrative method and the protagonist's constant subversion of Confucian orthodoxy: "Li Yu presents [the choices of male characters] as the ineluctable outcome of their karmic fates, using against the patriarchal norm, even queering, a Buddhism that, in complex ways, shored up patriarchal familial arrangements in this time."[5]

Indeed, the homoerotic contents of the novel are as explicit as the heterosexual ones. After leaving his wife, Wei meets a stranger who would eventually become his close friend, Sai Kunlun (賽崑崙). Spending a night together, naked, Wei insists that Sai share stories of his past sexual encounters with women. Sai accepts the request, and his stories fulfill Wei's desires: "At this point, it is as if the voice of a promiscuous woman emanates from someone next to Wei, causing his body to tremble. He suddenly ejaculates a dose of semen that he has kept to himself for too long. Unless he is asked otherwise, it is unquestionable what has just happened."[6] Similar to the kind of male–male intimacy that Eve Sedgwick uncovers in English literature, Wei's homosocial desire for Sai becomes intelligible by being routed through an implicit triangular relation involving women.[7] And before he acquires a hugely expanding dog's penis through surgery, Wei makes love to his sixteen-year-old boy servant one last time.[8]

Neither the implicitly homoerotic nor the explicitly homosexual scene appears in any of the twentieth-century adaptations of the story. Despite their prominence and wide circulation in contemporary popular culture, the modern versions of *The Carnal Prayer Mat* and *The Plum in the Golden Vase* in film and other media are notorious for being consistently marketed as commodities fulfilling the heteronormative desires of men. If one treats these "texts" as immediate historical evidence of sexuality across time, one might be inclined to conclude that homoeroticism "disappeared" in the twentieth century. Or more specifically, the juxtaposition between the seventeenth-century novels (with their frank and open homoerotic depictions) and their modern, more conservative variations seems to suggest a neat discrepancy between the *presence* of same-sex sexuality before its twentieth-century *absence*. It is perhaps more accurate to conclude that the afterlife and proliferation of these pornographic texts in the contemporary period rely on an indirect censorship of their homoerotic content. This censorship exemplifies what Sedgwick has called an "epistemological privilege of unknowing," a successful concealment of certain ways of thinking within the broader structures of knowledge.[9] In Sedgwick's words, "many of the major modes of thought and knowledge in twentieth-century Western culture as a whole are structured—indeed, fractured—by a chronic, now endemic crisis of homo/heterosexual definition, indicatively male, dating from the end of the nineteenth century."[10]

Similarly, we can interpret the evolving cultural representation of novels such as *The Carnal Prayer Mat* and *The Plum in the Golden Vase* through the lens of this "endemic crisis of homo/heterosexual definition." By spotlighting the rise of sexology in the 1920s as a pivotal turning point in the history of sexuality in China, this chapter offers an alternative explanation for the disappearance of homoerotic representations in their modern appropriations. After all, what the trajectory of this historical evolution reveals is not so much the coincidental disappearance of homosexuality but its very emergence. With the removal of their homoerotic contents, Ming-Qing erotic texts have essentially become heterosexualized in today's mass culture. The heteronormalization of *The Carnal Prayer Mat* therefore points to something more fundamental to the conceptual transformation of sex in the twentieth century: the emergence of its scientific designation as the subject of desire.

Epistemic Modernity

In the previous chapter, I show the ways in which Republican-era biologists and popular science writers translated the epistemological authority of natural science through the production of anatomical, morphological, and chromosomal images of sexual difference. These images affirm a certain form of distance from the viewer, making it possible to decipher truth's relation to nature through their means of visual objectivation. This chapter explores a different kind of relationship between truth and nature and a different rendition of distance between the subject and object of knowledge. By the 1920s, biological sex had become a commonsense in the popular imagination. With a supporting cast of social commentators, iconoclastic intellectuals began to contend that the hidden nature of erotic preference could also be discovered, deciphered, and known. Sex, they argued, was no longer something only to be seen; it was something to be desired, documented, and diagnosed as well. They participated in a new concerted effort, though not without friction, to emulate the European sexological sciences. Their translation and appropriation of Western sexological texts, concepts, methodologies, and styles of reasoning provided a crucial historical condition under which, and the means through which, *sexuality* emerged as an object of empirical knowledge. The disciplinary formation of Chinese sexology in the Republican period, therefore, added a new element of carnality to the scientific meaning of sex.

In the aftermath of the New Culture movement (1915–1919), an entire generation of cultural critics promoted sex education and sexological studies in an unprecedented, systematic fashion. Among the famous May Fourth iconoclastic intellectuals, some not only translated texts and adopted methodological rigor from European sexology but also developed their own theories of human sexual behavior and desire. They frequently engaged in heated debates over the meaning, principles, and boundaries of a science of sexuality. In the 1920s and 1930s, they greeted high-profile European sexologists, including Magnus Hirschfeld (1868–1935) and Margaret Sanger (1879–1966), in major cities such as Beijing and Shanghai.[11] During his visit, for instance, Hirschfeld lectured on the development of sexology as a field of scientific research as well as its significance for the practice of medicine, the promotion of feminist causes, and the implementation of sex education (figure 3.2).[12] Questions of competence, credentials, expertise, and authority preoccupied those of the early twentieth-century urban intelligentsia who spoke seriously about sex in public. By 1935 disparate efforts and conversations converged in the founding of such monthly periodicals

Figure 3.2 Magnus Hirschfeld's lecture with the Chinese Women's Association in Shanghai (1931).
Source: Zhi 1931, 298.

as *Sex Science* (性科學, *Xingkexue*). For the first time in China, sexuality was accorded a primacy of scientific "truthfulness."[13]

This chapter centers on the intellectual journey of two vital figures in this rich tradition of Republican Chinese sexology: Zhang Jingsheng (張競生) and Pan Guangdan (潘光旦). Historians have considered Zhang's prescription of proper heterosexual conduct as a hallmark of his sexological enterprise, especially as it involved his controversial theory of the "third kind of water."[14] Meanwhile, studies of Pan's contribution to Chinese sexology have typically focused on his annotated translation of Havelock Ellis's *Psychology of Sex*, which grew out of his lifelong interest in promoting eugenics in China.[15] Less well studied, however, is their discussion of same-sex desire.[16] From the early 1920s on, Zhang and Pan also debated vociferously about each other's legitimacy as a scientist of sex. Frequently joined by an extended cohort of sex educators and other self-proclaimed experts, such debates reflected the complexity of their sexological maneuver. Moving away from the heteronormative and eugenic emphases of their work, I draw from these examples a snapshot of the broader epistemic context in which the concept of homosexuality emerged as a meaningful point of referencing human difference and cultural identity in twentieth-century China.

Therefore, before we explore scientific discussions of sex change in detail (the topic of next chapter), it is useful to attend to a process of genealogical titration in which the concept of same-sex desire came to be extricated from the broader rubric of gender inversion. The emphasis on homosexuality and the relevant stakes of scientific disciplinarity revises the limited scholarly literature on the history of Chinese sexology. In his earlier study of the medico-scientific constructions of sex, Frank Dikötter argues that early twentieth-century Chinese modernizing elites did not fully grasp or reproduce European concepts of sexual "perversions," including homosexuality.[17] Similarly, Joanna McMillan asserts that, while "sexological studies of perversions were widespread in European medial circles, the literature in Republican China remained almost entirely silent on these enquiries."[18] More recently, in response to Dikötter's thesis, other scholars such as Tze-lan D. Sang and Wenqing Kang have exposed the ways in which selected May Fourth intellectuals—through various debates in the urban press—actually contributed to the increasing awareness of foreign categorizations of human sexuality in early twentieth-century Chinese mass culture.[19]

Taken together, these studies tend to depict Republican-era Chinese sexology as a unified field that treated homosexuality merely as a social, rather than a personal, problem.[20] According to Kang, for example, "Whereas in the West, sexological knowledge pathologized homosexuality as socially deviant, thus reducing it to an individual psychological problem, in China sexology as a form of modern knowledge was used more to diagnose social and national problems. . . . As Chinese writers and thinkers introduced Western sexology to China, male same-sex relations were stigmatized more as a disruptive social deviance than a personal medical condition."[21] Sang's analysis, too, seems to support the claim that no effect similar to the European "individualization" of homosexuality took place in Republican China. In the context of the May Fourth era, Sang observes, "*tongxing ai* ['same-sex love'] is primarily signified as a modality of love or an intersubjective rapport rather than as a category of personhood, that is, an identity."[22]

In this chapter, I suggest that this interpretation is an oversimplification. The view that homosexuality was only a social problem was not consistently shared by leading sexologists from the period, such as Zhang Jingsheng and Pan Guangdan. In the process of establishing sexuality as an appropriate object of scientific inquiry, they deliberated different opinions about the etiology, prevention, and significance of same-sex love. They even disagreed on the fundamental principles of sexological research. Given the multiple perspectives competing at the time, it is perhaps more compelling to suggest that homosexuality appeared to Chinese experts and popular audiences as much a personal problem as it was a social one—an explicit issue of personhood, subjectivity, and identity. Open communications between "sexperts," their readers, and other sexperts further enriched this incitement of a discourse that found truth in sex. To borrow Michel Foucault's insight on the incitement to speak about sex in modern bourgeois society, "Whether in the form of a subtle confession in confidence or an authoritarian interrogation, sex—be it refined or rustic—had to be put into words."[23] Sexology in Republican China was indeed a new system of knowledge in which, literally, new subjects were made.

Ultimately, participants of this new discourse established for China what Foucault has called *scientia sexualis*, which first distinguished itself in nineteenth-century Europe: a new regime of truth that relocated the discursive technology of the sexual self from the theological sphere of pastoral confession to the secular discourses of science and medicine.[24] I argue that from the 1920s through the 1940s, the conceptual space for articulating a Western-derived homosexual identity emerged in China precisely from the new regime of truth oriented by the introduction of European-type sexology. Moreover, whereas social scientists Dennis Altman, Lisa Rofel, and Judith Farquhar have respectively claimed that "gay identity" and *scientia sexualis* first appeared on the China scene only by the post-socialist era, my historicization suggests that both have deeper roots that can be traced to an earlier turning point—in the Republican period.[25]

Part of my divergence from these previous studies seems to stem from the absence of a theoretical vocabulary that fully registers the complexity of sexological claims in this period. Chinese sexologists' conviction that Western science held the key to effective modernization suggests that claims about tradition and modernity were embedded within claims of sexual knowledge. Though distinct, these two layers of the production of sexual truth are somewhat confounded in the analyses of Dikötter, Sang, and Kang: for them, sexological research on homosexuality in the Republican period itself marked a condition of modernization rather than a condition that permitted further referential points of argumentation about the authenticity, traditionality, and modernity of Chinese culture. This conflation rests on the assumption that broader trajectories of historical change—such as modernization and nationalization—are taken for granted and more immediately relevant to the emergence of a discourse of sexology in Republican China. But what if the stakes of the formation of such a discourse depended as much on these broader processes of historical change as on its internal disciplinary tensions and epistemic frictions? As generations of science studies scholars have shown, such conflicts and dissonances are crucial to the consolidation of any kind of scientific valuation.[26]

In order to differentiate the two levels of truth production on which sexological claims operated, this chapter proposes and develops the analytic rubric of epistemic modernity. My application of epistemic modernity in the following analysis refers to an apparatus in the Foucauldian sense that characterizes a historical moment during which a new science of sexuality became epistemologically rooted in Chinese culture. In the next section, I make even more explicit the historiographical rationale for implementing this theoretical neologism, including an operational definition appropriate for the purpose of this study. The core of this chapter consists of three interrelated sections, each featuring an aspect of epistemic modernity. Together they help to reveal a macro, multidimensional picture of East Asian *scientia sexualis*: the creation of a public of truth in which the authority of truth could be contested, translated across culture, and reinforced through new organizational efforts constitutes the social-epistemic foundation for the establishment of sexology in Republican China. I conclude by coming back to the central issue of how homosexuality emerged as a meaningful category of experience in this context. Its comprehensibility, I argue, depends on a new nationalistic style of argumentation that arose from the interplay between the introduction of a foreign sexological concept and the displacement of an indigenous understanding of same-sex desire.

Thresholds of Scientificity

Talking about (male) homoeroticism in China, one first thinks of its rich cultural history prior to the twentieth century, an ongoing topic of in-depth scholarly discussion.[27] This history, however, is not static but dynamic: over the years, the social significance of same-sex relations in dynastic China evolved according to the relevant historical factors. As Matthew Sommer's work on Chinese legal history has shown, sodomy appeared as a formal legislation in China only by the late imperial period. During the eighteenth-century Yongzheng reign (1723–1735), male same-sex practice was for the first time directly "assimilated" to heterosexual practice under the rubric of "illicit sex." This Qing innovation fundamentally reoriented the organizing principle for the regulation of sexuality in China: a universal order of "appropriate" gender roles and attributes was granted some foundational value over the previous status-oriented paradigm, in which different status groups were expected to hold unique standards of familial and sexual morality.[28] But whether someone who engaged in same-sex behavior was criminalized due to his disruption of a social order organized around status or gender performance, the world of imperial China never viewed the experience of homosexuality as a separate problem.[29] The question was never homosexuality per se but whether one's sexual behavior would potentially reverse the dominant script of social order. If we want to isolate the problem of homosexuality in China, we must jump to the first half of the twentieth century to find it.

The relationship between forms of experience and systems of knowledge thus occupies a central role in this historical problem, if only because what we have come to call "sexuality" is a relatively recent product of a system of medico-scientific knowledge that has its own unique style of reasoning and argumentation.[30] In the European context, sexuality emerged from the new conceptual space conditioned by the nineteenth-century shift from an anatomical to a psychiatric style of medical reasoning. "Before the second half of the nineteenth century," according to philosopher Arnold I. Davidson, "Anatomical sex exhausted one's sexual identity" because "the anatomical style of reasoning took sex as its object of investigation and concerned itself with diseases of structural abnormality." Hence, "as little as 150 years ago, psychiatric theories of sexual identity disorders were not false, but rather were not even possible candidates of truth-or-falsehood. Only with the birth of a psychiatric style of reasoning were there categories of evidence, verification, explanation, and so on, that allowed such theories to be true-or-false." "Indeed," Davidson claims, "sexuality itself is a product of the psychiatric style of reasoning."[31] The historical specificity and uniqueness of sexual concepts cannot be overstated, especially since our modern formulation of homosexuality, as the classicist David Halperin reminds us, does not anchor on a notion of object choice, orientation, or behavior alone but "seems to depend on the unstable conjunction of all three."[32]

If understanding the historical relationship between sexuality and knowledge claims in the Western context requires such careful historicism, zooming in on East Asia entails at least one additional layer of consideration. Since the mid-nineteenth century, the medical landscape in China had been characterized by an increasingly conspicuous struggle to reconcile the existing canon of indigenous Chinese medicine with foreign Western biomedical knowledge. For instance, as the last chapter has shown, Benjamin Hobson's anatomical drawings represented a radical epistemological departure from conventional theories of the sexual body in Chinese medicine. The heterogeneous efforts to bring together two coexisting but oftentimes competing systems of medical epistemology were overwhelmingly articulated within a larger sociopolitical project conceived in terms of nationalism.[33] Ideas and practices of nation making would come to assume the center stage in Chinese political and cultural discourses, especially following the First Sino-Japanese War (1894–1895). To ask the very least, why did modernizing thinkers like Zhang Jingsheng, Pan Guangdan, and other sex researchers use Western sexological ideas rather than traditional Chinese medical theory to purport a style of reasoning that stigmatized same-sex desire? What are the deeper historical implications behind such intellectual priorities? The relationship between systems of knowledge and notions of modernity in East Asia demands problematization as we historicize the concept of homosexuality—or, for that matter, sexuality—itself. In order to carefully account for the historical condition under which homosexuality became a meaningful Chinese category, we need a more potent historiographical framework for analyzing the relationship between science and sexuality in twentieth-century China.

To that end, I find what I call epistemic modernity, which builds on historian Prasenjit Duara's notion of "the East Asian modern," particularly useful. When proposing the idea of the East Asian modern in his groundbreaking study of Manchukuo, Duara aims to address two concomitant registers of historical production: how "the past is repeatedly re-signified and mobilized to serve future projects" and the transnationality of "the circulation of practices and signifiers evoking historical authenticity in the region." The concept allows Duara to treat "the modern" as a "hegemonic" project, "a set of temporal practices and discourses that is imposed or instituted by modernizers . . . rather than a preconstituted period or a given condition."[34] The emergence of homosexuality in early twentieth-century China adduces a parallel moment of contingent historicity. The analytic lens of epistemic modernity allows us to see homosexuality not as a strictly "modern" category but as a by-product of a contested historical process yielding specific cultural associations with the traditional, the modern, and the authentic.

Focusing on similar aspects of the transnational processes, flows, and interactions of regimes of cultural temporality and specificity in East Asia, my notion of epistemic modernity refers to a discursive apparatus of knowledge production that also governs implicit claims of traditionality, authenticity, and modernity: it essentially defines the index of imbrication in people's simultaneous preoccupation with the epistemology of scientific deliberation and the procedural determination of what counts as traditional, authentic, or modern. The analytic rubric enables a perspective on the historical question of, to cite Tani Barlow from a different context, "how our mutual present came to take its apparent shape" in "a complex field of relationships or threads of material that connect multiply in space-time and can be surveyed from specific sites."[35] As such, epistemic modernity does not merely denote a system of knowledge; rather, it features a set of ongoing practices and discourses that mediates the relationship between systems of knowledge (e.g., Chinese or Western medicine) and modalities of power (e.g., biopower) in yielding specific forms of experience (e.g., sexuality) or shaping new categories of subjectivity (e.g., homosexual identity). Modernity, to borrow the words of cultural critic Kuan-Hsing Chen, is therefore "not the normative drive to become modern, but an analytical concept that attempts to capture the effectiveness of modernizing forces as they negotiate and mix with local history and culture."[36]

By treating traditionality and authenticity as not ontologically given but constructed as such through the ongoing modernizing technologies of nationalistic measures, I thus attempt to offer sharper insights concerning the regional mediation of globally circulating discourses, categories, and practices in twentieth-century East Asia. The history of homosexuality in China, based on this model, is a history of how globally circulating categories, discourses, and practices were mediated within that particular geobody we call "China." A major aim of this chapter is to show that, in the context of early twentieth-century China, homosexuality was precisely one of these categories; sexology exemplified this kind of discourse; and the articulation of a Western psychiatric style of reasoning about sexuality represented one such practice. A relevant case in point is Ruth Rogaski's study of "hygienic modernity," for one can understand the hygiene–public health nexus as an exemplary model of how globally circulating discourses (of hygiene) and practices (as promulgated by public health campaigns and state interventions) were mediated by the discursive apparatus of epistemic modernity in the historical formation of national Republican China.[37]

Whether our analytic prism is sexuality or hygiene, epistemic modernity affords an opportunity to take the growing global hegemony of Western conceptions of health and diseases seriously without necessitating a full-blown self- or re-Orientalization. By that I mean an intentional project that continually defers an "alternative modernity" and essentializes non-Westernness (including Chineseness) by assuming that the genealogical status of that derivative copy of an "original" Western modernity is somehow always already hermeneutically sealed from the historical apparatus of Westernization.[38] Now that studies in the history of sexuality in non-Western regions have begun to mature, historians should be even more (not less) cautious of any effort to view the broader historical processes of epistemic homogenization as having any lesser bearing than forms of local (or "Oriental") resistance.[39] The idea that "local" configurations of gender and sexuality cannot be overridden by modern Western taxonomies of sexual identity is by now a standard interpretation of both the historical record and the cultural archive of non-Western sexualities. But a variant of this interpretation has already generated vehement repercussions in the field of Middle Eastern studies. Consider intellectual historian Joseph

Massad's controversial claim that all social significations of homosexuality, including internal gay rights activism, reflect the growing infiltration of Western cultural imperialism: "The categories of gay and lesbian are not universal at all and can only be universalized by the epistemic, ethical, and political violence unleashed on the rest of the world by the very international human rights advocates whose aim is to defend the very people their intervention is creating."[40] It bears striking similarity, however ironically and uncomfortably, to Lisa Rofel's adamant critique of a "globalized gay identity."[41] Whether the target of critique is global gay or global sex, post-Orientalist critical thinking should not deter the historian's interest in the condition of the translatability of sexual concepts, especially since the very same concepts have been invoked repeatedly by historical actors themselves.

To redress these analytical conundrums concerning the relationship between transnationalism and sexuality from a strong historicist viewpoint, what I am concerned with, then, is not a social history of homosexuals in China "from below," but an epistemological history in the Foucauldian sense that "is situated at the threshold of scientificity."[42] In other words, this is a study of "how a concept [like homosexuality]—still overlaid with [earlier] metaphors or imaginary contents—was purified, and accorded the status and function of a scientific concept. To discover how a region of experience [such as same-sex intimacy] that has already been mapped, already partially articulated, but is still overlaid with immediate practical uses or values related to those uses, was constituted as a scientific domain."[43] The rest of this chapter is devoted to examining closely the historical conditions under which the concept of same-sex desire came to fall within the realm of Chinese scientific thinking. Each of the following sections features an aspect of the cultural apparatus that I call epistemic modernity: a public of truth, a contested terrain of authority, and an intellectual landscape of disciplinarity. Each distinguishes the two levels of truth production on which sexological claims operated: one concerning explicit claims about the object of scientific knowledge (e.g., sexuality), and another concerning implicit claims about cultural indicators of traditionality, authenticity, and modernity (e.g., ways of narrating sex). Operating together within the governing apparatus of epistemic modernity, they anchored the ways in which same-sex sexuality crossed the threshold of scientificity and the very foundations upon which a *scientia sexualis* grappled the cultural context of Republican China.

Making Truth Public

No other point of departure serves the purpose of our inquiry better than the sex-education campaign that began to acquire some formality in the 1920s. In order to turn sex into a legitimate object of scientific inquiry and education, a notable segment of the urban intelligentsia drew on ideas from Western natural and social sciences. These discussions occurred in university lecture rooms, health care settings, public debates, and both the mainstream press and the vernacular print culture, including the newly established periodicals that featured explicit coverage of sex-related matters, such as *New Women* (新女性, *Xinnüxing*), *New Culture* (新文化, *Xinwenhua*), *Ladies Journal*, *Sex Magazine* (性雜誌, *Xingzazhi*), *Sex Science*, and *West Wind* (西風, *Xifeng*) (figures 3.3 and 3.4). In these forums, pedagogues, doctors, scientists, social reformers, cultural critics, and other public intellectuals taught people how to think about sexuality in scientific terms. In the years following the Xinhai Revolution and surrounding the New Culture movement, they viewed open talk about sexual behavior and desire as a sign of liberation. Or, to borrow the term from D. W. Y. Kwok's classic study, they squarely situated this frankness in the outlook of a new "scientism"—defined as "that view which places all reality within a natural order and deems all aspects of this order, be they biological, social,

Figure 3.3 Front cover of
Sex Magazine 1, no. 2 (1927)
Source: Xing zazhi (性雜誌)
[Sex magazine] 1, no. 2 (1927).

Figure 3.4 Front cover of *Sex Science* 1, no. 6 (1936).
Source: Xing kexue (性科學) [Sex science] 1, no. 6 (1936).

physical, or psychological, to be knowable only by the methods of
science"—that characterized Chinese culture in the first half of the twentieth century.[44]

In the 1920s many sex pedagogues considered adequate sex education
vital for unveiling the puzzling phenomenon of homosexuality. In a lead
essay in the inaugural issue of *Sex Magazine*, "The Necessity of Sex Education" (1927), Wu Ruishu (吳瑞書) argued that sexuality (性慾, *xingyu*)

should be a topic of weighty discussion because it represented one of the most fundamental human urges that helped to "protect the race" and "ensure human survival." According to Wu, if the subject of sexual desire was prohibited from bold and direct dialogue, it would exacerbate public confusion over such social problems as women's criminal behavior and homosexuality, the latter being rampant in school settings.[45] A more extreme, yet rather common, view of the role of sex education in disseminating knowledge about homosexuality can be found in an article titled, "A Paramount Problem in Sex Education: Some Reflections on Homosexuality" (1922), by Li Zongwu (李宗武). Li contended that sex education should caution young people of the danger of homosexuality, presenting it as a form of "sexual aberration" (性的畸形發達, xingde jixing fada) and the root of many unnecessary tragedies in life. He encouraged jettisoning the value placed on gender segregation in traditional Chinese society. According to him, separating the two sexes cultivated homosociality, which would in turn strengthen "unnatural" homosexual desires and weaken "natural" heterosexual bonds. By raising public awareness about the undesirability of homosexuality and by promoting mixed-sex interactions among young adults, he believed that the occurrence of homosexual love—which he contrasted with "platonic love" (純潔愛, chunjieai)—could be curtailed.[46] In 1926 Qiu Jun (丘畯), a biologist at the Nationalist-funded Guangdong University, identified many examples of homosexuality in the animal kingdom, a phenomenon he attributed to the outnumbering of xiong (male) by ci (female) animals in general. Echoing Li's sex-education agenda, Qiu believed that the best way to prevent animal homosexuality was to provide ci and xiong animals a congenial environment for their sexual expression and to breed a higher number of xiong animals.

In the context of growing support for sex education, foreign scientific ideas about sex and sexuality met with considerable enthusiasm. Public spokesmen who took the initiative to translate and disseminate Western psychobiological concepts typically received their advanced degrees at European, American, or Japanese institutions. Upon returning from abroad, many of them shared the conviction that adequate sex education was important for the strengthening of the nation, a belief intimately linked to the broader cultural ambience of the May Fourth movement. According to Frank Dikötter's observation of this period, "For the modernizing élites in Republican China, individual sexual desire had to be disciplined and evil habits eliminated, and couples were to regulate their sexual behaviour strictly to help bring about the revival of the nation."[47] By setting up the British sexologist Havelock Ellis as a role model, many of these modernizing elites singled out his seven-volume encyclopedic *Studies in the Psychology of Sex* as the epitome of scientific research on human sexuality. One of the foremost modernizing thinkers who emulated the empirical impulse of Ellis's work was China's own "Dr. Sex" (性博士, *xingbuoshi*), Zhang Jingsheng.

A university professor and a sex educator, Zhang Jingsheng treated his own sexological treatise, *Sex Histories* (性史, *Xingshi*), as a Chinese counterpart to Ellis's *Studies*. After earning his doctorate in philosophy from Université de Lyon, Zhang returned to China in 1920 and initially taught at the Jingshan Middle School in Guangzhou. For being educated abroad, Zhang was very much part of the work-study movement promoted by the French and Chinese governments in the 1910s. Although part of the initial rationale for this "work-study programme" was to popularize education and dissociate it from cultural elitism, by the end of the decade, the program was soon associated only with those who were anxious to study abroad. Not surprisingly, many of these individuals actually came from a family background that was fairly well-off. In the late nineteenth and early twentieth century, however, most students studying overseas actually went to either the United States or Japan.[48] Zhang's decision to study in France allowed him to maintain close ties with important figures such as Wang Jingwei (汪精衛, 1883–1944), Wu Yuzhang (吳玉章, 1878–1966), Cai Yuanpei (蔡元培, 1868–1940), and Li Shizeng. With these anarchists of the Nationalist party, Zhang participated in the founding of the Sino-French Education Association, branches of which, by 1919, could be found in Shanghai, Guangzhou, Chengdu, Hunan, Shandong, and Fujian.[49]

Zhang's participation in the association and the early work-study movement significantly shaped his intellectual orientation. When he was forced to resign from his post at the Jingshan Middle School in 1921, Cai Yuanpei offered him a teaching position at Peking University, the epicenter of the May Fourth movement. Throughout the second half of the 1910s, the Sino-French Education Association actively promoted the view that overseas study in France offered a rare opportunity for Chinese people to learn European science and humanist thinking without entirely relying on Japan. Adopting this vision, Zhang saw in Cai's offer to teach at Peking University at the peak of the May Fourth a unique opportunity to enlighten the Chinese public about sex. His first two books, *A Way of Life Based on Beauty* (美的人生觀, *Meide renshengguan*; 1924) and *Organizational Principles of a*

Society Based on Beauty (美的社會組織法, *Meide shehui zuzhifa*; 1925), expressed his conviction that the Chinese nation should be strengthened by learning from Europe, the United States, and Japan, especially on the topics of economic structure and military organization. Championing positive eugenics, Zhang even encouraged interracial marriage (and procreation) between Chinese people and those races that possessed strength where the Chinese race was weak, including the Europeans, Americans, Russians, and even the Japanese.[50]

Following these two well-received books, Zhang's publication of *Sex Histories* in 1926 earned him the popular title "Dr. Sex." In fact, he had been lecturing on human sexuality regularly at Peking University since September 1923.[51] *Sex Histories* comprised seven life histories written in the form of first-person narrative by those who responded to Zhang's "call for stories," which was originally published in the supplemental section of the *Capital Newspaper* (京報, *Jingbao*) in early 1926. This "call for stories" was titled "A Way to Kill Time for the Winter Vacation," and it asked young people to contribute stories and any other relevant, however mundane, information about their sex lives. It also indicated that these stories would be "psychoanalyzed" and serve the purpose of a "hygienic" intervention.[52] Zhang studied these life histories carefully and provided commentaries after each of the stories he included in *Sex Histories*. Therefore, Zhang's book adopted a case-study format similar to the way Western sexologists typically organized and presented their research finding.

Indeed, when Zhang published *Sex Histories*, he assiduously advertised the book as "a piece of science, because it documents facts."[53] In his view, there was nothing obscene or inappropriate about his effort to compile a volume based on people's accounts of their sexual thoughts and behaviors. After all, this documentation method had preoccupied European psychiatrists and other forensic doctors for decades already, although their focus had been primarily on deviant sexual expressions.[54] "To keep a strict record of how things happened in the way they did is the kind of mindset that any scientist should have," Zhang insisted.[55] He ended the book with a reprint of the "call for stories" entry, which also solicited collaborators for a project that he had envisaged on translating Ellis's *Studies*.[56] In a word, Zhang considered what he was doing in China to resemble what the European sexologists were doing on the other side of the world.[57]

Zhang's appropriation of Western sexological empiricism—as exemplified by the effort to collect case studies and "document facts"—illustrates a straightforward example of epistemic modernity: implicit in his self-proclaimed expertise on human sexuality lies a claim of another sort concerning referential points of tradition and modernity in Chinese culture. In Zhang's sexological project, knowledge about sexuality involved a modern phenomenon of narrating one's life history in a *truthful* manner. Whereas literature (e.g., fiction, poetry) had been the traditional vehicle for the cultural expression of love and intimacy (including homoeroticism) in late imperial China, according to Zhang's sexology, this mode of representation was no longer appropriate in the twentieth century.[58] His empirical methodology posited a new way of confessing one's erotic experience in the name of science, the domain of modernity in which the truthfulness of sexual desires was to be recorded, investigated, and explained. Similar to the ways in which "sex was constituted as a problem of truth" in nineteenth-century Europe, the procedure for producing sexual knowledge promulgated by Zhang transformed personal desire into scientific data: "sex was not only a matter of sensation and pleasure, of law and taboo, but also of truth and falsehood."[59]

By encouraging people to talk about their sexual experiences in a new order of knowledge that conformed to the "norms of scientific regularity," Zhang hoped to achieve more than just archiving "the facts of life."[60] As the "call for stories" makes clear, narrators who were brave enough to speak out and report their sex life were rewarded with the unparalleled opinion of a sexpert, who, according to the entry, possessed the kind of enlightening scientific knowledge about sexuality from which laypersons could learn and benefit. So, drawing on his academic training in philosophy and the empirical approach of European sexologists, Zhang framed the modernism of his sexological science with another epistemological tool: theoretical innovation. He did this by developing a coherent set of guiding principles in human sexual conduct based on concepts of Western bioscience.

His theory of a "third kind of water" is perhaps the most famous and controversial example. According to this theory, the female body produces three kinds of water inside the vagina: one by the labia, another by the clitoris, and a third from the Bartholin glands. The release of all three kinds of water, especially the "third kind," during sex would benefit the health and pleasure of both partners. Reflecting its eugenics underpinning, the theory claims that the release of this "third kind of water" at the right moment, which normally means twenty to thirty minutes into sexual intercourse as both partners achieve simultaneous orgasm, is crucial to the conception of an intelligent, fit, and healthy baby.[61] At least one other self-proclaimed sexpert, Chai Fuyuan, author of *ABC of Sexology*, supported Zhang's idea of female ejaculation.[62]

Interestingly, apart from construing women as active agents in heterosexual intercourse (e.g., by asking them to perform "vaginal breathing"), Zhang also held them responsible for reducing male homosexual behavior in China.[63] In *Sex Histories*, for instance, Zhang reasoned that since the anus lacked "momentum" and any kind of "electrolytic *qi*," it could not compete with the vagina, which was filled with "lively *qi*." As long as women took good care of their vagina and used it properly for sex, such as by complying to his theory of the "third kind of water," the "perverted," "malodorous," "meaningless," and "inhumane" behavior of anal intercourse among men could be ultimately eliminated.[64] This example powerfully illustrates the subtle ways in which male same-sex practice came to be discussed in the language of biological science: although not the direct cause of homosexuality per se, according to Zhang's theory, the properties, quality, and physiological mechanism of female reproductive anatomy were nonetheless understood as a key determinant of the prevalence of male homosexual conduct. Meanwhile, in prioritizing Western biology as a modernist discourse for the cultural appreciation of female heterosexuality, his theoretical project construed Daoist alchemy as a symbol of tradition in conceptions of sexual health in Chinese culture.

Zhang, above all, sought to create a new public of truth about sex. By privileging the scientific public as the ultimate site for sexual understanding and narration, his effort made unproblematic a discourse based on *reason* to speak of sex. The autobiographical narratives that he collected in *Sex Histories* strictly cohered around this vision. Additionally, in his capacity as the founding editor of the popular magazine *New Culture*, he published translations of excerpts from Ellis's *Studies in the Psychology of Sex*. The periodical soon became a venue for other kindred spirits to present the science of sexology to a popular audience and to establish their own "sexpertise." But most importantly, *New Culture* was not a forum devoted exclusively to the voice of experts; it published readers' responses to not only its most controversial essays but also any contemporary issue that seemed relevant to the scope of the magazine, including sex-related subjects. In the pages of *New Culture*, "the speaking subject [was] also the subject of the statement."[65]

Readers, presumably many of whom resided in urban areas where the mass circulated print publications were most readily accessible, seized the opportunity to respond to Zhang's provocative writings. Some felt the need to confirm the scientific value of his work. One reader, for example, interpreted *Sex Histories* as an "outstanding scientific piece of 'sex research.'"[66] Another reader urged him to publish more sexological treatises like *Sex Histories* by asking "why have you published only one volume of *Sex Histories*? Have you met your goal with this singular contribution?"[67] Others similarly maintained that *Sex Histories* "definitely cannot be viewed as a pornographic piece of writing. Its contents are all valid research material on sexual activities."[68]

Some readers did not find it necessary to justify the scientific nature of Dr. Sex's advice. From the outset many took for granted that his words already constituted science. One woman wrote to Zhang:

> There is one part of your advice that said "the female partner should try to become excited, so that there will be a great amount of water released in the vagina. The male partner could then gradually insert his penis into her vagina . . . and rub it back and forth smoothly and easily." This part, I think, is a little too idealistic. In fact, it cannot be accomplished: although I am a woman who has been married for over a year, if I follow your suggestion, I think it certainly will not work. This is because people who are impatient, men or women, would quickly lose sexual interest in the process. As for those who prefer to take their time, they probably would start getting tired and annoyed of the process, and this might even have a negative effect on two persons' love for each other. What do you think?[69]

Although disagreeing with Zhang's initial advice, this reader still considered him the ultimate authority on matters pertaining to sex. In fact, the letter squarely conveyed her desire to contribute to Dr. Sex's science by providing a personal perspective, which bore a similar empirical value to the case studies collected in *Sex Histories*. Another reader named Xu Jingzai (徐敬仔) even offered Zhang his own insight concerning the proper way of "sexual breathing."[70] Others similarly respected what Zhang had to offer but either wanted to learn more about his theory of the "third kind of water" from the perspective of men or expressed frustration with its impracticality based on their own experience in the bedroom.[71]

A number of readers directly responded to Dr. Sex's brief discussion of homosexuality. Supporting Zhang's effort in promoting sex education on scientific grounds, a lady named Su Ya (素雅) argued that the prevalence of undesirable sexual practice would decrease once adequate sex education became common in China. Su wrote to Zhang by echoing the ambition of many sex educators: "As long as sex education continues to be promoted and advanced, all the illegal sexual behaviors, such as rape, homosexuality, illegal sex, masturbation, etc., could be eliminated."[72] Miss Qin Xin (芹心), however, disagreed: "Homosexuality is not a natural sexual lifestyle. It is a kind of perversion and derailment in human sexuality, so it does not have a proper place in sex education."[73] Another reader asked, "It seems that homosexuality exists among both men and women, but could these people's 'sexual happiness' be identical to the kind of enjoyment experienced in sexual activities with the opposite sex?" Zhang answered with a blatant no: "Other than being a personal hobby, homosexuality cannot compare to the kind of happiness one achieves in heterosexual intercourse. Since on the physical level it cannot generate the kind of electric *qi* found in heterosexual mutual attraction, homosexuality also does not provide real satisfaction on the psychological level."[74] Zhang's response thus reminded his readers the importance of knowing and practicing the correct form of heterosexual intercourse. It implied the paramount significance of following his theory of the "third kind of water," which defined women's befitting sexual performance, attitude, and responsibility.

Together the guidelines that Zhang offered in *Sex Histories* and his communications with readers in *New Culture* shed light on the grounding of sexological science in Republican China through the means of expert intervention. To borrow Foucault's insight on this matter again, it was a technology of power in which "one had to speak of sex; one had to speak publicly and in a manner that was not determined by the division between licit and illicit . . . one had to speak of it as of a thing to be not simply condemned or tolerated but *managed*, inserted into systems of utility, regulated for the greater good of all, made to function according to an optimum. Sex was not something one simply judged; it was a thing one administered."[75] Starting in the 1920s, under the influence of Dr. Sex, some Chinese urbanites began to treat heterosexuality and homosexuality as scientific categories of discussion and sexology as a serious discourse of expertise knowledge. In 1927 one individual who worked for the Fine Arts Research Society (美術研究會, *Meishu yanjiuhui*) observed that "due to the recent progress in academia, there is a new independent scientific field of study that surprises people. What kind of science is it? It's called sexology."[76] In particular, Zhang Jingsheng's theory of the "third kind of water" both biologized and psychologized sex. It biologized sex because it discussed people's erotic drives and motivations in terms of the somatic functions of male and female reproductive anatomy. The theory psychologized sex by explaining people's sexual behavior and activities in terms of what they thought and how they felt.

The case studies approach advocated by Dr. Sex marked a decisive turning point in Chinese sexology. This empirical signature gradually came to characterize the entire scientific discourse of homosexuality in the Republican period. This was especially true for writings on female same-sex intimacy. In the scattered commentaries on the subject that appeared in the urban press before the time of Dr. Sex, the case studies format was notably absent. In one of the earliest Republican-era accounts of female homosexuality, for example, the author Shan Zai (善哉) introduced the concept of "sexual inversion" (情慾之顛倒, qingyu zhi diandao) and commented on its prevalence as documented in the historical record worldwide. However, she never mentioned a single concrete case, leaving the tone of her overview general and abstract.[77] In 1923 a contributor to the Ladies Journal explained homosexuality as a consequence of gender segregation in education and military settings. Again, the author did not discuss in depth an actual homosexual affair in these unisex institutions. Referencing Western psychological theories of marriage, the author merely upheld the value of mixed-sex education.[78] The translation of a Japanese article on the implication of homosexuality for women's education appeared two years later in the same journal. Although the article brought up Oscar Wilde's trial in England, specific cases of homosexuality in contemporary East Asian society appeared nowhere in this piece. Interestingly, unlike most of the essays printed at the time, this article challenged a stigmatized understanding of homosexuality and pushed for a more sentimental definition by drawing on the theories of the British sexologist Edward Carpenter (1844–1929).[79]

After Zhang published Sex Histories, the majority of popular writings on female homosexuality included actual "cases" as examples. These publications were often billed with an "objective" voice, imitating the sexological authority of Dr. Sex, and with the intention to cast light on new anecdotes of homosexual relations between women. An article in New Ladies Monthly, titled "Homosexuality" (1946), opened with two stories of lesbian love. The first story involved two students who always spent time together in their daily routines, from having meals together to sitting next to each other in classrooms. As their relationship grew, one of them became an object of affection to a third classmate, leading the other to cultivate a strong sense of jealousy. The envious girl therefore attempted to stab her lover although, as it turned out, without success. The second story was about the intense romance between a masculine "Miss S" and a feminine "Miss Y." Given their attachment, the sexual nature of their relationship was unambiguous to their peers. Eventually, Miss S acquainted herself with a male partner, and, depressed in consequence, Miss Y tried to convince Miss S to commit suicide with her. This tragedy was avoided when a friend introduced Miss Y to another male date, and both Miss S and Miss Y got married separately. These two specific cases provided the empirical basis for the author of the article to use scientific theories of sexual psychobiology to explain this "perverse sexual desire" (變態性慾, biantai xingyu). The narratives thereby anchored a broader discussion of "the prevalence of these kinds of story in Chinese unisex schools before the Japanese occupation period" and how the popularity of free love made the cultural context at the time especially congenial for developing homosexual tendencies.[80]

Other writers contributed to the empirical understanding of lesbianism by proposing that it was regionally specific—a situation endemic to the South. In 1934 Jian Yun (澗雲) documented the few cases of lesbianism that she personally witnessed in a two-part essay published in the Shanghai-based Choumou Monthly (綢繆月刊). Specifically, she substantiated two stories in detail. The first story occurred in 1922, the year when she arrived in Hong Kong to stay with a female friend, W, who cohabitated with another lady, S. Jian Yun compared their residence to the kind of living arrangement typical of a conventional married couple. Although she slept on the same bed with W and S for the first two nights, Jian Yun decided to leave them on the third day and went to stay with her cousin instead. Years later, she discovered that both W and S got very sick and were hospitalized for several weeks. S eventually married a man, and the couple moved to the Philippines. W, on the other hand, refused to compromise and so still clung onto the past.[81] The second, much more convoluted story featured her coworker, L from Nanhai (a district of Foshan, Guangdong), and L's mother. Both L and L's mother, so Jian Yun's report suggests, were enticed into being romantically involved with other women, causing both to experience serious illness involving menstrual irregularities and eventually death.[82] Jian Yun presented these stories as examples of the kind of lesbianism recurrent in Southern China, especially in Hong Kong, Guang-zhou, Nanhai, among other regions.[83]

Additional stories came from lesbians themselves. These tended to be couched in a more "subjective," self-confessional style. For example, two women who met in school valued their mutual feelings for one another so much that they declared their "same-sex marriage" in a newspaper in 1934.[84] In 1941 another woman, Li (莉), recounted her past romantic rela-tionship with a girlfriend, Lin, in an article titled "Sister Lin, Please For-give Me: Narrating My Homosexual Life." What prompted Li to compose this article was an anonymous commentary that she came across earlier in the year. Departing from the dominant depictions of homosexuality as unnatural and sick, the published commentary broached the similarities between same-sex and opposite-sex attractions and suggested that people "do not always have to turn their back on homosexual relations."[85] This message moved Li and inspired her to restore the details of her relationship with Lin, who was three years elder. They had met in junior high and became lovers immediately. However, in subsequent years, Li fell in love with new incoming girls and abandoned her relationship with Lin. After writing numerous letters to Li, Lin fell ill one day. At that point, Li was still not sufficiently motivated to pay Lin a visit. When Lin finally passed away, Li gathered a considerable measure of guilt and regret. Her article, in other words, detailed more than just her feelings for Lin. This documented "proof" of a lesbian relationship was also an ad hoc letter asking for forgiveness.

The case studies format that brought these stories to light was consistent with the empirical approach of fin de siècle Western sexology. Among the field's founding figures, Havelock Ellis, Sigmund Freud, Magnus Hirschfeld, Richard von Krafft-Ebing (1840–1902), Iwan Bloch (1872–1922), and Max Marcuse (1877–1963) all discussed, classified, understood, theorized, and, in essence, imparted knowledge claims about human sexuality by collecting and studying individual life histories. This approach bore little resemblance to the sociological-statistical method adopted later by Alfred Kinsey (1894–1956), the American sexologist who would assume an international repu-tation by the midcentury.[86] As reflected in their correspondences, the Chinese Dr. Sex and his readers faithfully believed that sexuality—hetero or homo—was something to be known scientifically, and that both the experts and non-experts mutually relied on one another for valuable infor-mation. The intersubjective dynamic between the Chinese sexperts and their readers mirrored the reciprocal dialogue between medical doctors and their patients in European and American *scientia sexualis*. As medical historian Harry Oosterhuis has claimed, "the new ways of understanding sexuality emerged out of a confrontation and intertwining of professional medical thinking and patients' self-definition."[87] Foucault's observation, again, is germane here: "It is no longer a question of saying what was done—the sexual act—and how it was done; but of reconstructing, in and around the act, the thoughts that recapitulated it, the obsessions that accompanied it, the images, desires, modulations, and quality of the pleasure that animated it. For the first time no doubt, a society has taken upon itself to solicit and hear the imparting of individual pleasures."[88] In his attempt to enlighten the public with reliable and "accurate" knowledge about proper heterosexual behavior, Zhang's sexological project gave true or false statements of homosexuality an unprecedented scope of comprehensibility in China. It is worth reemphasizing that what *scientia sexualis* produced was not so much homosexual experience per se than the historical conditions under which it became a conceptual possibility—a system of truth and falsehood that structures identity along the axis of a heterosexual–homosexual polarity.

Competing Authorities of Truth

The public dissemination of scientific knowledge about sexuality was a hallmark of Zhang Jingsheng's "utopian project," to borrow the phrase from Leon Rocha.[89] In pushing for the public circulation of private sexual histories, Zhang's sexological enterprise simultaneously defined certain aspects of China's sexual culture as traditional or modern, whether in terms of modes of narration (literary versus scientific) or knowledge foundations (Daoist alchemy versus Western biology). In this new public of truth, the nature of human desire and passion was openly debated by experts and their readers. But the cast in these debates included other public contenders as well. This section of the chapter highlights another aspect of epistemic modernity crucial to the development of *scientia sexualis* in Republican China: a public platform on which authorities of truth competed.

Whereas a considerable mass of urban acolytes extolled Zhang by calling him *the* Dr. Sex, other adepts publicly gainsaid his teaching. These critics ridiculed Zhang's sexological work mainly for its lack of scientific proficiency. A contributor to the periodical *Sex Magazine*, Han (瀚), called Zhang's sexology "fraudulent science [偽科學, *weikexue*]." In describing the specific type of pleasant odor emitted by women during sexual intercourse, Zhang cited the example of Lin Daiyu, the principle female protagonist of the Qing classic *Dream of the Red Chamber*. Han criticized this reference on the ground of its historical and factual inaccuracy: "Is Lin Daiyu your relative? Were you physically next to Jia Baoyu [Lin's lover in the story]? Please do not make such preposterous statements. Speaking from a fact-based perspective, I would like to invite you to collect these pleasant-smelling gases that you describe and subject them to chemical testing." The author also attacked Zhang for suggesting that male genitals produced *yang* electrolytic *qi* and female genitals produced *ying* electrolytic *qi*: "to demonstrate that he is deeply familiar with the traditional depiction of *yang* as masculine and *ying* as feminine in Chinese culture, [Zhang] inevitably imposed the notions of *ying* and *yang* onto the concept of electrolytic *qi*." Above all, Han was dismayed by the various kinds of female water that Zhang described. If liquids were indeed released during sexual intercourse, Zhang's suggestion that electrolytic *qi* could be produced concomitantly appeared unconvincing. For Han, this only demonstrated that Zhang "does not even understand the most basic principles of electric generation."[90]

In the same issue of *Sex Magazine*, another writer, Qian Qian (倩倩), composed a piece called "Research on the Third Kind of Water." Targeting the theory of the third kind of water, the article began with a statement that actually attested to the ecumenical significance of Dr. Sex's work: "Since the publication of Zhang Jingsheng's *Sex Histories*, scholarly understandings of sex have multiplied in unexpected ways. At the same time, it has also initiated an increasing number of individuals studying sex and a growing wave of popular fascination with sex research." However, precisely because Zhang's theory of the third kind of water had stirred up a revolution of some sorts in popular culture, Qian Qian set out to clarify "what the third kind of water is" for non-experts. In twelve pages, Qian Qian explained in detail three main types of fluids that could be found in female genitalia: menstrual fluid, the secretion of the Bartholin's gland, and vaginal fluid. "Professor Zhang's understanding of the third kind of water as the secretion of the Bartholin glands is evidently incorrect," Qian Qian concluded. "The significant quantity of female discharge during orgasm" that Zhang described "was nothing other than vaginal fluid." Refuting

Zhang, Qian Qian added, "the secretion of vaginal fluid is a continual process, though with growing intensity in intercourse, and does not happen only during orgasm as Professor Zhang understood it."[91]

The writings of Dr. Sex drew fire from other more mainstream scholars. Even though Zhou Jianren, author of numerous popular life-science books and an editor at the Shanghai Commercial Press, had praised Zhang's first two books for their sound philosophical argument, he, too, attacked Zhang's theory of the third kind of water immediately following the publication of *Sex Histories*. Zhou argued that Zhang's theory did not correctly account for the biological process of ovulation in women's menstrual cycle. Zhou noted that if the female body produces an ovum only on a periodic basis, Zhang's advice for women to voluntarily release an egg and the "third kind of water" in each sexual intercourse was evidently "pseudo-scientific" at best. Another sex educator, Yang Guanxiong (楊冠雄), even described Zhang as a public figure destructive to the entire sex-education movement. For regular interlocutors in sex education like Zhou and Yang, who kept up with the latest developments in the natural sciences, the most problematic aspect of Zhang Jingsheng's sexology was its inaccurate grounding in human biology.[92] Some writers used the unscientific foundation of Zhang's writings as a benchmark for distinguishing domestic, second-rate from foreign, "serious and respectable" sex research.[93] Other high-profile intellectuals, including Zhou Zuoren, Chen Cunren, Zhang Taiyan (章太炎, 1869–1936), and Hu Shih, weighed in on the debate over the literary value and (in)decency of *Sex Histories*.[94]

Out of the many critics of Zhang, the most vociferous was probably Pan Guangdan, the famous Chinese eugenicist who presented himself as a loyal devotee of Havelock Ellis's sexological oeuvre. Pan described his first impression of Zhang as a "literary freak" (文妖, *wenyao*) and Zhang's work as "fake science" (假科學, *jiakexue*), "fake art" (假藝術, *jiayishu*), and "not philosophy" (非哲學, *feizhexue*).[95] Pan received his bachelor's and master's degrees in biological science at Dartmouth College in 1924 and Columbia University in 1926, respectively. In light of his high academic performance, Pan was conducted into the Phi Beta Kappa honor society upon his graduation from Dartmouth.[96] His educational experience in New York coincided with the peak of the American eugenics movement, the center of which was located in the upper-class resort area of Cold Spring Harbor on Long Island. In 1904 the Station for the Experimental Evolution was established there under the directorship of Charles Davenport with funds

Figure 3.5 Pan Guangdan (1899–1967).
Source: http://tupian.baike.com/doc/潘光旦/ao_68_64_0130000032862212319164981o 937_jpg.html?prd=citiao_tuce_zhengwen.

from the Carnegie Institution of Washington. In the summer of 1923 and between his undergraduate and graduate studies, Pan visited Davenport's Eugenics Record Office (founded in 1910) to learn more about human heredity research.

After returning to China in 1926, Pan did not conduct experimental research in biology (given his interest in eugenics, experimentation with human breeding was of course not an option). Instead, like most European and American eugenicists, he spent most of his time studying the ethnosocial implications of sex by constructing extended family pedigrees and collecting other forms of inheritance data.[97] His *Research on the Pedigrees of Chinese Actors* (中國伶人血緣之研究, *Zhongguo linren xueyuan zhi yanjiu;* 1941) is an example of this.[98] Similar to the Anglo-American eugenicists whom he tried to emulate, Pan prioritized the making of an "eugenic-minded" public.[99] He did this by delivering numerous lectures around the country and publishing extensively in both academic journals and the popular press to promote his positive vision of eugenics.[100] The Chinese public in general viewed him as a trustworthy intellectual in light of his impressive academic credentials. Through Pan, "eugenics" (優生學, *youshengxue*) quickly became a household term in China in the 1920s and 1930s.[101]

Sharing the same intellectual worries with Zhou Jianren, Pan depicted Zhang Jingsheng's writings on human sexuality as "unscientific." Pan was particularly disdainful of anything Zhang had to say about the relationship between sex and eugenics because he despised Zhang's lack of formal training in biology. Even though Zhou, like Zhang, had a background in philosophy, his writings on evolution proved his erudition in the modern life sciences. On the contrary, in Pan's view, Zhang's ideas about human sexuality demonstrated a fatal failure in communicating the basic principles of human biology. In 1927, Pan responded to Zhang's theory of female ejaculation:

> [Zhang] claims that he has discovered a "third kind of water," but we do not know what it is. He has indicated that it simply refers to the secretion of the Bartholin glands. If that is the case, then it is really nothing new to any educated person who is familiar with the physiology of sex. . . . One of the functions of the Bartholin secretions is to decrease resistance during sexual intercourse. The amount of secretion increases as the female partner becomes more aroused, so the quantity of secretion depends entirely on the intensity of her sexual desire and arousal. . . . Since this function is present in most females, one wonders on what statistical basis does [Zhang] claim that women in our nation usually do not release this third kind of water. When he claims that this kind of water is more typically released in the body of European urban women, one is equally suspicious about the statistical evidence on which he relies, if there is any at all. If he has none yet still speaks so unabashedly in these words, the intention behind making these unsupported claims is dubious.[102]

Pan attacked Zhang's understanding of eugenics by citing the statistical data collected by Charles Davenport and Francis Galton. Pan even accused Zhang for having overlooked Galton's work completely: "Since the Englishman Francis Galton published his *Hereditary Genius* in 1869, the book has proven to be immensely useful; and the recent developments in intelligent testing have grown exponentially. Why doesn't [Zhang] consult these works more substantially? He probably is not even aware of the existence of these studies; one really cannot understand why someone would speak about eugenics so elaborately without showing some basic familiarity with this body of scholarship."[103]

In his reply, Zhang showed no acquiescence. He pointed out that Pan's comments "have in fact proven the scientific content of my theory. The third kind of water is, of course, something present in every woman. . . . I am merely bringing this kind of water to people's attention and teaching them the ways in which it can be discharged." Zhang even construed Pan's recourse to the work of Galton as evidence of poor research and understanding of eugenics: "In terms of heredity and eugenics, [Pan's] knowledge in these subjects is even more limited. He is familiar with Francis Galton's work, but Galton's theory does not seem well-grounded. . . . Three years ago, I had already indicated in my book, *A Way of Life Based on Beauty*, that Galton's eugenic theory is not real science, but what we want is real science. . . . Please allow me to invite [Pan] to study my work more carefully in addition to Galton's."[104] To Zhang, Pan was the one who lacked scientific and scholarly integrity. In response to other critics, Zhang emphasized his foreign language ability (which allowed him to distill Galton's work) as a key credential that set him apart from the numerous "quack" doctors who dismissed *Sex Histories*.[105] In fact, he repeatedly pitched his sexological treatise as marking a pivotal turning point in the larger intellectual endeavor to move away from "pornography" (淫書, *yinshu*) and toward "scientific sexology" (科學的性學, *kexue de xingxue*).[106]

This public correspondence between Pan and Zhang offers a window onto the ways in which, in the 1920s and 1930s, experts defined and debated the boundaries of a scientific discourse of sexuality. An important aspect was the mutual contestation of the credibility and validity of expertise, a regular facet of any scientific discipline. For Pan, formal training in the biological sciences represented a crucial feature of sexological credibility. Even if an expert lacked this credential, sexological competence could still be achieved by acquiring Western scientific knowledge faithfully and refraining from making unsubstantiated empirical claims about sex. This is why he regarded Zhou Jianren as a better equipped sex educator and a more respectable scientist than Zhang Jingsheng. To Zhang, Pan had obviously misinterpreted what he was trying to do. In fact, Pan's oversight of Zhang's earlier scholarly output implied a defect in Pan's research and scholarship. In turn, Zhang even encouraged Pan to study his earlier writings more carefully in addition to the work of foreign scientists like Galton.

Since he had already built a foundation of sexological expertise, Dr. Sex believed that this foundation should be consulted, or at least acknowledged, by new incomers to the field, including Pan.

Although Zhang's theory of female ejaculation formed the most controversial part of his work, his sparse discussion of same-sex sexuality often served as a springboard for rich critique and elaboration. In 1929, for example, Yang Youtian (楊憂天) authored a lengthy essay in *Beixin* (北新) titled "The Problem of Homosexuality" (同性愛的問題, *tongxing'ai de wenti*). The opening paragraph squarely framed Yang's essay as a response to Dr. Sex:

> There are many mysteries in the human world, but perhaps sex tops the list like no other topic. In the sexual realm, perhaps the most surprising phenomenon is none other than homosexuality. Dr. Sex, Zhang Jingsheng, left behind [Auguste] Comte's positivist philosophy, as well as his professorial position at Peking University, in order to explore sexology. This invariably casts him as one odd creature. Yet, Dr. Zhang's discussion always revolves around male-female sexual relations and is really about nothing more than the different techniques of heterosexual intercourse. In the end, Zhang is still ridiculed by many. . . .
>
> I do not hold a doctorate in sexology. I am not an education expert, a psychologist, or a medical professional. However, I am keen to raise the issue of homosexuality for public discussion precisely because I want to unravel this mystery and expose its secrecy.[107]

As Wenqing Kang has noted, although we know very little about the biographical details of Yang Youtian, the significance of this essay can be seen in the debate on homosexuality it generated among sexologists in the late 1920s and early 1930s.[108] Specifically, Yang's essay prompted Hu Qiuyuan (胡秋原), a leftist intellectual and translator of Edward Carpenter's work, to write a rebuttal, "Research on Homosexuality" (同性愛的研究, *tongxing'ai de yanjiu*).[109] Both essays were collected in a book titled *Essays on the Problem of Homosexuality* (同性愛問題討論集, *Tongxing'ai wenti taolunji*) published by the Shanghai-based Beixin shuju (北新書局) in 1930.[110]

In contrast to Zhang's scant attention to the topic of homosexuality, Yang's essay enriched the depth of Chinese sexological understanding in substantial ways. In fact, Yang's weighty discussion of same-sex sexuality both reflected and foreshadowed the diversification of Chinese sexology in this period. First, Yang extended existing debates on the congenital versus acquired nature of homosexual desire. While most writers simply brought up these two opposing visions of etiological cause, Yang explored their respective genealogies in detail and with impressive nuance. In discussing the intellectual foundations of these two etiological perspectives, he provided rare information on how Western sexologists related these perspectives to the ideas of latent homosexuality, biological and psychosexual hermaphroditism, gender inversion, and bisexuality. This discussion enabled Yang to tap into familiar terrains in the conclusion, focusing on the prevention of homosexuality. Second, while most Chinese sex researchers looked up to British sexologists such as Havelock Ellis, Edward Carpenter, and Marie Stopes as intellectual predecessors, Yang's balanced coverage drew attention to the significance and contributions of continental European sexological scientists.[111] He gave equal weight to the competing theories of homosexuality articulated variously by Carl Westphal (1833–1890), Richard von Krafft-Ebing, Albert Moll (1862–1939), and Karl Heinrich Ulrichs (1825–1895), among others. This impressed Chinese readers with a more sophisticated spectrum in the historical roots of sexological taxonomy. Third, in presenting a more comprehensive coverage of European sexology, Yang used this opportunity to introduce other categories of sexual perversion, such as sadism, masochism, and fetishism. "Homosexuality is closely connected to these other types of abnormal sexuality," Yang explained, "so it is quite common to find a homosexual who is inflicted with one or all three of them."[112] Yang's sexological contribution, although rooted in the often-pejorative language of Western sexual science of the period, thus greatly amplified the scope of sex research in the aftermath of Dr. Sex.

The debates between Zhang and his critics reveal the broader evolving context in which homosexuality became a matter of scientific discussion. This contested terrain of authority denotes a public platform on which self-proclaimed experts in sexology competed and challenged each other's scientific legitimacy. By scientific "legitimacy," they considered a host of criteria, including academic credentials (whether someone was trained in the humanities or sciences and in what discipline), methodological approach, accuracy in understanding and communicating the specific contents of Western scientific knowledge (entailing foreign language ability), evidence of candid research experience (including grasp of previous scholarship), and acquaintance with the intricate genealogies of Western sexual taxonomy, among others. In this regard, East Asian sexology, as a regionalized but globally circulating discourse marked by the trends and currents of epistemic modernity, reflected the broader stakes of scientific disciplinarity looming over Chinese culture at the time.[113] Similar to the famous 1923 "science versus metaphysics" controversy, debates over sexual knowledge contributed to the increasingly hegemonic intellectual agenda in which the interrogation of the very meaning of science became a pre-occupation unique to the early Republican period. In a bifurcated direction, the growing currency of debates on scientism—itself a new symbol of modernity—contextualized the gradual process by which the category of homosexuality absorbed the dominant frame of thinking about same-sex desire in the twentieth century.[114] In China and abroad, the competing authorizations of truth established sexology as a historically specific configuration of contested knowledge. Deriving particular pleasures from the scientific analysis of sexual confession, the diverging opinions and claims to disciplinary propriety created a heterogeneous—and perhaps even ambiguous—space in which, echoing Foucault's own suspicion, the distinction between *ars erotica* (under which Foucault himself initially categorized China) and *scientia sexualis* eventually dissolved.[115]

Disciplinary Consolidation

In addition to the invention of a new public of truth and a contested terrain of authority, the grounding of *scientia sexualis* in Republican China involved a third endeavor: the consolidation of its disciplinarity through the translation and reinforcement of specialized authority across culture. The novelty of Zhang Jingsheng's *Sex Histories* was highlighted in its incitement of a new Chinese discourse in which the truth of people's sexual experience was negotiated in public; but the book's cultural legacy and significance was even more pronounced in the way it reproduced the social dynamics between the observer (the sexologist) and the observed (sexual desire and behavior) that characterized Western sexual science. The criticisms leveled against Zhang, by Pan Guangdan and others, broadened the purview of such dynamics and relations of power. They made public not only people's sex life but also each other's (in)competence to speak about the scientific nature of sex. By the 1930s, through translating, reinforcing, and recontextualizing the cultural authority of sexology, Chinese sex scientists accomplished beyond disclosing sexual truths and the contested nature of their "sexpertise" in public: they introduced, on the level of epistemology, a new style of reasoning about sexuality and, in the domain of social sphere, an unprecedented forum for intellectual debates that defined their project as culturally urgent, socially legitimate, and disciplinarily independent.

By the time that *Sex Histories* had undergone numerous reprints and could be found in almost every corner of urban cities such as Shanghai, Tianjin, Beijing, and Guangzhou, it seemed pressing to sex educators that the study of sexuality required a more rigorous scientific grounding.[116] This drew the line between Dr. Sex, who was primarily concerned with popularizing his theory of the third kind of water, and his critics, who increasingly viewed his work as narrow, unscientific, and arcane. In 1933 Zhou Zuoren compared China's Dr. Sex to Havelock Ellis and Magnus Hirschfeld. Zhou praised the "scientific" and "naturalist" endeavors of these European sexologists, who advocated a more benign interpretation of sexual abnormalities that was critically absent in the writings of China's Dr. Sex.[117] Again, the degree of scientific difference was embodied most notably in the contrast between Zhang Jingsheng and Pan Guangdan in their approach to the empirical study of sex.

Despite their shared interest in emulating Havelock Ellis, Pan has been considered by many as a more pivotal figure than Zhang in pioneering the introduction of Western sexology to China. For one, Zhang rarely offered insights concerning human sexuality other than heterosexual intercourse. In 1929 Yang Youtian wrote his piece on homosexuality to illuminate "the most unimaginable secret of sex," since "Dr. Zhang's discussion always revolves around male–female sexual relations and is really about nothing more than the different techniques of heterosexual intercourse."[118] In contrast, Pan often discussed a wide range of "deviant" sexual practices in writing and lectures. For critics of Dr. Sex, investigation into diverse topics of human sexuality not limited to "normal" heterosexual practice was a cornerstone of European sexology that he had obviously neglected. Zhang's work, in short, pushed for the normalization of heterosexual monogamy; the work of other sexologists including Pan achieved "the setting apart of the 'unnatural' as a specific dimension in the field of sexuality."[119] Diverging from Dr. Sex, their writings made room "for [marginal] figures, scarcely noticed in the past, to step forward and speak, to make the difficult confession of what they were."[120]

In terms of quantity, Pan also translated more Western sexological texts. Although claiming that the facts and autobiographical narratives he solicited from readers formed the scientific basis of his sexological writing, Zhang translated relatively few foreign sexological works into Chinese. And even though Zhang frequently cited Ellis, Pan translated at least three monograph-length studies by Ellis, including the entire manuscript of *Psychology of Sex: A Manual for Students*.[121] Pan was so intrigued by Ellis's discussion of sexual inversion that at the end of his annotated translation of *Psychology of Sex*, he even included an appendix on "Examples of Homosexuality in Chinese Literature."[122] For the Ming-Qing period, Pan listed twelve cases of male homosexuality and one case of female homosexuality.[123] The work of other English sexologists similarly intrigued Chinese readership. In 1929 *New Women* published a translation of Edward Carpenter's "The Homogenic Attachment," the third chapter of his *The Intermediate Sex* (1908).[124] As it is well known, both Carpenter and Ellis were friends with John Addington Symonds (1840–1893), with whom Ellis collaborated on their controversial volume on *Sexual Inversion* (1897).[125] Classics by other prominent turn-of-the-century European sexologists such as Marie Stopes, Auguste Forel, and Solomon Herbert were also translated into Chinese, and they provoked similar public interests in the topic of same-sex affect.[126] Meanwhile, Chinese readers became increasingly conscious of the varying extent to which the original versions of these sexological texts were censored in Europe and the United States. Zhou Zuoren, for instance, discussed the Nazis' attack of the Institute for Sexual Research directed by Hirschfeld and the burning of its books and archives in 1933.[127] The banning of Radclyffe Hall's *The Well of Loneliness* (1928) in England in the late 1920s invited heated discussions in China.[128] All of these endeavors fell beyond the intellectual concerns of Dr. Sex.

Apart from topical diversity and the wide range of translated texts, Chinese sex scientists also valued the role of historical information in the cultural authority of sexology. If the signature of Dr. Sex's sexology encompassed the empirical understanding of sexual behavior through compiling and collecting actual life histories, it also involved, for Pan and others, the rendition of historical data on sexual variations so as to better illumine their relevance in contemporary Chinese society. Elsewhere, throughout the 1920s and 1930s, other writers followed Pan in looking back at same-sex practice in ancient societies (most notably Greek) and discussing its implications for the modernization of the Chinese nation.[129] Although both valued empiricism, Zhang and Pan adopted contrasting approaches to emulating Havelock Ellis: whereas Zhang was more concerned with collecting and responding to the contemporary "stories" or "cases" that people had provided him about their sexual experience, Pan devoted more effort to translating Ellis's work, a project supplemented by his own historical, sociological, and ethnological insights.

Although many Chinese sex researchers considered Ellis the preeminent European sexologist whom they tried to emulate, he was not the only role model. For example, Pan also introduced Freud's view of human sexuality to the Chinese public. If American eugenicists like Davenport paid no attention to Freud, Pan certainly embraced Freudian psychoanalysis wholeheartedly and used it as a legitimate scientific theory to explain sexual desire.[130] In his psychobiographical study of the late Ming poetess Feng Xiaoqing (馮小青, 1595–1612), Pan psychoanalyzed Feng's writings and concluded that she had narcissistic tendencies.[131] Other sinologists have viewed this effort as an early example of how psychoanalysis was transferred to China in the early twentieth century.[132] According to literary critic Haiyan Lee, "In [the hands of Western-educated May Fourth intellectuals], psychoanalysis was divorced from its clinical setting and retooled as a critical hermeneutic strategy. It served the enlightenment agenda of displacing both the Confucian moral discourse of sex/lust and the cultivational discourse of health/generativity with a scientific discourse of sexuality."[133] In addition to narcissism, Freud's theory of infantile sexuality also acquired some traction. For example, Sex Science published an article in 1936 titled, "Preparations for the Scientific Study of Sex," which postulated that "sex education should not stop at the level of genital anatomy and reproductive health, as infants between the age of two or three already develop their sexual roots."[134]

Indeed, Pan systematically used psychoanalysis in his writings as a modernizing scientific tool for diagnosing the sexual problems of Chinese society. In his annotated translation of the chapter "Sexual Education" from Ellis's Sex in Relation to Society (the sixth volume of Studies in the Psychology of Sex), Pan, in a footnote, recapitulated the idea of a five-stage progression of psychosexual maturation that he first articulated in his psychobiographical study of Feng: "'primary identification between mother and son,' 'maternal desire,' 'narcissism,' 'homosexuality,' and 'heterosexuality.'"[135] Two years later, in an article titled "Sexuality Today," Pan reiterated an identical pathway of psychosexual development: "it is necessary for the development of sexual desire to go through several stages: (1) primary identification, (2) the objectification of the mother's body and the desire for her,

(3) the realization of self-awareness and narcissism, (4) homosexuality as a result of the expansion of narcissism, and (5) heterosexuality as the result of the maturation of sexual physiology and sexual psychology."[136] When his translation of Ellis's *Psychology of Sex* appeared in 1946, he would refer to this process of psychosexual development again to explain the single case of female homosexuality he included in the appendix.[137]

In his 1910 revision of *Three Essays on the Theory of Sexuality* (1905), Freud added the following footnote on homosexuality: "In all the cases [of sexual inversion] we have examined we have established the fact that the future inverts, in the earliest years of their childhood, pass through a phase of very intense but short-lived fixation to a woman (usually their mother), and that, after leaving this behind, they identify themselves with a woman and take *themselves* as their sexual object. That is to say, they proceed from a narcissistic basis, and look for a young man who resembles themselves and whom they may love as their mother loved them."[138] Therefore, it seems that, from the 1920s to the late 1940s, Pan had endorsed Freud's explanation of homosexuality faithfully. Pan insisted throughout his publications that psychosexual maturation "is like a stream of water, and two changes could occur in the middle of this process: arrested or reversed development."[139] Readers who subscribed to Pan's psychoanalytic explanations would thus interpret same-sex desire in Freudian terms as an arrested or reversed phase of sexual maturation and as an inadequately developed psychological condition due to early childhood disturbances. Henceforth, the absorption of the sociocultural meaning of "same-sex desire" by the scientific category of "homosexuality" was in part enabled by the new epistemological framework of psychoanalysis.

Other medical and scientific experts shared this psychopathological view. In 1936, after returning from her psychiatric training at Johns Hopkins University, the practicing gynecologist Gui Zhiliang (桂質良) wrote in her widely read *The Life of a Woman* that "homosexuality is a kind of intermediate or preparatory stage to heterosexuality; it is necessary for people to go through it." According to Gui, those who develop "normally" would "pass" (過度, *guodu*) the stage of homosexuality, but others would "get blocked" (阻礙, *zuai*) or "stop" (停止, *tingzhi*) in the process and express "abnormal homosexuality" (不普通的同性戀愛, *buputong de tongxing lian'ai*). Similar to what Freud argued as early as in 1903, Gui did not think that homosexuality was necessarily "treatable" or "correctable."[140] In expressing strong sympathy for "homosexual persons" (同性戀患者, *tongxinglian huanzhe*), another writer, Wang Xianli (王顯理), defined the age range between twelve and sixteen years old as "the period of homosexual love" (同性愛時期, *tongxing'ai shiqi*). According to Wang, homosexual affection during this phase of growing up was a "normal" but "temporary" experience for teenagers. A normal young adult would get over this phase eventually and enter "the period of heterosexual love" (異性愛時期, *yixing'ai shiqi*).[141]

Unlike Zhang Jingsheng's somatic-oriented statements about sexuality, the transmission of Freudian psychoanalysis in the 1920s and 1930s offered a strictly psychogenic way of explaining same-sex desire. Sexologists such as Gui and Pan were no longer "concerned solely with what the subject wished to hide," which defined Dr. Sex's preoccupation, "but with what was hidden from himself." And by implication, they maintained their power—constituted not in advance, of course, but in a hermeneutic function, or "a discourse of truth on the basis of its decipherment"—to verify the truth of sex by rendering the subject's confession incomplete.[142] The making of a new science of sexuality both produced and relied on the psychoanalytic codification of their authority as the inducer, interpreter, interrogator, and ultimate arbiter of truth about people's libidinal drive. Serving as a new conceptualizing and modernizing tool, psychoanalysis emerged as an important cultural technology of the Republican period that made homosexuality a serious candidate of scientific thinking, a subject whose truth and falsehood became debatable among doctors and scientists of sex.

Out of the many debates on homosexuality, the most popular in this period concerned the question of whether the condition could be treated or cured. Besides Gui, many interlocutors of the debate, who had either translated foreign (Western or Japanese) sexological texts into Chinese or written about sex from a "scientific" viewpoint themselves, did not consider homosexuality necessarily curable. In an article that appeared in the journal *Sex Science* in 1936, for instance, the translator Chang Hong (長虹) defined "sexual perversion" as "those expressions of sexual desire that neither accompanied male-female love nor established procreation as its ultimate goal." The author presented homosexuality as one among the many existing types of sexual perversion (others include bestiality, fetishism, sadism, and masochism), and remarked that "if a man expresses both feminine and homosexual tendencies, no natural treatment is effective. At the same time, there is no pharmaceutical cure for this kind of situation."[143] Despite this explicit acknowledgment that no effective treatment of homosexuality was available, the article still construed same-sex desire and behavior as unsavory, especially by underscoring its categorical affinity to other kinds of sexual perversion like sadism, fetishism, and bestiality.

Chang's translated piece represented just one among the many perspectives circulating in a thematic issue of *Sex Science* devoted to the topic of homosexuality (figure 3.6). Another translated article in this issue with the title "Can True Homosexuality Be Cured?" advocated a less stigmatizing position. The author claimed that "recent scholars have come to believe that the nature of homosexuality is inborn, congenital, and immutable. The only situation in which an individual's homosexual desire could be changed is if it is an 'acquired' or 'fake' homosexuality. I agree with this perspective."[144] Elsewhere in the same issue, treatment methods for homosexuality such as surgical castration or psychological hypnosis were cast in a highly suspicious light.[145]

By and large, however, the essays in this thematic issue of *Sex Science* emphasized the likelihood for homosexuality to be acquired. While acknowledging that most experts had agreed on the inborn nature of homosexual tendencies, they nonetheless paid more attention to the prevalence of homosexual behavior in unisex settings, such as in schools, dormitories, factories, military units, prisons, and so on.[146] Yang Kai (楊開), a doctor who earned a medical degree at the University of Hamburg in Germany, noted that the number of homosexuals "among female students, employees, and workers is especially large in the present time." Although he suspected that the main cause of this "perversion" may be "inherited," Yang still attributed the high frequency of homosexual practice to "habits and the environment."[147] This was consistent with the impression one would get from reading the popular handbook *ABC of Sexology*, in which the author, Chai Fuyuan, noted that male homosexuality was more prevalent in schools, the military, and temples and that lesbianism occurred most frequently in the work place and factories.[148] According to another lengthy article in this special issue of *Sex Science*, "The Study and Prevention of Homosexuality," "the question of how homosexuality can be prevented is an empty question. Since homosexuality is widely recognized as a congenital condition, preventive methods are certainly very ineffective. But a hygienic social environment could suppress the occurrence of acquired, immature, or temporary homosexuality. Schools should be the primarily targets of hygienic intervention, because this could prevent the spreading of homosexuality on campuses."[149]

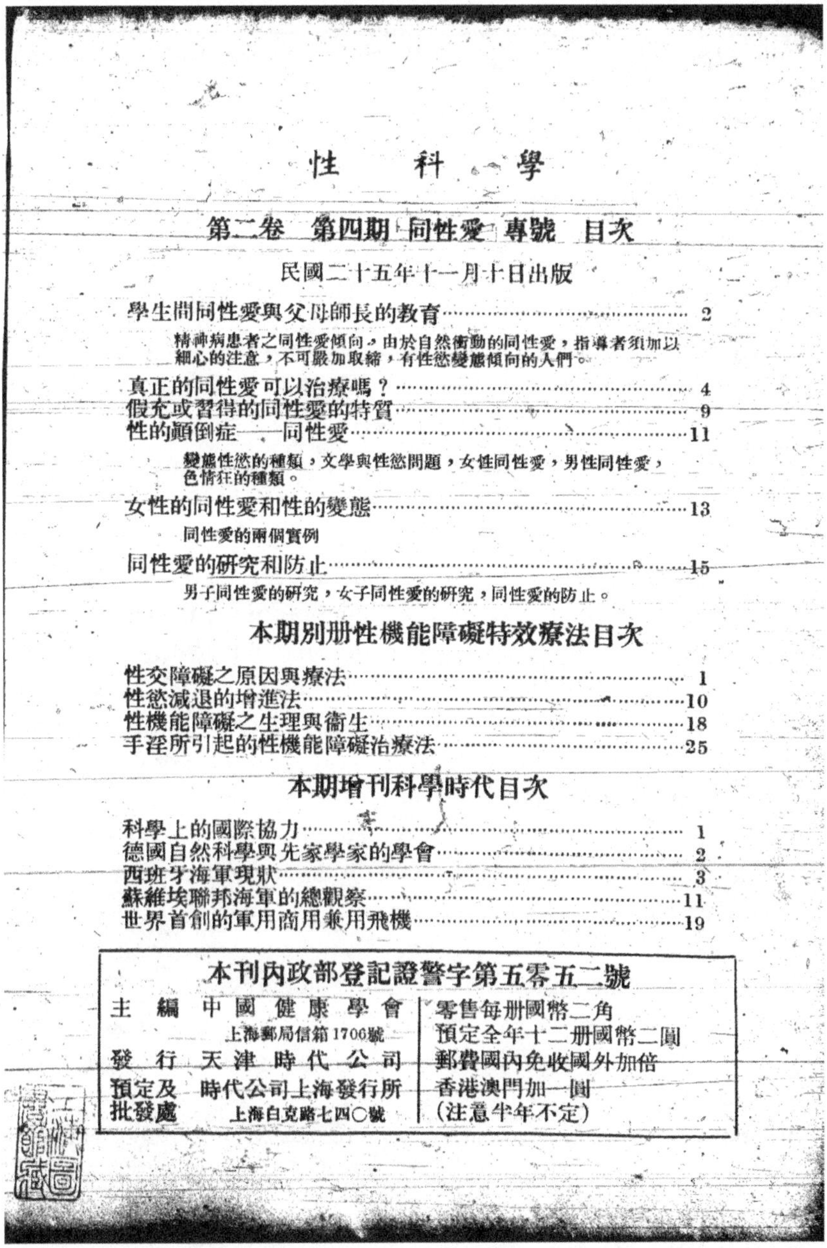
性　科　學

第二卷　第四期　同性愛　專號　目次

民國二十五年十一月十日出版

本期別册性機能障礙特效療法目次

本期增刊科學時代目次

本刊內政部登記證警字第五零五二號

主　編　中國健康學會
上海郵局信箱1706號
發　行　天津時代公司
預定及批發處　時代公司上海發行所
上海白克路七四〇號

零售每册國幣二角
預定全年十二册國幣二圓
郵費國內免收國外加倍
香港澳門加一圓
(注意半年不定)

Figure 3.6 The table of contents of the *Sex Science* special issue on homosexuality (1936). *Source: Xing kexue* (性科學) [Sex science] 2, no. 4 (1936).

But this must be done with great caution, as the opening essay of the forum warned its reader: if the surveillance policies of school dormitories are too strict and rigid, students might become "overly sensitive to sexual stimuli," and this would lead to a situation in which students are actually "more likely to engage in masturbation and homosexuality."[150] Hence, most of the articles in this special issue of *Sex Science* encouraged opportunities for mixed-sex socialization as a way to prevent homosexuality, implying that most same-sex erotic behaviors are perhaps more correctable than assumed.[151]

Correctable or untreatable, inborn or acquired, same-sex desire was now discussed in the Western psychiatric style of reasoning. The acquisition and articulation of this novel style of reasoning gave same-sex desire a new scientific meaning in twentieth-century China. In 1932 Gui Zhiliang stated in her book *Modern Psychopathology* that "some experts in psychopathology claim that homosexuality is the cause of paranoia . . . but although homosexuality could possibly induce paranoia, it does not have to be the sole cause of it."[152] Gui's allusion to the famous Freudian association of male homosexuality with paranoid delusions revealed that the Western psychiatric style of reasoning exhausted the semantic meaning and comprehensibility of same-sex eroticism in the context of this knowledge claim by the early 1930s. In the 1940s medical authorities began to foreground a distinctively psychopathological definition of "homosexuality" and criticize an overgeneralized usage of the term outside clinical or in quotidian contexts.[153] When twentieth-century Chinese commentators used "homosexuality" as a conceptual blueprint for understanding same-sex relations, they had completely displaced any of its nonpathological connotations in the premodern context. What they translated was not merely the vocabulary of homosexuality itself but a whole new style of reasoning descending from Western psychiatric thought about sexual perversion and psychopathology.

It should be stressed again that sex was not new to conceptions of health in traditional Chinese medicine. Concerns about the dangers of undisciplined sexual activities can be found in the very opening chapter of the *Inner Canon's Basic Questions*:

> The people of archaic times who understood the Way modeled [their lives] on [the rhythms of] yin and yang, and accorded with the regularities imposed by disciplines [of self-cultivation]. Their eating and

drinking were controlled, their activity and rest were regular, and they did not exhaust themselves capriciously. . . .

People of our times are not like that. Wine is their drink, caprice their norm. Drunken they enter the chamber of love, through lust using up their seminal essence (*jing*), through desire dispersing their inborn vitality (*zhenqi*). . . . Devoted to the pleasures of the heart and mind, they reject the bliss that accompanies cultivation of the vital forces.[154]

Unlike the Western psychiatric style of reasoning about sexual disorders, this passage makes it evident that traditional Chinese medical thinking conceptualized sexual desire and activity in quantitative terms, conveying a general rubric of "sexual economy."[155] This economy of sex follows the idea of an orderly life, stressed by medical scholars since the first millennium, that requires strict moral self-regulation and a spiritual life lived in harmony with the environment. In this cosmically ordered world of imperial China, as Charlotte Furth reminds us, "no *kind* of sex act or object of desire was singled out in medical literature as pathological."[156] To paraphrase Davidson, then, we can confidently say that as little as one hundred years ago, Western psychiatric notions of sexual identity (e.g., homosexuality) were not false in China but rather were not even possible candidates of truth or falsehood. Only after the translation and introduction of a psychiatric style of reasoning by the modernizing thinkers in the 1920s and beyond were there ways of arguing, verifying, explaining, proving, and so on that allowed such notions to be true or false.

The translation, mediation, and introduction of this new psychiatric style of reasoning hinged on an intellectual landscape of sexological disciplinarity. Although priding itself as a symbol of modernity, Zhang Jingsheng's *Sex Histories* soon triggered an opposite reaction. His critics perceived his sexological project as unscientific and attempted to move beyond its limitations. The scope of Pan's sexology, for example, included a broader range of topics not limited to "normal" heterosexual intercourse, translated a significantly higher quantity of foreign sexological literature, sought and drew on historical data for valuable insights concerning contemporary sexual problems, introduced a purely psychological account of human sexuality in the language of Freudian psychoanalysis, and thereby enabled debates on the etiology and prevention of "deviant" sexual practices. The convergence of all these efforts formed the necessary pillars upon which sexology came to be established as an independent scientific discipline. For feminist critics like Yan Shi (晏始), sexology (性慾學, *xingyuxue*) played an instrumental role for expanding public discussions about the "woman question."[157] Others stressed the importance of sexological research for distinguishing sexuality from love and for enumerating both scientific and artistic dimensions of sex.[158] The consolidation of sexology, in turn, provided sufficient grounds for bringing a foreign psychiatric style of reasoning into comprehensibility in Chinese culture. In depicting Zhang's sexological enterprise as hopelessly out of date, sex educators and scientists used it as a foil against which new measures of being "scientific," "modern," and, by extension, "traditional" could be juxtaposed.

No other example illustrates the outcome of this epistemic modernity better than the existence of an academic periodical called *Sex Science* in 1930s China. The Chinese Academy of Health (中國健康學會, *Zhongguo jiankang xuehuui*) was named as its official editorial governing board on the front page of each issue, and another, the Shanghai Sexological Society (上海性學研究社, *Shanghai xingxue yanjiushe*), was listed as the editorial collective of *Sex Magazine*.[159] According to one journalistic account, a closed "sexological association" (性學會, *xingxuehui*) ran weekly and monthly events, featuring lectures on abortions and other domains of sexual knowledge delivered by accredited experts, and it involved more than 130 regular members by 1931.[160] Although there is no doubt that many modernizing intellectuals at the time viewed human sexuality through the lens of social problems, these learned societies and disciplinary journals set the stage for sexual problems to be considered topics worthy of serious investigation in their own right.[161] In addition to providing a more focused venue for the translation of foreign sexological literature, *Sex Science* offered Chinese sexperts an unique opportunity to publish original contributions and sophisticated opinion pieces in direct dialogue with one another, foreign sexologists, and their readers. Their debates on the congenital or acquired nature of homosexuality reflected the gradual spread of psychoanalytic thinking about sexual latency and the psychosexual self, which in turn cast Dr. Sex's earlier explanation of heterogenital contact in terms of electrolytic *qi* as overly simple, insufficient, and passé. Similar to its Western counterparts such as the *Journal of Sexual Science* in Germany and *Sexology* in the United States, *Sex Science* functioned as a textual database reinforcing the specialized authority of sexology across culture. Its founding and circulation thus marked an important episode in the intellectual translation and disciplinary consolidation of *scientia sexualis* in Republican China.

Historicism Uncontested

If Foucault was correct in asserting that Western civilization was "the only civilization to practice a *scientia sexualis*," such practice had certainly proliferated to the East Asian world by the early twentieth century like never before.[162] But this chapter has also attempted to show that the historical significance of this proliferation rested on a level deeper than the superficial transfer of ideas across cultural divides. The unfolding of *scientia sexualis* in Republican China was governed by a discursive apparatus that I call epistemic modernity, in which explicit claims of sexual knowledge were imbricated with implicit claims about cultural indicators of traditionality, authenticity, and modernity.

In the context of Zhang Jingsheng's sexology, whether it is the dualism between literary representations of love versus scientific truthfulness of sex or the juxtaposition between Daoist cultivational ideas in Chinese medicine versus the biopsychological language of Western biomedicine, the analytic of epistemic modernity delineates the two registers of truth production on which sexological claims operated reciprocally: one concerning explicit claims about the object of scientific knowledge (human sexuality) and another concerning implicit claims about cultural markers of traditionality, authenticity, and modernity (modes of narrating sex, theoretical foundations of medicine, etc.). But Zhang's project quickly turned into the antithesis of science and modernity in the eyes of his contemporaries. Moving beyond the limitations of his work, they aimed to establish an independent discipline more akin to the richness of European sexological science. By the mid-1930s, disparate efforts in making sexuality a legitimate subject of scientific discussion and mass education culminated in such projects of disciplinary consolidation as the founding of *Sex Science* and the Shanghai Sexological Society. These unprecedented achievements gave rise to a radical reorganization of the meaning of same-sex desire in Chinese culture around a new psychiatric style of reasoning.

In the politically volatile context of Republican China, the introduction of Western sexology often reframed same-sex desire as an indication of national backwardness. In his *Sexological Science*, after documenting the prevalence of homosexual practice in different Western societies, Zhang Minyun (張敏筠) concluded that "the main social cause for the existence of homosexuality is upper-class sexual decadence and the sexual thirst of the lower-class people."[163] And this, according to Zhang, should help shed light on "the relationship between homosexuality and nationality."[164] "For the purpose of social improvement," according to another concerned writer, "the increasing prevention of homosexuality is now a pressing task."[165] Pan Guangdan expressed a similar nationalistic hostility toward the boy actors of traditional Peking opera: since they often participated in sexual relationships with their male literati patrons, Pan described them as "abnormal" and detrimental to the social fibers of morality. He explained that their lower social status prevented them from participating in the civil examination system, implying that a modernizing nation in the twentieth century certainly has no place for them.[166] The physician Wang Yang (汪洋), known for his expertise in human sexuality and reproduction, went so far to identify homosexuality as "a kind of disease that eliminates a nation and its races."[167]

Therefore, if we take the insight of Lydia Liu and others seriously, the apparatus I call epistemic modernity that mediated the transmission of *scientia sexualis* to China ultimately characterizes a *productive* historical moment.[168] When Republican Chinese sexologists viewed the *dan* (旦) actors and other cultural expressions of homoeroticism as signs of national backwardness, they had in essence domesticated the Western psychiatric style of reasoning and turned it into a new nationalistic style of argumentation about same-sex desire.[169] In addition to staging certain elements of the Peking opera field as being out of time and place, epistemic modernity occasioned an entrenched nationalistic platform on which other aspects of this cultural entertainment also functioned as a powerful symbol of quintessential Chinese tradition and authenticity. Rendered as a prototypical exemplar of the modern homosexual, the twentieth-century *dan* actor became a historic figure signifying a hybrid embodiment of the traditionality and what Prasenjit Duara aptly calls "the regime of authenticity" of Chinese culture.[170]

It is therefore possible to contrast this new nationalistic style of argumentation with the culturalistic style of argumentation that underpinned the comprehensibility of same-sex desire in the late imperial period.[171] For this purpose, we can turn to the late Ming essayist and social commentator Zhang Dai (張岱), who reflects on his friend Qi Zhixiang's fondness for a young man named Abao in his *Tao'an mengyi* (Dream reminiscence of Tao'an). Tao'an is Zhang's pen name, and this collection of miscellaneous notes serves as a good window onto the literati lifestyle circa the Ming-Qing transition, since Zhang is often considered as an exemplar of literati taste of the time. An example from the late Ming is also most apt because the period is infamous for marking the peak of a flourishing "male love" (男色, *nanse*) homoerotic culture in late imperial China. The title of this passage is "The Obsession of Qi Zhixiang," and because it places seventeenth-century male same-sex love in the context of multiple desires, it is worth quoting in full:

If someone does not have an obsession (*pi*), they cannot make a good companion for they have no deep passions; if a person does not show some flaw, they also cannot make a good companion since they have no genuine spirit. My friend Qi Zhixiang has obsessions with calligraphy and painting, football, drums and cymbals, ghost plays, and opera. In 1642, when I arrived in the southern capital, Zhixiang brought Abao out to show me. I remarked, "This is a divine and sweet voiced bird from [the paradise of] the western regions, how did he fall into your hands?" Abao's beauty was as fresh as a pure maiden's. He still had no care for decorum, was haughty, and kept others at a distance. The feeling was just like eating an olive, at first bitter and a little rough, but the charm is in the aftertaste. Like wine and tobacco, the first mouthful is a little repulsive, producing a state of tipsy lightness; yet once the initial disgust passes the flavor soon fills your mind. Zhixiang was a master of music and prosody, fastidious in his composition of melodies and lyrics, and personally instructing [his boy actors] phrase by phrase. Those of Abao's ilk were able to realize what he had in mind. In the year of 1645, the southern capital fell, and Zhixiang fled from the city to his hometown. En route they ran across some bandits. Face to face with death, his own life would have been expendable, but not his treasure, Abao. In the year of 1646, he followed the imperial guards to camp at Taizhou. A lawless rabble plundered the camp, and Zhixiang lost all his valuables. Abao charmed his master by singing on the road. After they returned, within half a month, Qi again took a journey with Abao. Leaving his wife and children was for Zhixiang as easy as removing a shoe, but a young brat was as dear to him as his own life. This sums up his obsession.[172]

This passage also sums up what a man's interest in young males meant in the seventeenth century remarkably well: it was perceived as just one of the many different "obsessions" that a male literatus could have—a symbol of his refinement. For Zhang, a man's taste in male lovers was as important as his "obsessions" in other arenas of life, without which this person "cannot make a good companion." Despite all the hardship, the romantic ties between Qi and Abao still survived, and perhaps even surpassed Qi's relationship with his wife and children.

Let us now compare this to a twentieth-century example. For the most part, there was a distinct absence of discussion about same-sex sexuality in the numerous sex-education pamphlets published throughout the late 1940s and 1950s.[173] But in the few instances where homosexuality assumed exceptional presence, the way it was described and the specific context in which it was brought up would appear strange and foreign to Ming-Qing commentators on the subject. In a sex-education booklet for adolescents published in 1955, the author wrote: "Certainly, sometimes 'same-sex desire' is only psychological and not physical. For example, a girl might be very fond of a same-sex classmate, to the extent that she even falls in 'love' with her. Their relationship could be quite intimate in that they might even share the same bed and touch each other, but there is actually nothing beyond that. For this type of same-sex love/desire, it is easily curable. As long as they get married separately, whatever happened could be easily forgotten."[174]

The author, Lu Huaxin (陸華信), went on to describe a symmetrical situation for those adolescent boys who have developed a similar kind of affection for classmates of the same sex. But Lu insisted that "as long as [these] teenager[s] get married, the pathological feelings will disappear." Only for certain teenagers whose "lifestyle has become decadent" and who "really starts pursuing abnormal sexual gratifications," Lu continued, "their brain then really needs to be treated. Because their thoughts are unhealthy and filthy; they have been infected by a pornographic virus. If an individual of this type is identified, friends should encourage everyone to offer help and assistance."[175]

By the mid-twentieth century, same-sex desire had acquired a set of social meaning and cultural significance radically different from the way it was conceived before the onset of epistemic modernity. For one, the relationship between same-sex desire and heterosexual marriage is viewed as incommensurable or incompatible, even antithetical. One could not possibly be married to an opposite sex while still passionately desiring someone of the same sex.[176] In fact, according to Lu, heterosexual marriage is precisely the most useful "cure" of same-sex desire. Same-sex desire now also means a pathological—and not just abnormal—tendency, based on which an autonomous relationship between two persons of the same sex is conceivable regardless of their social status. Lu located the seat of this deviant subjectivity inside the brain, via a vague notion of viral infection, which underscores the "pathological" or "unhealthy" nature of its psychological status. Again, as same-sex desire now represents something that is "curable," heterosexual marriage could serve that function most powerfully. No longer understood simply as one of the many "tastes" or "obsessions" that a man of high cultural status could have, erotic preference for someone of the same sex becomes something that could be eliminated with the help of friends, as opposed to something that could be appreciated by them.

To assess the epistemological transformation of same-sex desire in Chinese culture from an indigenous historical perspective, then, we can begin to reconstruct some of the polarized concepts that constitute two opposed styles of argumentation. We are presented, for instance, with the polarities between literati taste and sick perversion, refined obsession and pathological behavior, cultural superiority and psychological abnormality, markers of elite status and signs of national backwardness. The first of each of these pairs of concepts partially makes up the culturalistic style of argumentation about same-sex desire while the second of each of these pairs helps to constitute the nationalistic style of argumentation. These polarities therefore characterize two distinct intellectual modes of representation, two conceptual spaces, two different kinds of deep epistemological structure. It follows that the discursive apparatus of epistemic modernity has not merely mediated the introduction of the foreign sexological concept of homosexuality (thereby pushing the concept of same-sex desire over the threshold of scientificity), but, in doing so, it has simultaneously catalyzed an internal shift in the conceptual paradigm of Chinese same-sex desire.

According to Ari Larissa Heinrich, in the nineteenth century, China metamorphosed from being identified as "the Cradle of Smallpox" to a pathological empire labeled with growing intensity as "the Sick Man of Asia."[177] My analysis suggests that this transformation took another turn in the early Republican period as China emerged from its image as a "castrated civilization." After the introduction of European *scientia sexualis* in the 1920s, the Chinese body could no longer be conceived in mere anatomical terms. It became rather appropriate, and perhaps even necessary, to conceptualize the Chinese body as explicitly *sexual* in nature. Chinese corporeality is now always linked to implicit claims of psychiatric reasoning and nationalistic significance. Put differently, a distinct problem in modern Chinese historiography has been the question of why, starting in the Republican period, Chinese modernizers began to view earlier expressions of same-sex eroticism and gender transgression as domestic indicators of cultural deficiency. And what I am suggesting is that, much like how the gradual acceptance of an intrinsically pathological view of China helped the reception of Western-style anatomy in nineteenth-century medicine, the epistemic alignment of prenationalistic homoeroticism with the foreign notion of homosexuality precisely undergirded the appropriation of a new science of Western-style sexology in twentieth-century China.

In light of the prevailing criticisms of Foucauldian genealogy, many historians of sexuality have refrained from advancing a claim about the occasioning of an epistemological break in the Republican era by showing that earlier concepts associated with male same-sex sexual practice (e.g., *nanse* or *pi*) jostled alongside and informed the new sexology discourse.[178] However, it has been my intention to show that the congruency between earlier and later understandings of same-sex practice is *itself* a cultural phenomenon unique to the Republican period and not before. Wenqing Kang, for example, has argued that preexisting Chinese ideas about male favorites and *pi* "laid the ground for acceptance of the modern Western definition of homo/heterosexuality during [the Republican] period in China." His first explanation is that "both the Chinese concept *pi* (obsession) and Western sexology tended to understand same-sex relations as pathological."[179] He then relies on Eve Sedgwick's model of the overlapping "universalizing discourse of acts" and "minoritizing discourse of persons" to suggest that indigenous Chinese understandings shared a comparable internal contradiction in the conceptualization of male same-sex desire.[180] In his words, "The concept *pi* which Ming literati used to characterize men who enjoyed sex with other men, on the one hand implied that men who had this kind of passion were a special type of people, and on the other hand, presumed that the obsession could happen to anyone."[181]

My reading of Zhang Dai's passage on *pi* suggests that isolating both a pathological meaning and this internal conceptual contradiction of *pi* represents an anachronistic effort that reads homosexuality into earlier modes of thought. Zhang's remark precisely reveals the multiplicity of the meaning and cultural significance of *pi* that cannot be comprehended through a single definition of pathology or an independent lens of same-sex relations decontextualized from other types of refined human desire. Kang therefore seems to forget that the very semblance between what he calls "the internal contradictions within the Chinese indigenous understanding of male same-sex relations" and "those within the Western modern homosexual/heterosexual definition" was made possible and meaningful only in contemporaneity with the emergence of homosexuality as a concept in China.[182] In this regard, the following statement confuses his interpretation of historical sources with the very colonial landscape it claims to exceed: "When Western modern sexology was introduced to China in the first half of the twentieth century, the Chinese understanding of male same-sex relations as *pi* (obsession) was very much alive, as evidenced in the writings of the time. It was precisely because of the similarity between the two sets of understandings that Western modern sexology could gain footing in China."[183] The claim is confusing because the similarity Kang points to would not have made much sense in a context without the epistemological salience of the very concept of homosexuality itself, that is, before the twentieth century. Treating the discursive nature of discourse seriously requires us to pay closer attention to how old words take on a new meaning (and life) in a different historical context rather than imposing later familiar notions onto earlier concepts.[184] A distinct problem with Kang's reading remains the way he turns a blind eye to the hierarchical nature of the invocation of *pi* in literati discourses. It might be useful to rephrase this problem by borrowing David Halperin's remark: "Of course, evidence of conscious erotic preferences does exist in abundance throughout the surviving documentary record, but it tends to be found in the context of discourses linked to the senior partners in hierarchical relations of pederasty or sodomy. It therefore points not to the existence of 'gay sexuality' per se but to one particular discourse and set of practices constituting one aspect of what counts as gay sexuality nowadays."[185]

Despite how Pan Guangdan's condemnation of the homosexuality of boy actors (and, by implication, their patrons) was informed by the long-standing and still-continuing practices of male prostitution, his condemnation was made possible—and comprehensible—only by the arrival of a psychiatric style of reasoning that construed same-sex relations in negative terms. In their study of nineteenth-century "flower guides" (*huapu*), cultural historians Wu Cuncun and Mark Stevenson have probed the many social taboos surrounding this literary genre that extolled the beauty of boy actors, including "rules about money and taste and passion and lust, and also rules about the representation of social competition." They conclude that "none of these were concerned with fears of same-sex desire or of stigma through connection to the world of Beijing's homoerotic nightlife."[186] The scientific reasoning of desire that gained rapid momentum in the 1920s, on the other hand, ushered in a new era of the social stigmatization of male same-sex relations. Pan and other sexologists isolated homosexuality as a conceptual blueprint for individual psychology independent of hierarchical indexes of power relations, social status, class subjectivity, and so on, but it was a concept that, unlike heterosexuality, carried a pathological connotation and linked to notable cultural signifiers of traditionality contributing to, according to them, China's growing national deficiency. It was in this context that homosexuality came to set itself apart from gender transgression as two distinct nodes of conceptualization in modern Chinese culture.

Therefore, the twelve cases of male homosexuality and the one case of female homosexuality that Pan enumerated in his annotated translation of *Psychology of Sex* should be understood less as historical evidence of homosexual experience in the Ming and Qing dynasties than as a reflection of how the epistemological reorientations brought about by a new psychiatric style of reasoning crystallized the condition of their comprehensibility. Here is where I part company with the literary scholar Giovanni Vitiello, who interprets Pan's effort "as if to provide a Chinese perspective on an experience inadequately represented in the Western book. These negotiation attempts remind us that the transformation of sexual culture in twentieth-century China cannot be read simply as the replacement of one model with another."[187] Two major assumptions are embedded in Vitiello's statement: first, that the internal coherence of an unified structure of homoerotic sentiment had *always already* existed in China before the Western concept of homosexuality, and, second, that the alignment between the former and the latter structures of knowledge was inevitable and unproblematic.

The heart of the matter does not concern the question of whether the contested process of translation is itself fraught with the possibility of "losing" or "adding" new dimensions of knowledge (of course it is). But what escapes Vitiello's reading is the way in which both the internal coherency of an indigenous structure of knowledge on which the foreign model of homosexuality could be easily mapped *and* the condition of possibility of this mapping were themselves historically contingent on—even historically produced by—the very process of translating "homosexuality" into a comprehensible Chinese concept. Likewise, when Pan and other sexologists used examples from ancient Greece to render the modern category of homosexuality intelligible, the result was a similar moment of epistemic refiguration in the establishment of *scientia sexualis* in China. Their debates on "true" or "fake," "inborn" or "acquired," "natural" or "curable" homosexuality in the pages of *Sex Science* already took for granted the new psychiatric style of reasoning and so treated sexuality and its attendant disorders, such as homosexuality, as if they were naturally given and carrying broader implications for the modern nation. Simply put, the epistemic continuity established by Chinese sexologists between the foreign concept of homosexuality and earlier examples of homoeroticism do not undermine the kind of Foucauldian epistemological rupture I have been proposing but actually exemplify it. Before the rupture, according to the normative definition of desire in male spectatorship and connoisseurship, the possibility of having the same (homo)sexuality as either the *dan* actor or the male favorite would appall the literati gentleman.

Epistemic modernity, then, is more than just an example of "translated modernity"; rather, it refers to a series of ongoing practices and discourses that could generate new ways of cultural comprehension and conceptual engagement, allowing for possible intersecting transformations in history and epistemology. If we ever wonder how to make sense of the prevalence of same-sex sexual practice in China before the rise of an East Asian *scientia sexualis*, as so vividly captured in *The Carnal Prayer Mat*, we only need to remind ourselves that as little as a century ago the question of sexual identity did not even fall within the possible parameters of Chinese thinking. For in China there is no such thing as homosexuality outside epistemic modernity.

CHAPTER IV

Mercurial Matter

A Malleable Essence

I n an era when science carried an unrivaled measure of cultural prestige, the comprehensibility of the concept of sex was crystallized through the intersection of its three epistemological coordinates—as the object of observation, the subject of desire, and a malleable essence of the human body. The last two chapters have explored the unilateral labors of visual persuasion, newly invented words, narrative techniques, methodological innovation, expertise friction, professional politics, and contested claims of modernity that anchored the development of new structures of knowledge around the visuality and carnality of sex. The aim of this chapter is to investigate its third epistemic linchpin, mutability, and the vibrant discourse of "sex change" that saturated the mass circulation press from the 1920s to the 1940s.

In the second quarter of the twentieth century, the urban intelligentsia began to envision a more fluid definition of humanity. They no longer drew on the limited language of anatomy to talk about two different but equal sexes; rather, they started to think of men and women as two versions of a universal human body. In this period, Chinese sexologists came to embrace the plausibility of sex transformation based on the vocabulary of sex endocrinology, famous sex-reversal experiments on animals, and a new scientific theory of universal bisexuality (referring to a form of biological sex constitution rather than sexual orientation). Indigenous Chinese frameworks for understanding reproductive anomalies, such as eunuchs and hermaphrodites, provided scientific writers a conceptual anchor for communicating new and foreign ideas about sex. After mapping these cross-cultural shifts, the second half of this chapter reconstructs a highly sensationalized case of female-to-male transformation in mid-1930s Shanghai, the story of Yao Jinping (姚錦屏), and assesses its impact on the wider awareness of human sex change. The tensions of these epistemological reorientations nourished the authoring of a science fiction short story called "Sex Change" (1940) by the pedagogical writer Gu Junzheng (顧均正). The chapter concludes by highlighting the 1940s as a new era in which people began to consider sex-reassignment opportunities through the possibility of surgical intervention.

I argue that over the course of this period, as scientific ideas were transmitted and disseminated into popular culture and the pledge to the value of medical science deepened beyond its professional parameters, accounts of "sex change" gradually loosened its association with animal experiments and human reproductive defects. These accounts turned into sensationalized stories of bodily change that, in the decades before the concept of "transsexuality" was available, introduced Chinese readers to the possibility of transforming physical sex. As stories about the relative ease of sex metamorphosis flooded the press, they made the body seem more mercurial than previously assumed.

A New Hormonal Model of Sex

In the mid-1920s Chinese sexologists gradually withdrew from an anatomical framework and shifted their definition of sex to one based on chemical secretions. Informed by European endocrine sciences, they began to view sex as a variable of specific chemical substances found in the bloodstream.[1] Previously, the anatomical register of the human body proved to be a useful guide for deciphering the biological difference between male and female.[2] As the May Fourth fervor began to wane, however, Chinese writers no longer looked to the structural underpinnings of male and female reproductive organs as the natural arbiter of sexual difference. Rather, they became invested in the idea that gonadal secretions—specifically, the chemical substance produced by testes or ovaries—were the actual determining agents of human sex difference. Whereas eggs and sperm occupied the center stage in an earlier scientific discourse of sex, *hormones*—the name that was available for the chemical messengers that control sexual maturity and development—constituted the focus of discussions on sex by Chinese modernizing thinkers throughout the 1920s and 1930s.[3]

In *The Internal Secretions* (1924), Gu Shoubai (顧壽白, 1893–1982) expressed this new view of sex in unambiguous terms. A Japan-educated doctor who authored numerous books on biology and medicine for the Shanghai Commercial Press, Gu posited that "in addition to sperm cells," testes "produce a kind of stimulating substance [刺激素, *cijisu*] that gives the physical body a uniquely male quality [男性特有之發育, *nanxing teyou zhi fayu*]." Similarly, for the female sex, "besides eggs, ovaries produce a kind of stimulating substance, whose clinical presence has been experimentally proven by researchers." The surgical removal of ovaries would thus result in an unfeminine physical and mental state: "Specifically, [a woman's] body becomes larger and stronger; she lacks gentleness; her genital develops inadequately; she lacks sexual desire; psychologically, she does not show the kind of characteristics and temperaments typically associated with women."[4]

Other sexologists spoke more nebulously about the effects of the internal secretions. "Other than producing eggs," Chai Fuyuan explained in his widely read *ABC of Sexology* (1928), "the ovary, like testicles, plays a functional role in the internal secretions. It excretes a fluid with an unpleasant odor in the bloodstream that promotes the development of femaleness [形成女性, *xingcheng nüxing*]." Whereas Gu described the chemical secretion of the sex glands in a more cautious way (by calling it a "stimulating substance"), Chai simply called it a liquid. His reader was thus led to believe inaccurately that hormones are actually fluids. Less ambiguous was Chai's effort in holding these chemical substances responsible for determining one's biological manhood and womanhood. He boldly declared that "a woman is a woman only because of this fluid," which, according to him, "has three main effects on the female sex: first, it increases sexual desire and the intensity of orgasm when the body comes into contact with men. Second, it stimulates secondary sex characteristics [次性特徵, *cixing tezheng*], including the enlargement of the pelvis, the scarcity of body hair, the smoothness and paleness of the skin, etc. Third, it nourishes the body, strengthens the mind, and increases memory capacity and the ability to imagine." Notably, two of the three effects identified by Chai correspond to those found in Gu's discussion above. And lest any reader felt unsure about what to conclude from all of this, Chai pronounced, "without the internal secretions, a woman is not a woman."[5]

According to the new vision of sex as articulated by Gu and Chai, the degree of maleness and femaleness depended less on the presence of gonads than on the quantity of chemical agents that Gu called "stimulating substance," a term he evidently used to translate the Western concept of hormone. It would be at least another half a decade before male and female sex hormones were structurally discovered, isolated, and synthesized by scientists in Europe and the United States.[6] But in the 1920s, the Chinese lay public was already introduced to a quantitative definition of sex. As Wang Yang, the head of Zhongxi Hospital (中西醫院, *Zhongxi yiyuan*), corroborated in 1927, "the development of sexual differences between men and women is entirely contingent on the secretions of the reproductive glands."[7] In this way, the natural construction of manhood and womanhood appeared more malleable than previously thought. Earlier discussions on the subject by May Fourth feminists tended to ground social gender equality in the biological development of dual anatomical sexes. With new ideas from endocrinology, the nature of sex no longer relied on the structural outgrowth of genital ontology but was directly governed by the invisible chemical messengers circulating in the bloodstream.

As a substance whose natural effects depended crucially on its quantity, the internal secretions for the first time solved the mystery of an age-old practice in China: urine consumption. According to an article published in *Sex Science*, the practices of "replenishing yang with yin" (採陰補陽術, *caiying buyangshu*) and drinking the urine of young boys represented solid evidence for serious alertness to the endocrine system in ancient China. The article chronicled a history of the endocrine sciences, detailing the work of such crucial pioneers as the German physiologist Arnold Berthold (1803–1861), who experimented on the role of gonads in the development of secondary sex characteristics, and Ernst Laqueur (1880–1947), who coined the term "testosterone" with the Organon research group in the Netherlands. By depicting the male gonad and urine as the key source of male sex hormones, the author claimed that "the reason for the Chinese to consume the urine of young boys stems from an awareness that it contains internal secretions; although the quantity is not much, the advantages are telling if consumed on a long-term basis." The article described female sex hormones as bearing a functional similarity that boosts vitality and metabolism in women.[8] In fact, ever since the French physiologist Claude-Édouard

Brown-Séquard (1817–1894) claimed success in increasing physical and mental strength by injecting himself with animal testicular extracts in 1889, scientists and intrigued laypersons all over the world had praised what came to be known as the "Brown-Séquard Elixir."[9] When this "discovery" reached the Chinese press, the presumed rejuvenating properties of testosterones distinguished the new biochemical vision of sex and the perception of hormones as the quintessence of life.[10]

Hormones straightforwardly linked primary to secondary sexual characteristics.[11] They helped to explain, for instance, the positive correlation frequently found in male bodies between penis and testicles, on the one hand, and masculine physical traits such as muscular strength, larger bone structure, deeper voice, and so forth on the other. In a sexological textbook called *The Principle of Sex*, translated from Japanese into Chinese by Wang Jueming in 1926, "hormone" was referred to as "something without which the development of secondary sex characteristics cannot happen." The author qualified that "adequate growth of all secondary sex characteristics begins only with the full maturity of the sex glands."[12] The new language of endocrinology was not available to an earlier cohort of reformers and nationalists who also advocated gender equality. From the 1930s on, the growing popularity of endocrinology helped to illustrate the connection between anatomical sex and secondary sex characteristics— those bodily traits typically considered the keystone of masculinity and femininity. To a new generation of Chinese commentators on gender, hormone authorized an enticing biological lexicon for naturalizing the social interpretation of sexed bodies.

A highlight in this transitional period was a special issue of *Sex Science* devoted to the theme "Sex Endocrinology" (性內分泌科學, *xing neifenmi kexue*) published in 1936. This issue represented the first sustained treatment of sex hormonal science in a professional serial publication in Chinese. Billed as a "focused section," the entire issue was in fact penned by a single author, Li Yongnian (李永年), who held a medical degree from the University of Berlin. Li opened his piece by stating that one of the most powerful contributions of modern science had been the discovery of the internal secretions and their role in shaping sexual difference. After going over the familiar numerical calculations of the autosome-to-sex chromosome ratio in humans and animals, Li's discussion stressed the sex-defining nature of the sperm: because it carried the Y chromosome, "the sperm plays a causal role in the differentiation between male and female sexes."[13]

Aware of the ongoing debates about the clinical importance of endocrinology, Li devoted a significant portion of his essay to addressing the issue of homosexuality. His overarching claim was that "homosexuality still cannot be explained by today's scientific understanding of the endocrine glands." Under the section titled "*nanse*" (男色, male lust), Li observed that most "passive" male homosexuals achieved mature sexual development with ease, but their manners tended to be overtly feminine. Since this phenomenon "resembles that of the *dan* actors in Peking opera," he concluded that the effeminacy of male homosexuals "has resulted entirely from behavioral habit." For Li, it was not useful to explain homosexuality through glandular science. After speculating a number of hypotheses concerning the causation of homosexuality, including a troubled marriage prospect and the lack of an appropriate partner for channeling one's erotic urges in certain social contexts, he endorsed a sociological perspective, rather than one rooted in biology, to explain this "social vice." This conceptual gap between the science of sex hormones and homosexuality evident in Li's discussion demonstrates that even as some Chinese sexologists began to imagine the sexed body as more pliable than anatomically predetermined, they still distinguished the biological basis of sex from the subjective realm of sexuality.[14]

Animal Sex Transformations

In the early twentieth century, the idea that masculinity and femininity were alterable through biochemical agents soon triggered an avalanche of publicity about sex transformation. For decades, European sexologists had produced a vast quantity of clinical literature on "aberrant" gender identification and sexual inclinations. Late Victorian sex scientists often conflated a range of different gender and sexual expressions in their writings, usually under the rubric of "sexual inversion."[15] At the same time, they devised an impressive taxonomy to classify these diverse orientations. As many historians have shown, this vast array of sexological vocabulary emerged from the intervention of people with unconventional gender or sexual identification in that they were not merely passive agents objectified by the medical profession.[16] It was only by the mid-twentieth century, nonetheless, that such crucial sexological categories as homosexuality, bisexuality, and transsexuality came to be distinguished more cogently in the medical literature.[17] But when this process was just beginning to unfold, scientists in Europe and, to a lesser extent, America were already broaching the broader significance of sex-change surgeries on a frequent basis.

In the 1910s Vienna stood at the forefront of sex-change experiments. The Austrian physiologist Eugen Steinach (1861–1944) attracted international acclaim for his "transplantation" experiments on rats and guinea pigs. In 1912 and 1913, respectively, he published "Arbitrary Transformation of Male Mammals into Animals with Pronounced Female Sex Characteristics and Feminine Psyche" and "Feminization of Males and Masculinization of Females." The articles soon became scientific classics, and the experiments on which they were based led Steinach to place his research in the larger turn-of-the-century scientific project that attempted to locate the biological essence of sex in gonadal secretions. These groundbreaking experiments also suggested the possibility of medically transforming sex. As he put it, "the implantation of the gonad of the opposite sex transforms the original sex of an animal."[18] His work directly influenced Magnus Hirschfeld, Harry Benjamin, and other sexologists who participated in the delineation of the concept of "transsexuality" around the mid-twentieth century.[19]

Word of the sex-change experiments conducted by Western biologists reached China primarily through the mass circulating sexological literature in the 1920s. Some Chinese sexologists placed Steinach's "transplantation" studies in a broader discussion of the relationship between secondary sex characteristics and heredity. As early as 1924 Fei Hongnian (費鴻年), a professor of biology at Beijing Agricultural University, introduced Steinach's work in his *New Treatise on Life* (新生命論, *Xin shengminglun*), which enumerated the effect of "transplantation" surgeries on generation. Fei first described the results obtained by the German scientist Johannes Meisenheimer (1873–1933), who claimed to have inserted ovaries into male moths and testes into female moths, with the outcome that the transplanted organs remained functional and grew without impeding the process of metamorphosis.[20] Steinach's work was then pointed out as another example of the success of gonadal transplantations without producing detrimental effects on animal vitality. Finally, Fei mentioned American physiologist Charles Claude Guthrie's (1880–1963) findings from grafting ovaries of black hens into white hens and ovaries of white hens into black hens. The change in the color of eggs as a result of ovarian transplantation provided evidence for the effect of transplantation on heredity.[21]

In the 1920s Chinese scientists considered physical sex transformation the most intriguing aspect of these transplantation experiments. The scientific reports from Europe and America allowed some Chinese to entertain the possibility of sex reversal at least in animals. In *The Internal Secretions*, for example, Gu Shoubai offered a more sustained discussion of Steinach's studies under the sections called "The Feminization of Males" and "The Masculinization of Females."[22] He began by stipulating the recent discovery that gonadal secretions bear a causal relationship to male and female traits, both psychological and physical. "According to this line of reasoning," Gu wrote, "if a male organism's testes are removed and replaced with ovaries before puberty, he can turn into female [男性當可化為女性, *nanxing dangke huawei nüxing*]." It was with the intention of providing this statement a scientific basis that Gu presented Steinach's findings in remarkable detail.[23]

Gu began by describing Steinach's laboratory method as succinctly as possible: "Steinach removed the testes from male animals and transplanted ovaries into their body as an attempt to feminize them both physically and psychologically [使其肉體精神均為女性化, *shiqi routi jingsheng junwei nüxinghua*]." Gu then described Steinach's findings in great detail. After three to four weeks, Gu noted, Steinach made the following observations about castrated male rats with implanted ovaries: their implanted ovaries developed normally and even produced eggs; their original penis shrank and degenerated (退化萎縮, *tuihua weisuo*); the size of their enlarged breasts was similar to the size of breasts found in regular female rats, and they even exhibited "maternalistic" tendencies; unlike the thicker body hair found in normal male rats, these animals had finer and smoother hair; they accumulated more body fat; their bone structures were smaller than normal males; and they displayed more "female-like" qualities, including a softer and more gentle physique. But for Gu, the key evidence that suggested the feminization of male animals was the psychological changes induced by the surgeries: the laboratory animals "displayed no male psychological traits"; they were "not passionate, not stimulated, and not excited" when put in contact with female animals; and in contrast, when they were acquainted with other male peers, they "suddenly displayed manners that are uniquely female, including raising the posterior end of their body to seduce male animals. . . . They basically exhibited any trait typically associated with female animals."[24]

In the section titled "The Masculinization of Females," Gu offered a symmetrical description of Steinach's experiments on female animals. Steinach inserted testicles into infant female rats whose ovaries had been eliminated. According to Gu, Steinach made the following observations about these transplanted animals: their implanted testicles developed normally; their original vulva degenerated and all or parts of their vaginal opening shrank significantly; their breasts could not grow into the size of regular female breasts; their hair became as thick as regular male body hair; they did not accumulate fat in a way that would give them a regular female physical appearance; their bone structure developed into a manly size and shape; and psychologically they became as competitive as their male counterpart. After presenting Steinach's experiments on rats, Gu expressed a considerable measure of optimism toward the prospect of similar sex-change phenomenon in humans: "Although the two kinds of sex transformation described above are experiments conducted on animals with success, we do not yet have formal reports of similar procedures tested on humans. Theoretically speaking, though, it is reasonable to entertain the feasibility [of human sex transformation]."[25]

In the early twentieth century, Steinach's sex-change experiments soon became classics in not only Western but also East Asian sexological circles. European-trained scientists such as Gu Shoubai and Fei Hongnian had the linguistic ability to introduce a range of European and American scientific research to Chinese readers in a firsthand manner. However, around the same time, more Chinese students on Qing government scholarships studied in Japan rather than Western countries. Indeed, a significant portion of the Chinese public acquired familiarity with Western sexology from reading translations of Japanese sexological literature. Steinach's work was mentioned, for instance, in Wang Jueming's translation of *The Principle of Sex* (1926), originally written in Japanese. The book pitched Steinach's studies, along with other transplantation experiments, as evidence for the direct influence of glandular secretions on the development of secondary sex characteristics. After briefly summarizing Steinach's experimental procedures, the author was convinced that "secondary sex characteristics can easily switch between the two sexes."[26]

Beside the Steinach experiments, scientific journals brought to light other reports of animal sex change. In 1924 Lingnan Agricultural University's *Agricultural Monthly* introduced the writings of Francis Albert Eley Crew (1886–1973), a professor of animal genetics based in the Animal

Breeding Research Department at the University of Edinburgh. An article titled "The Sex Change of a Hen" (雌雞之變性, *ciji zhi bianxing*) impressed its readers with a story of how a hen, as witnessed by Crew, turned into a rooster at the age of three and half. Upon dissection, Crew found "a rounded growth next to the uterus" that was "roughly seven centimeters long" and "blue in color and covered with yellow spots." Next to the growth, he came across something that "did not look all that different from a testicle" (與睪丸無異, *yu gaowan wuyi*) with an identical counterpart located symmetrically opposite to it. The seminal epithelium connected to the testes seemed "identical to what can be found in the reproductive organ of normal roosters," and apparently they even had the capacity to produce sperm cells, though scarce in quantity. The main difference between the genital structure of this sex-changed hen and that of the normal roosters, according to Crew's observation, was that the testicles seemed "smaller."[27] Crew's work was later picked up by a Chinese journalist in the 1930s to give weight to the thesis that "among all animals, chickens are the ones most frequently found to undergo sex transformation."[28]

Whereas biologists such as Zhu Xi had begun to interpret hermaphroditism through genetic theories, others who espoused the endocrinological model investigated intersexuality in a new light. For example, the zoologist He Qi (何琦), a graduate of Yenching University and a biology instructor at Cheeloo University in Shandong, published an article in 1928 titled "Intersexuality and Hormones" (中間性與內分泌, *zhongjianxing yu neifenmi*). He began with a seemingly banal insight that the scientific study of animal intersexuality had become increasingly popular in the last decade or so. His discussion focused on the findings of Frank Rattray Lillie (1870–1947), an American embryologist best known for his work on freemartins—the infertile, masculinized female cotwin of a male calf. In 1917 Lillie published his now-classic study in which he showed that these infertile female mammals, though born with both male reproductive organs and nonfunctioning ovaries, were in fact genetically female, but their development had been altered by the hormones of their twin brother, resulting in masculine behavior and development. He Qi highlighted Lillie's findings to confirm the importance of sex endocrinology: "the sex of animals is determined at the moment of conception, but the sexual organs depend largely on the gonadal secretions for development." In other words, rather than reinforcing the division between the genetic view of sex and the hormonal view, Lillie's work helped to reconcile this tension by proving that the two worked in concert.[29]

Foreign animal sex-change experiments continued to attract the interest of Chinese readers into the 1930s and 1940s. In June 1934 Guo Shunpin (郭舜平), the translator of Maurice Cornforth's *Dialectical Materialism* (1952), reviewed the startling hypothesis postulated by the Russian cell biologist Nicholas K. Koltzoff that "electric current determines sex." Koltzoff founded the Institute of Experimental Biology in Moscow in 1917 and was widely recognized by his peers as the first to consider a chromosome as "a giant molecule" and, later, a critic of Trofim Lysenko (1898–1976). He put forth a theory of cellular mechanism involving the existence of an electric "force field" in the cell and, building on the assumption that the Y chromosome played a sex-determining function, suggested that the sperm cell may be responsive to electric stimulation.[30] His hypothesis acquired international acclaim because it carried the potential of enabling "stock raisers to breed for male or female as they wish" and, in Guo's words, even the "artificial manipulation of human sex."[31] According to *Popular Science Monthly*, "[Koltzoff's] discovery now is being tested on an elaborate scale at government farms in Russia," and he may have "found the answer to the age-old riddle of determining sex."[32] Seven years later the sex-change experiments on swordtail fishes conducted by two biologists at the University of Southern California, E. M. Baldwin and H. S. Goldin, proved earlier understandings of sex hormone functions. Baldwin and Goldin injected male sex hormones into female swordtails and observed the latter's eventual development of a sword, a characteristic unique to male swordtail fishes. Yun (雲), who wrote about the experiment in Chinese, attributed this sex reversal to "the transformation of the reproductive gland by way of male hormonal injections."[33] Stories of animal sex metamorphosis, induced experimentally or not, fascinated Chinese readers because they debunked a static framework of dual sexes that drove much of the earlier feminist discourse on gender equality.

The Theory of Universal Bisexuality

By the late 1920s the idea that maleness and femaleness were flexible fit nicely with the new endocrinological model of sex. If biological sex was "determined" not through gonadal presence or chromosomal makeup but through glandular secretions, scientists began to question a fixed and immutable definition of sex. In German-speaking circles, thinkers such as Sigmund Freud and Otto Weininger (1880–1903) vehemently challenged the Victorian notion of separate and opposite sexes.[34] Social context mattered too. In the late Qing and early Republican periods, Christian missionaries created a steadily increasing measure of educational opportunities for women, and after the 1911 revolution the Nationalist government recommended for the first time coeducation policies in the national educational system.[35] As more women pursued higher education, entered the labor force, and participated in social reform movements, Chinese leaders voiced the importance of granting women greater access to the public and political spheres. The new emphasis on gender equality construed men and women as more similar than different. At the same time, the influx of new, Western-derived categories like "feminism" and "homosexuality" called attention to masculine women and effeminate men.[36]

Against this backdrop, Chinese intellectuals who were drawn to Western natural sciences began to shift their vision of sex. As accounts of foreign research on sex change became available in the bourgeoning print culture, they started to cast doubt on the old notion of binary opposite sexes. Meanwhile, in the 1920s and 1930s Western biochemists learned to extract and detect hormones from the organs and urine of animals, and they soon discovered that men and women had both male and female hormones. Moreover, they found that the chemical compositions of male and female sex hormones highly resembled one another. It made sense in this social and intellectual context to consider all humans as having the potential of being both male and female. Early twentieth-century scientists, in China and abroad, gradually pushed for the argument that male and female were absolute forms that did not exist in reality. What the new wave of scientific findings showed, they said, was that everyone "fell somewhere between the two idealized poles."[37] All females had elements of the male; all males had elements of the female. By the 1930s a significant number of Chinese writers joined experimental scientists in Europe and America to biologize the human body as inherently two-sexed. As the editor of *Sex Science* proclaimed in 1936, "today, with an expanding corpus of empirical evidence, science tells us definitively that the distinction between the sexes is only a matter of relative difference (性的區別只是程度的差異, *xingde qubie zhishi chengdu de chayi*)."[38] News of surgical attempts at changing sex, along with the emergent hormonal model that conceptualized sex as a malleable construct, posited a new scientific theory of universal bisexuality.

From the start, the introduction of bisexual theory to China relied on the writings of Japanese sexologists.[39] Again, the Chinese translation of *The Principle of Sex* included an elaborate discussion of human innate bisexuality and reviewed the work of Western theorists who supported the view. The chapter on "Sex in Theory and Sex in Practice" listed numerous human conditions that blurred the biological boundaries of gender: men with overdeveloped breasts, women with flat chests; men without facial hair, women with mustaches; men with female-shaped pelvis, women with male-shaped pelvis; men with feminine throats, women with manly throats; women whose voice, facial appearance, temperament, and body hair became mannish after regular menstruations have stopped; and, most notably, women who were "conspicuously masculinized" because they have never conceived. Despite how exceptional these physical conditions may be, the author stressed, "even normal men and women actually possess latent aspects of the opposite sex in their body." The exceptional cases, then, were simply "extreme" occurrences of the universal bisexual condition. Such biological categories as pure male or pure female only existed in theory, the author insisted, as "they do not exist in reality."[40]

The names of renowned Western proponents of the bisexual theory found their way into the book, including Otto Weininger, Robert Müller, Rosa Mayreder, Solomon Herbert, and Edward Carpenter, among others. Out of this group, the name most frequently associated with the theory of bisexuality was Weininger.[41] Although Weininger's ideas anchored most of the discussions, the book closed by highlighting another influential study of the time, *The Sex Complex* by the British gynecologist William Blair-Bell (1871–1936).[42] In siding with Bell's clinical findings, the author drove home the theory of biological bisexuality: "Each individual has the inner qualities and external morphology of both sexes to varying degrees. All men and women are mixtures of the essential elements of both sexes."[43] What this new theory of sex challenged was the feasibility of discrete categories. To proponents of this view such as Weininger and Blair-Bell in Europe or the author and translator of *The Principle of Sex* in East Asia, average men were merely made up of a higher portion of "maleness," or traits typically associated with men, and a lower level of "femaleness," or qualities normally associated with women. Normal women, on the other hand, were the combination of predominant female elements and a subordinate composition of maleness. Everyone had the potential of expressing both ways. To quote from *The Principle of Sex* again, all men and women were simply variants of how certain traits "receded" to the background or "lay latent."[44] In the 1920s and 1930s scientific investigations of sex moved toward "an emphasis on individual variation in which categories blended into spectra or continua." The popular view of sex shifted, that is, "from the categorical to the scalar."[45]

Some Chinese sexologists went directly to the original English sources. In 1928 the Shanghai Commercial Press published a translation of Solomon Herbert's *The Physiology and Psychology of Sex*. Covering both the biology and psychology of sex, the book offered a comprehensive overview of the main intellectual currents in Western sexology. As discussed in the last chapter, many iconoclastic May Fourth intellectuals had already written extensively about European sexological ideas of homosexuality. By the time the translation of Herbert's work appeared, the notion of "sexual inversion" (性的顛倒, *xing de diandao*) invoked by Havelock Ellis, Sigmund Freud, and other sexologists frequently appeared in the vernacular lexicon of urban China. In 1929, for instance, a tabloid article identified a handful of foreign cross-dressers as individuals embodying "sexual inversion."[46] Meanwhile, Chinese writers often mentioned Edward Carpenter's idea of the "intermediate sexes" to interpret the existence of feminine men, mannish women, and other in-between types.[47] By the time that the Commercial Press published a Chinese edition of Herbert's book, the subtle distinction between somatic and psychological sex had already gained some footing in the popular consciousness.

From reading the Chinese translation of Herbert, one could easily relate contemporaneous findings in endocrinology to the new quantitative definition of sex and the theory of universal bisexuality. "Recent scholars have come to the consensus that rather than assuming sex as something determined by reproductive organs, it is more correct to say that sex is determined by the various combinations of the internal secretions. Any individual with one of the sex glands (testes or ovaries) simultaneously maintains the characteristics of the opposite sex [同時保有他方異性之特徵, *tongshi baoyou tafang yixing zhi tezheng*]."[48] These words prepared the reader for a fuller exposition of what the Western sexologists called "sexual inversion." In the same paragraph, bodily sex and psychological traits were carefully distinguished to challenge a dominant perspective of this clinical condition: "The general public tends to consider male sexual inverts as individuals with a male soma and a female psyche [肉體為男性而精神為女性, *routi wei nanxing er jingsheng wei nüxing*], but this view is too extreme and simplistic. In fact, the entire mental state [of the male sexual invert] is not female: only their sexual desire and emotions are female and the remaining parts [of their bodily constitution] remain normal."[49]

Unlike most European sexologists, Herbert did not use the soma/psyche distinction to support a straightforward explanation of sexual inversion. Influenced by the emergent perspective of sex as quantitative rather than qualitative, he found fault with the existing interpretation of "a female soul trapped inside a man's body." The theory of human bisexuality posited that everyone was a mixture of both sexes. Thus, the paragraph continued: "However, individuals with female characteristics are not a minority even among men, and yet most of them do not have perverted tendencies [變態的傾向, *biantai de qingxiang*]. Therefore, the difference between feminine men and male sexual invert is all the more difficult to discern based on a single criterion of the presence or absence of female sexual emotion [是否具有女性之性的感情之一點決定之, *shifou juyou nüxing zhixing de ganqing zhi yidian jueding zhi*]. Similarly, distinguishing sexual inverts from normal [men] is not an easy task."[50] So subtle and confusing was the distinction between normal and pathological individuals. The translation of *The Physiology and Psychology of Sex* allowed Chinese readers to rethink some of the fundamental issues underlying the subject of same-sex desire, scientific narratives about abnormality, and the nature of sex itself.

Overall, Chinese thinkers reacted to the theory of universal bisexuality in diverging ways. A small, more conservative group adhered to the idea of sexual dimorphism and construed it as an evolutionary benefit for ensuring procreation in higher-level organisms.[51] In contrast, a more substantial group of writers pursued the view that sex was a delicate balance of maleness and femaleness. In his 1929 essay, "The Problem of Homosexuality," Yang Youtian, for instance, asked, "If the male body contains latent degrees of femaleness and the female body contains dormant elements of maleness, what constitutes true male or female?"[52] His own response to the question was, "In fact, the categories of male and female are convenient shorthands for describing morphology," and both men and women should be more accurately understood as "compound sex" (復性, *fuxing*), meaning that each individual "embodies elements of both sexes."[53] Also writing about homosexuality, Wang Xianli agreed in 1944: "Recent scientific findings suggest that there is no clear boundary that can be drawn to differentiate sex in animals and humans. Female properties exist in male bodies; similarly, male properties can be found in female bodies. In other words, a 100 percent man or a 100 percent woman does not exist—there are no perfect or absolute sexes in the world. Scientists have experimented with glandular transplantations, and it is now possible to transform sex accordingly."[54]

Around the same time, a *Wanxiang* (萬象) article drew on the bisexual theory to explain the sex-change experience of the American Barbara Ann Richards, who was born Edward Price Richards and shared a dorm with Virginia Prince (born Arnold Lowman) at Pomona College in the 1930s.[55] The article described Richards's condition as an example of hormonal imbalance: "A doctor surmised that Richards was born dual sexed," but "as the level of female hormones in his body rose, they took precedence and suppressed his male hormones." Since "the human body comprises both sexes," the article cautioned, "anyone could follow a similar fate."[56]

Ironically, as the conceptualization of sex became less rigid, the fluidity of bisexual theory and the traction of the hormonal model led some proponents to argue that homosexuality may be "cured" with gonadal transplantation surgeries. The renowned German sexologist Magnus Hirschfeld, for instance, referred male homosexual patients to Eugen Steinach and Richard Mühsam (1872–1938), a Dresden gynecologist, for treatment via testicular implants.[57] In China, an article that appeared in *Sex Science* in September 1936 asserted that "we have a scientific way of correcting the constitutional bisexuality of homosexuals: by manipulating male and female hormonal levels, we can consolidate greater distinctions between the desired maleness or femaleness among the sufferers of this illness."[58] The theory of biological bisexuality, then, promoted a definition of sex that seemed more malleable than before, but precisely due to its flexibility, the theory also allowed for the buttressing of conservative attitudes toward sex and sexuality.

Castration and the Feminized Male Sex

However startling or luring, new discoveries in sex hormones, animal sex change, and universal bisexuality convinced some Chinese sexologists in part because they shared aspects of the existing *mentalité*. Or, to put it differently, the comprehensibility of these foreign ideas had to do with the specific intellectual and political agendas of Chinese modernizing thinkers and social commentators at the time. After the founding of the new Republic, long-standing corporeal practices in China such as footbinding and castration came to be denounced in elite and popular discourses as unfavorable reminders of the past. The first chapter of this book explored how the normative regime of eunuchism came to an end through the lens of the history of knowledge production about the castration operation itself. This section highlights another dimension of its social and cultural demise: how the body of eunuchs provided a concrete example from traditional Chinese culture that enabled Republican Chinese sexologists to focus their attention on and grasp new Western theories of sex. This regendering of eunuchs as feminized males signaled an epistemological departure from the cultural norms of Chinese castration according to which eunuchs retained a distinct masculine identity in late imperial political culture.[59]

Most of the key figures who subscribed to the theory of universal bisexuality brought up the example of eunuchs to elucidate the glandular model of sex. Again, Gu Shoubai played an important role in disseminating new findings in hormonal biology in this period. In *The Internal Secretions*, Gu noted that boys at a relatively young age do not have "reproductive desires" (生殖慾, *shengzhiyu*) and their bodily makeup "is similar to girls." The rapid development of male sexual characteristics, Gu pointed out, "begins only at the age of sixteen, when a boy enters a stage of human development that is typically called puberty." At this point, a boy develops "the desire for and fondness of the opposite sex."[60] All of this, however, can be altered by castration:

> The removal of testicles will bring obstacles to the physical and mental development of a boy. Even when he reaches puberty, his body will not undergo those physical changes that are uniquely male [男性特有之變化, *nanxing teyou zhi bianhua*]. Specifically, his muscles are less stringent; his strength is weaker; his body accumulates more fat; he has less body hair; the growth of his larynx stops and his vocal folds do not get any longer or thicker, and as a result his vocal production is similar to that of adolescents; he has no reproductive desire; his mental reaction is slower than normal; and he lacks both moral judgment and the will to compete.[61]

No other corporeal figures could exemplify these postcastration changes better than the "eunuchs in our country's past" and the "castrati in southern

Europe." For Gu, Chinese eunuchs (宦官, *huanguan*) and Italian castrati were "concrete human examples of castration—one created for the purpose of preventing promiscuity [防其淫亂, *fangqi yinluan*] and the other for the purpose of maintaining a beautiful voice [保其妙音, *baoqi miaoyin*]."[62]

By the 1930s, most scientists and doctors considered the biology of castration central to a solid understanding of the endocrine system. In his *Sex Science* essay "Sex Endocrinology," Li Yongnian provided a lengthy discussion of castration and its physical consequences before bringing up the testicular transplantation "sex-reversal" (性之交換, *xing zhi jiaohuan*) experiments. With reference to Chinese eunuchs, he corrected the popular view that the main goal of castration was to eliminate a man's vitality.

> Speaking of castration (去勢, *qushi*), one immediately thinks of the eunuchs of the despotic past (專制時代的宦官, *zhuanzhi shidai de huanguan*). Eunuchs serve in the imperial palace. In order to maintain order and avoid promiscuity inside the palace, where many women resided, castration operations were performed to eradicate the vitality of these men (使他們失去一切精力, *shi tamen shiqu yiqie jingli*). Yet in fact, serving the emperor, eunuchs often intervene in political affairs, sometimes even disrupting the selection of the heir of the imperial throne. . . . As such, far from lacking vitality, eunuchs possess a tremendous degree of energy that deserves our critical attention![63]

Li also alluded to the castrated slaves in ancient Rome, the Skoptsy (the self-castrated) sect in Tsarist Russia, and European castrati singers as comparable examples of a corporeal practice aiming to "maintain a pure body by way of severing the reproductive organ."[64] Since a tradition of castrating animals for agricultural purposes had long existed in China, Li went so far as to ponder: "Did China invent castration? Or was it first invented in Central Asia?"[65]

For endocrinological experts like Li, the discovery of gonadal secretions pulled people out of "barbaric" and superstitious beliefs and clarified the reasons for castration-induced bodily change. Testicles were not only the "source of sexual development" but also "a prominent variable in bodily growth." Such an understanding, for example, made it possible to link castration to sexual desire. According to Li, castration would "significantly reduce, but not remove entirely, the erotic interest of any higher-level species with intercourse experience." In contrast, eunuchs, "although castrated," would "still feel strong erotic urges by imagining a beautiful lady standing next to him" because "the sexual desire of humans differs from that of animals." Above all, the physical transformations of the body offered indisputable evidence for the ways in which hormonal stimulation mediated the effects of castration. "If one is castrated at a relatively young age," Li wrote, "his bodily and psychological developments will be stumped, physical traits will become feminized or intersexual, muscles will turn flabby, and body fat will accumulate significantly." "Based on the above descriptions," Li's point was simply that "castration is against nature, so it should not be executed without sufficient justification." Endocrine biology handed critics such as Li a new set of vocabularies for foraying into the question of eunuchs' sexual urges and, relatedly, condemning the practice of castration as "not only cruel, but also harmful to individuals, families, society, and the country."[66]

Whatever the perceived value of eunuchs' existence in imperial China, their image as "emasculated" or "effeminate" living creatures remains pervasive in and out of China.[67] If eunuchs were to be recognized as historical agents with some degree of masculinity at all, they were often cast as feminized men with abhorring bodily traits. The discovery of chemical messengers that linked the sex glands to conventional gender morphologies unlocked the secret, for many, of those physical and psychological attributes that observers typically identified with eunuchs. Zhang Luqi (張祿祺), an expert on human development, thus observed in 1941: "The eunuchs of ancient China, the monks of medieval Europe, and the individuals sterilized for their deep-rooted depravity or hereditary diseases in Europe and the United States are all known as castrated persons [閹人, *yanren*]. After the operation, castrated men develop abnormally and become androgynous, so that they have a gentler temperament, a bulkier physique, a squeakier voice, more body fat, less facial hair, and brighter and more tender skin."[68]

These feminizing effects, according to Zhang, precipitated from testicular elimination. Following this logic, the European glandular transplantation experiments "unambiguously confirmed the [sex-]dictating status of hormones."[69] Some scientists such as Chen Yucang, a Japan- and German-educated doctor and proponent of the body-as-machine metaphor, emphasized the "psychological consequences" (精神上的變化, *jingshengshang de bianhua*) of castration: "most eunuchs are warm and gentle without any degree of masculine temperament, and this is because a particular type of endocrine secretion is missing from their bodies."[70] In contrast to Li Yong-nian's view, Chen depicted "our country's eunuchs in the past" as lacking "mental sharpness," "the ability to discern the good from the evil," and "competitive mindset."[71] Figuratively and materially, the eunuch's body became an engine of transcultural exchange, a token of historic signifier, and a conduit for the material epistemology of knowledge that allowed scientific thinkers in China to grasp and comprehend the rapidly evolving science of the internal secretions.

Symmetrical discussions about the castrated male body can be found in Chinese translations of foreign sexological texts. In Wang Jueming's translation of *The Principle of Sex*, an analysis of the effects of human castration followed the section on Steinach's sex transplantation surgeries. Since Steinach experimented only on animals, the implications of his studies for humans seemed worth expounding upon: "eunuchs who are castrated at youth" have "underdeveloped genitals." Their "vocal folds do not elongate," so "the pitch of their voice is high like women's." Their body "undergoes greater fact accumulation" and "grows lesser facial and body hair." Their pelvis would "not grow properly, just like that of a child." Their psychological condition rendered them "as gentle and sweet as a virgin youth [溫順如處子, *wenshun ru chuzi*]."[72] Similar statements appeared in the Chinese edition of Herbert's *Physiology and Psychology of Sex*: "eunuchs who are castrated at youth [幼年去勢之宦官, *younian qushi zhi huanguan*] experience no change in 'vocal production,' as they still maintain a voice with high pitch like the voice of young children."[73] In 1936 the Chinese translation of an essay by a British physician listed the following qualities to describe the effects of castration: "the degeneration of external and internal reproductive organs," "scrotum retraction," "impotence," "penis shrinkage," "lack of facial and body hair," "voluminous hair growth," "undetectable view of the larynx," "body fat accumulation," "abnormal pelvis growth," and "breast enlargement." The essay argued that the "feminization of males" (男子變成女性, *nanzi biancheng nüxing*) and the "masculinization of females" (女子變成男性, *nüzi biancheng nanxing*) were accompanied by not only physical changes but also "psychological sex reversals" (精神異性化, *jingsheng yixinghua*) following the surgical removal of gonads.[74]

Although these quotes were presumably taken from works originally written in Japanese or English, the translators' cautious word choice in translating "eunuchs" into Chinese—*yanhuan* (閹宦) and *huanguan* (宦

官)—unravels a long-standing cultural lexicon deeply rooted in the norms of dynastic Chinese social life. But the invocation of these terms in a conceptual horizon shaped by the contours of Western natural science also points to something more: a new consideration of castration as a scientific procedure and eunuchism as a bodily state that defied the fixed nature of sex binarism. This conceptual reorientation would easily alienate Chinese commentators on the subject before the twentieth century (see chapter 1). In the modern period, the body of eunuchs became a "text" whose corporeal feminization suggested that men could become more female. As much as the new biochemical model of sex may have helped to explain the effects of human castration, the castrated male body conferred a concrete material basis for the circulation, transaction, and articulation of new truth claims about sex and its transmutability/embodiment in the transition from late imperial to national Republican China.

Hermaphroditism as a Natural Anomaly

Escalated publicity on surgical attempts at changing sex, on top of a novel biochemical interpretation of the castrated body, led to a renewed interest in natural reproductive anomalies. To readers of the modern sexological literature, castration was an unambiguous case in point of human-induced alteration of sex. For centuries, however, Chinese physicians endorsed various perspectives on individuals born with ambiguous genitalia. The first systematic medical categorization of intersexed bodies appeared in the late Ming, with Li Shizhen's listing of five "non-males" and five "non-females" in his compendium of material medica, *Bencao gangmu* (1596).[75] As Charlotte Furth has suggested, Qing physicians for the most part adhered to, or at least systematically referenced, Li's classification.[76] As chapter two has shown, the situation started to change with the growing popularity of Western-style anatomical texts since the mid-nineteenth century. By the Republican period, Chinese life scientists readily filtered the competing perspectives and experimental findings on hermaphroditism—even on the microscopic scale of genes and chromosomes—from Europe and the United States.[77]

In the 1920s, though, some popularizers of sexology boiled down complex scientific theories for a general readership. Toward the end of his widely read *ABC of Sexology*, Chai Fuyuan included a chapter on "Abnormal Sexual Lifestyle" (畸形性生活, *jixing xingshenghuo*). The chapter focused on four topics in particular: "incomplete male growths" (男性發育不全, *nanxing fayu buquan*), "incomplete female growths" (女性發育不全, *nüxing fayu buquan*), "ambiguous genital sex" (男女性別不明, *nannü xingbie buming*), and "homosexuality" (同性戀愛, *tongxing lian'ai*).[78] In the sections on incomplete male and female growths, Chai borrowed from Li Shizhen's categorization to explain human reproductive anomalies in Western anatomical terms. On "incomplete male growths," Chai wrote:

> The incomplete growth of male reproductive organs is a phenomenon commonly known as "natural castration" (天閹, *tianyan*). There are several types of natural castration. The first type is characterized by the incomplete development of external genitalia. Even with fully functional testicles and the biological capacity to produce sperm, people with this condition cannot mate due to the small size of their penis. The second type is exactly the opposite: although people with this condition are physically capable of mating, they lack sexual motivation and the physiological ability to produce sperm. The last type is a combination of both: people with this condition have incomplete internal and external sexual organs. Internally, their bodies do not possess functional testes; externally, they have an immature penis.[79]

Chai laid out a commensurate discussion of "incomplete female growths":

> Women with an incomplete set of reproductive organs are typically called "stone maidens" (石女, *shinü*). The original meaning of "stone maiden" refers to women with genital anomalies of the sort that would make sexual penetration impossible. Eventually, the term was broadened to be associated more generally with incomplete female growths. With the external type, the genital organ is completely in an impenetrable state. It may be that the hymen is too thick so that it covers the vagina, or the labia is too thick and the vaginal opening too small making sexual penetration impossible. With the internal type, the individual either lacks ovaries or lacks ones that are functional, resulting in the absence of sexual motivation and the reproductive organs without proper female functions.[80]

Interestingly, none of the physical conditions Chai described here would be sufficiently considered "hermaphroditic" by modern standards in Western biomedicine; rather, he merely reiterated a long-standing concern in traditional Chinese medicine for the generational capacity of individuals.[81] For Chai, persons with the kind of physical symptoms and sexual experience outlined here would still be treated as reproductively anomalous even if their genital sex per se was not necessarily ambiguous.

But Chai went further and enriched his catalog with modern scientific linings. In his discussion of the congenital malformations of the reproductive system, he included an additional section on "ambiguous genital sex."

"Ambiguous Genital Sex" refers to people with external male genitalia and internal female reproductive system or with external female genitalia and internal male reproductive system, also known as "half *yin-yang* persons." If a man has a penis with a slight vaginal opening, he is called "male half *yin-yang*" (男性半陰陽, *nanxing banyinyang*). If a woman has an enlarged clitoris with reduced vagina and labia, she is called "female half *yin-yang*" (女性半陰陽, *nüxing banyinyang*). If both ovaries and testes are internally present and a penis, labia, and a vagina are externally present, if an individual is physically capable of engaging in sexual intercourses with both men and women and experience organism from them, and if both the male and the female sex appear in the same body, this condition is called "bisexual half *yin-yang*" (兩性半陰陽, *liangxing banyinyang*).[82]

Using Chinese words with which lay readers felt more comfortable (i.e., *nan* and *nü*, *yin* and *yang*) than entirely foreign medical terms, Chai implicitly distinguished pseudo-hermaphrodites from true human hermaphrodites. His classification provided a way in which people could understand those new modern anatomical concepts, such as penis, vagina, testes, and ovaries based on traditional ideas about human reproduction.[83] The category of "male half *yin-yang*" was endorsed by such reputable physicians as Ding Fubao (丁福保, 1874–1952), who later published a case study of a patient with both a 3.5-cm-long penis and a vaginal opening with hidden testicles.[84] The biological implications of such categories as "male half *yin-yang*," "female half *yin-yang*," and "bisexual half *yin-yang*" blended nicely with the emergent theory of constitutional bisexuality. At least in these rare cases of reproductive anomalies, the physical makeup of the human body was portrayed as innately dual sexed.

More often, Chinese urban intelligentsia assumed a more rigorous approach: to avoid simplifying scientific information. In this spirit, the self-proclaimed natural scientist Liu Piji (劉丕基) treated the topic of human hermaphroditism with finer detail in his *Common Misinterpretations of Biology* (人間誤解的生物, *Renjian wujie de shengwu*). The Shanghai Commercial Press published *Common Misinterpretations* in 1928, and, as the title of the book suggests, it was written to inform a popular audience about general misunderstandings of problems in biology. The motivation behind publishing the book squarely reflected the normative ethos of middlebrow print culture in the aftermath of the May Fourth: that it was important for educated Chinese to move toward a greater appreciation of the epistemological value of Western natural scientific knowledge and away from Confucian philosophy or misguided superstitions (迷信, *mixin*). Liu authored *Common Misinterpretations* in order to furnish greater public interest in Western science, such as by straightening out those puzzles of everyday life for which modern biology seemed to offer the most adequate and reliable answers.

For Liu, misconceptions of human sexual oddity reflected a crucial oversight in Chinese knowledge about life. Popular errors and unfounded myths about sexually ambiguous bodies were pervasive, but Liu insisted that they could be displaced only with accurate interpretations of modern biological knowledge. In Chai Fuyuan's *ABC of Sexology*, for example, earlier typologies of reproductive anomalies (from traditional Chinese medicine) were invoked to render the modern scientific category of hermaphroditism meaningful. Chai never articulated a neat distinction between "true" and "pseudo-" hermaphroditism, but his differentiation of "male half *yin-yang*" and "female half *yin-yang*" from "bisexual half *yin-yang*" was implicitly informed by it. In contrast, Liu was keen to translate the "true" and "pseudo-" distinction more explicitly into Chinese. He began by collapsing various Chinese labels for human hermaphrodites: "*Ci-xiong* humans (雌雄人, *cixiongren*) are also known as *yin-yang* humans (陰陽人, *yinyangren*), dual-shaped (二形, *erxing*), or bisexual abnormality (兩性畸形, *liangxing jixing*)." Despite this variety, Liu asserted that they all designated a similar biological condition, which could be classified into two main categories: "true *ci-xiong* humans or true half *yin-yang*" and "pseudo-*ci-xiong* humans or pseudo-half *yin-yang*."[85]

Liu further divided pseudo-*ci-xiong* humans into two subtypes: "male *ci-xiong* humans" (男性雌雄人, *nanxing cixiongren*) and "female *ci-xiong* humans" (女性雌雄人, *nüxing cixiongren*). According to Liu's definition, the former label referred to individuals who had "internal male sex glands (with testes)," but, externally, his "genital appearance resembles the female sex"; the second subtype referred to those who had ovaries but with male external genitalia. The physical appearance of male *ci-xiong* humans resembled a woman and the physical appearance of female *ci-xiong* humans, a man. This left "true *ci-xiong* humans" for people born with "both types of male and female sex glands (meaning, having both testes and ovaries)." This unique condition, Liu hastened to add, was referred to as "a man yet a woman, a woman yet also a man" (值男即女值女即男, *zhinan jinü zhinü jinan*) in the late Ming materia medica *Bencao gangmu* by Li Shizhen.[86] To illustrate his point, Liu reproduced a hand-drawn image of the genial area of female *ci-xiong* humans (figure 4.1) and a picture of Marie-Madeleine Lefort (1799–1864), a famous female pseudo-hermaphrodite (figure 4.2).[87]

More so than this classification scheme, Liu devoted a significant part of his discussion to the question of why these congenital malformations occurred in nature. He drew on embryological knowledge to explain the existence of these rare human conditions. Supporting the theory of innate bisexuality, he located their cause in irregular embryonic development. In his words, "the sex glands (referring to testes and ovaries) of the human embryo are identical for men and women. . . . Sexual differentiation begins only during the second to the third month of fetal development, when the sex glands mature into a finer differentiation between the two sexes. This is also the time when [testicular and ovarian cells] are formed."[88] He noted the "disappearance" of the Müllerian duct in normal male embryonic development and the "disappearance" of the Wolfferian duct in the maturation of the female fetus. Typically, the former embryonic duct developed into uterus, vagina, and fallopian tubes in women, whereas the latter would be transformed into seminal vesicle, epididymis, and vas deferens in men. "The concurrent growth of both ducts," Liu explained, "would thus produce a true *ci-xiong* individual."[89] On the other hand, the external genitals of male pseudo-*ci-xiong* humans were "insufficiently developed," thus "appearing not as a penis but as a clitoris" (不成陰莖而像女的陰核, *bucheng yinjing erxiang nüde yinhe*), whereas female pseudo-*ci-xiong* persons had clitorises that were "irregularly developed" and "appear like a penis" (外觀像是陰莖, *waiguan xiangshi yinjing*).[90]

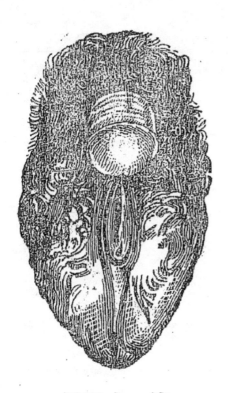

圖二十三第

女性假半陰陽之陰部

Figure 4.1 "The Genital Area of Female Pseudo-Hermaphrodites" (1928). *Source:* Liu (1928) 1935, 84.

The scientific classification and explanation put forth by Liu Piji featured none of the overriding concerns with reproductive potential in traditional medicine. In introducing the modern biomedical divide between "true" and "pseudo-" hermaphrodites, he reworked Li Shizhen's naturalist observations in the new lexicon of glandular and embryological sciences. Chai's allusion to *Bencao gangmu* in his *ABC of Sexology*, on the other hand, foregrounded those issues of generational capacity that was foundational to the naturalist's original typology. But whether Liu was more "accurate" than Chai is perhaps less important here. The significance of their writings lies in the fact that they provided Chinese readers the first sustained contact with Western biological interpretations of sexually ambiguous bodies. In the 1920s the hermaphroditic body, like the castrated male body, became a "text" whose corporeal significance helped to anchor a new vision of sex. If the embodied experience of eunuchs exemplified

Figure 4.2 "The Portrait of the Female Pseudo-Hermaphrodite, Nagdalena Lefort (65 Years of Age)" (1928). *Source:* Liu (1928) 1935, 87.

第二十四圖

女性假牛陰陽 Nagdalena Lefort 之肖像

（時年六十五歲）

the potential transformation of a man into a feminine subject, true and pseudo-hermaphrodites were the most basic and natural examples of the universal bisexual condition.

Despite the multiplicity of modernizing attitudes surrounding a new vision of sex, early twentieth-century sexologists remained oblivious to the possibility of complete sex transformation in humans. Even though new findings in endocrinology, accompanied by the biological theory of bisexuality, boosted the notion that men could become female and women could become male, for the most part, Chinese scientists spoke with greater confidence (and ease) of sex metamorphosis in animals only. When they discussed explicit examples of human sexual defects, they focused on eunuchs and hermaphrodites. Indeed, Liu Piji intended his scientific exposition of intersexuality to alleviate any misconceptions about *renyao* (人妖), for which the best English translations would be "freak," "fairy," or "human prodigy."[91] Specifically, he wanted to dismiss the validity of this traditional concept, which had been used in Chinese discourses to describe a diverse spectrum of individuals in ambiguous and, according to him, unscientific ways.

After listing the various categories of human hermaphrodites, Liu remarked that "our country has a long history of calling these individuals 'freaks' [怪胎, *guaitai*] or 'human prodigy' [人妖]. Many of them were tortured to death, but even for the fortunate minority who survived, they were never treated as decent human beings."[92] Liu's message was clear: the long-standing popular rendition of *renyao* lacked a scientific basis, especially as it led to, for centuries, the ostracizing of natural variants in the human population. Popular errors and myths about the figure of *renyao*, he argued, should be replaced with modern biological accounts of human intersexuality. Liu concluded his chapter, entitled "*Ci-Xiong* Humans Misunderstood as Human Prodigies," with the following remark:

> With pseudo-hermaphrodites, it is possible for the male sex to change into a female and the female into a male. His/her inner physiology is usually without any defect, but the outer part is not completely formed. As a result, the body undergoes many changes at puberty, when the outer part fully develops and reveals itself in its true appearance. Traditionally, people did not understand the reasons for these changes and considered men who become women and women who become men demonic. Consequently, records of such phenomena in official histories and popular gazetteers have been ambiguous and lacked specificity. In reality, it is nothing but a very ordinary phenomenon; what is there to be surprised about?[93]

In Liu's formulation, then, men and women who turned into the opposite sex were basically pseudo-hermaphrodites and nothing more.

Liu restated this view in *Scientific China*, one of the leading popular science magazines of the Republican period. In 1934 the magazine featured a question-and-answer section on "What Is the Explanation for Female-to-Male Transformation?" Zi Yin, a reader from Shanghai, had learned about the sudden transformation of a sixteen-year-old French girl named Alice Henriette Acces into a boy. According to Acces's doctor, orthogonadist Robert Minne, "Henriette Acces has become physiologically male," and "it is entirely possible, and even probable, that Henri can become a father."[94] *Scientific China* asked Liu to respond to this foreign incident of sex change that Zi Yin had brought up. Consistent with the reasons provided in his book, Liu described human sex transformation as a natural outcome of pseudo-hermaphroditism. Again, Liu insisted, "men who become women and women who become men are only due to their biological structural defect and should not be considered as freaks." For people with incomplete external genital formations, including Acces, they may switch sex around the age of fifteen or sixteen. Liu suggested that this was because bodily pathologies and defects usually surfaced at puberty. But how come some individuals with ambiguous genitalia never underwent sex transformation? These individuals, according to Liu, were true hermaphrodites who possessed both normal male and female genitals. To underscore his point, Liu concluded: "True hermaphrodites cannot experience sex change, a possibility limited to pseudo-hermaphrodites. Since Acces is a pseudo-hermaphrodite, she can go through the type of sex change that is also known as female-to-male transformation. Such bodily transformation merely reflects her original masculine trait and should not be deemed as a rare and repulsive event."[95]

In depicting hermaphroditism as an example of universal bisexuality and sex transformation as a natural occurrence, Liu bestowed a powerful script through which earlier life histories could be brought to light and reinterpreted. For example, in summer 1931, the Tianjin-based *Beiyang Pictorial* (北洋畫報, *Beiyang huabao*) published a short journalist account titled "Sex Change" (性變, *Xingbian*), in which the contributor Qu Xianguai (曲綫怪) recalled a story of a *yin-yang* person. In 1908 Zhao Tingfu (趙廷苻), a native of Deqingzhou (德慶州), appeared in public wearing government official boots (官靴, *guanxue*) but walked in a swishy (扭捏, *niunie*) way. As Zhao explained to the reporter, he was actually born a girl, required by his parents to pierce his ears and bind his feet. Around the age of ten, however, he became very ill one day with swelling genitals, which turned out to be a symptom of sex switch. Learning that their child had suddenly turned into a boy, his parents immediately unbound his feet, but he confessed that it was not easy to walk at first. By the time that he had assumed an official post, he was already fifty years old and even had grandchildren. According to Qu, Zhao's sex change could be attributed to the fact that he was a "female *yin-yang* person" whose "condition worsened" around the time approaching puberty. These "physiological" changes were "most certainly possible" and "factual," and one did not need to resort to "mythic" (神話, *shenhua*) explanations or to treat this incident as "fictional."[96]

This justification for sex change introduced a compelling distinction between hermaphroditic transformations and the feminizing effects of castration. Such a distinction was meaningful at least to Li Shilian (李士璉), a reader of the Shanghai-based *Saturday* (禮拜六, *Libailiu*)—a magazine that epitomized the evolving literary output of the so-called Mandarin Ducks and Butterflies genre (鴛鴦蝴蝶派, *yuanyang hudie pai*).[97] In 1934 Li wrote to the editor of *Saturday* seeking clarification on something he found confusing. Learning from Mendelian genetics, he was under the impression that sex determination

is established at the time of conception . . . and has nothing to do with the nurturing environment. However, the world is full of surprises. Recently, the news reported on a man with poor health, whose genital organs became severed from his body after falling ill at the age of 10 and consequently turned into a girl all of a sudden. At the time, only his mother and himself knew about his sex change. However, the truth was revealed later when he was accused of adultery and impregnating his alleged sexual partner. The accusation forced him to come clean with his biology, which made it impossible for him to commit the crime. This case demonstrates that sex is not determined through heredity but is in fact shaped by development. Could you please explain why this is so? Does this person have an abnormal set of reproductive organs? And why can he suddenly turn into a woman by way of genital dismemberment? If that is the case, how come eunuchs do not become the female sex after castration? I have also heard that there is a type of individuals that are known as "half man-woman." How does their biology develop over time? What do their reproductive organs look like?[98]

The responding editor noted that, although he lacked formal training in physiology and these questions were best left to experts, he had once come across a similar example from Denmark. Again, this case concerned a pseudo-hermaphrodite, who was originally thought to be a man but later, when his female reproductive features predominated, turned into a woman. "As for eunuchs," the editor added, "they cannot change into the female sex even after genital eradication because the internal makeup of their body remains completely male."[99] This correspondence goes a long way to reveal the growing awareness—among both scientific writers and their readers—of the difference between the genetic and hormonal definitions of sex as well as the distinction between the anatomical sex changes of hermaphrodites and the glandular feminization of eunuchs.

Perhaps the story that epitomized the possibility of sex transformation among intersexed individuals in the 1930s was none other than that of Lili Elbe.[100] An acclaimed artist, Elbe was born in Denmark as Einar Wegener and married to illustrator and painter Gerda Gottlieb in 1904. After traveling through Italy and France, the couple settled in Paris, where one day Wegener posed for Gottlieb in women's clothes and felt surprisingly comfortable in the clothing. The cross-dressed Wegener soon became Gottlieb's favorite model and adopted the persona "Lili Elbe." In 1930 Elbe traveled to Germany for a series of sex-reassignment surgeries under the supervision of Magnus Hirschfeld, who by this point had acquired an international reputation for overseeing similar operations at his Institute for Sexual Research in Berlin.[101] The book that recounts the transformation of Lili Elbe, *Man into Woman: The First Sex Change*, appeared in English in 1933, and it was translated into Chinese and serialized in the Shanghai-based magazine *Art Life* (美術生活, *Meishu shenghuo*) in 1935.[102] The translator, Wei Shifan, echoed Liu Piji's theory and used Elbe's intersex condition to justify her sex change.[103] Both Chen Yucang and Zhou Zuoren similarly drew on the theory of constitutional bisexuality to underscore the mundane nature of Elbe's transformation.[104] Other accounts in the Chinese press placed Elbe's experience within a broader framework of natural science, relating this episode to accounts of sex transformation in the Chinese historical record, animal sex-change experiments, and the story of female-to-male Yao Jinping, to which we will return in the next section.[105]

As embedded and echoed in these popular media discussions, the intention of sexologists like Chai Fuyuan and Liu Piji was, of course, not to generate novel scientific hypotheses about biological malformations but simply to demonstrate the power of modern science to throw light on all aspects of life. In hoping to correct the widespread tendency to demonize and marginalize the *renyao* figure, Liu in particular implied that human sex change was possible—but only among people born with intersexed conditions. His rendering of human hermaphrodites as normal and benign simultaneously articulated the *impossibility* of sex transformation in persons with no reproductive deficiencies. Liu's point was not that human sex reversal was impossible, but just that such a biological phenomenon could be explained with an adequate grasp of modern biological knowledge about natural genital defects. Ultimately, even modernizing voices such as Liu Piji's did not convey a keen message about the physical change of sex among ordinary (non-intersexed) adults. In the 1920s and early 1930s Chinese scientists entertained the possibility of human sex change and even offered scientific explanations for it, but they oftentimes retreated to biologically anomalous cases such as eunuchs and hermaphroditic subjects. They had yet to articulate a vision of individuals as agents capable of seeking surgical sex transformation for themselves.

A Lightning Strike

The idea that non-intersexed individuals could change their sex began to reach a wider public in the mid-1930s when the press reported on a lady from Tianjin named Yao Jinping, who allegedly turned into a man and changed her name to Yao Zhen (姚震) in 1934. On March 17, 1935, news of Yao's sex transformation appeared in major papers, including *Shenbao* (申報) and *Xinwenbao* (新聞報), and soon became the spotlight of urban public discourses in China. According to Yao's grandfather, the family lost contact with her father, Yao Yotang (姚有堂), after his army was defeated by the Japanese troops and retreated to Xinjiang. Yao cried day and night and would rarely get out of bed until one night in the late summer of 1934, when a lightning strike hit the roof of their house. Yao suddenly felt something different about her body. On the next morning, she reported her possible sex change to her grandmother, who felt Yao's covered genital area and was confident that Yao had turned into a man. Her bodily metamorphosis earned the uniform label *nühuanan* (女化男, woman-to-man) in the media.[106]

Although surprised by the transition at first, Yao's grandfather eventually decided to bring her to Yao Yotang's senior officer, General Li Du (李杜, 1880–1956), who played an important role in defending Northeast China from Japan in the 1930s. They explained the situation to him in Shanghai. On the day before Yao's news appeared in print, reporters met Yao in person but mainly spoke to her grandfather, who assumed the responsibility of communicating the details of her bodily change to the press. Her grandfather presented several pieces of "evidence" to prove Yao's former biological femininity, including a diploma indicating Yao's graduation from a female unisex elementary school in 1930. The most striking evidence that her grandfather showed the reporters were two photographs of Yao taken before and after her sex change (figure 4.3). These photos were printed and distributed in most, if not all, of the newspapers that featured the Yao story.[107] Apart from this crucial proof, journalists drew attention to other fragmented hints of femininity, including her pierced ears and slightly bound feet. Given these indications, the headline of Yao's account in *Shenbao* read "evidence points to the factual status [of sex change] and awaits examination by experienced physiologists."[108] *Xinwenbao* described the Yao story as "something similar to a fairy tale" and the evidence brought forth by her grandfather as "nothing like the *biji* [筆記] notes . . . but hard facts."[109]

On the second day of her publicity, Yao finally opened herself up and narrated her experience to the reporters, in part because they soon considered the details provided by her grandfather "inconsistent." Yao explained that during her childhood, her life resembled that of an ordinary girl. She underwent menarche at the age of fourteen. One day in the summer of 1934, Yao felt extreme physical discomfort, dizziness, and a notable lack of appetite. She stayed in bed throughout the day until upon hearing a

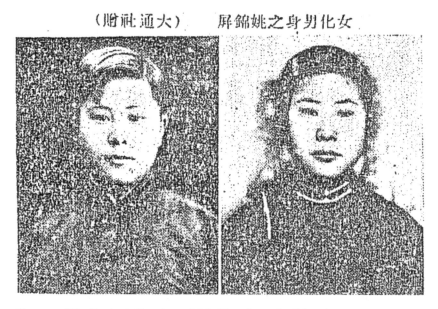

Figure 4.3 "Yao Jinping's Female-to-Male Transformation" (1935).
Source: *Shenbao* 1935a, 13.

lightning strike in the evening, when her genitals suddenly transformed into the opposite sex. After her grandmother confirmed Yao's physical change indirectly, Yao was kept in the house for an entire month. Over the course of her recovery, Yao's chest flattened so that her upper body looked more masculine, and a bulge appeared on her throat that resembled an Adam's apple. She turned into a man at the age of twenty. According to the account in *Shenbao*, "these are the biological changes that occurred following the transformation of [her] reproductive organ."[110] Leaking another piece of information about her past, *Xinwenbao* ran the exposé with the headline "Marriage Arranged Prior to Sex Change Now Canceled."[111]

Upon hearing Yao's own recollection, enthusiastic reporters jumped to ask about psychological changes. They were eager to find out whether Yao had begun to experience "sexual feelings toward women," especially in light of her decision to cancel her arranged marriage. Yao expressed unease upon hearing this question, so she refused to respond directly. Instead, she wrote on a piece of paper: "I am currently no different from a normal man. One hundred days after my physical sex change, I started to experience an admiration of some sorts toward other women." At this point, her grandfather stepped in and told the journalists that although Yao's genital area had masculinized, it remained underdeveloped. The reporters recommended Yao to undergo a physical examination. But her grandfather insisted that she still needed to rest and recuperate, only after which they may consider a full medical exam. Meanwhile, General Li Du, her father's senior official, promised to schedule a full physical checkup for Yao. He also expressed a strong willingness to support Yao financially so that she could eventually go back to school.[112]

On the same day, reporters asked a handful of medical experts to deliver professional opinions on Yao's case. The most important input came from the president of the National Shanghai Medical College (國立上海醫學院), Yan Fuqing (顏福慶, 1882–1970; figure 4.4). Yan received his medical degree at Yale University and a diploma from the Liverpool School of Tropical Medicine in 1909. Following the fate of many intellectuals in China, he eventually suffered and in fact died in Shanghai during the Cultural Revolution. Yan suggested that Yao's bodily changes were likely symptoms of a "*ci* becoming *xiong*" (雌孵雄, *cifuxiong*) condition, or pseudo-hermaphroditic female-to-male sex transformation. Nonetheless, he insisted, the truth behind Yao's case could only be confirmed with a thorough physical examination following strict scientific standards. Xu Naili (徐乃禮), the

Figure 4.4 Yan Fuqing (1882–1970).
Source: https://zh.wikipedia.org/wiki/颜
福庆#/media/File:Yan_Fuqing2.jpg.

acting chair of the Chinese Medical Association (醫師公會監委), con-
curred that although a case like Yao's was indeed rare, the facts remained
to be fully uncovered. A definitive diagnosis of Yao's condition must not
be formulated based on unfounded speculations.[113]

On the following day, other interlocutors from the medical profession
chimed in. The gynecologist Mao Wenjie (毛文杰) paid Yao a visit on the
morning of March 18 and asked to inspect her body carefully. Yao refused
to disrobe, so Mao proceeded with an assessment of her genital region in
a fully dressed situation, which was identical to the way her grandmother
had verified her transformation previously. Based on this indirect observation, Mao conjectured that Yao's condition was congruent with what doctors normally called "female pseudo-hermaphroditism," or what was more
commonly known in Chinese as "*ci* becoming *xiong*." Mao testified that
Yao's male organ remained underdeveloped because, although he could
sense a penis that was immediately erected upon physical contact, he could
not detect the presence of testicles. He also called attention to specific resid-
ual female traits, such as a large right breast (but a small left one) and a
significant amount of vaginal secretion that left a strong odor in her lower
body.[114]

Mao compared Yao to a similar case in 1930 of a man who experienced sex metamorphosis in Hangzhou. The twenty-one-year-old Shen Tianfang (沈天放) had had abdominal pain on a monthly basis since the age of sixteen. By July 1930 the periodic discomfort, which had troubled him for years, reached an unbearable stage, so his mother finally brought him to several doctors for treatment. While some physicians attributed Shen's condition to intestinal problems, others thought that Shen had contracted some type of venereal disease. Yet still, after a handful of consultations, Shen's condition only worsened. Finally, in August, Shen and his mother met Dr. Wang Jiren (王吉人) of the Tongren Hospital (同仁醫院) on Qingnian Road (青年路) in Shanghai. Wang specialized in the treatment of sexually transmitted diseases, with a secondary expertise in surgery. He considered Shen's reproductive organs symptomatic of a congenital defect rather than a venereal infection. Nor did Wang think that there was any problem with Shen's digestive system. Wang found no testes inside Shen's scrotum and deemed his enlarged chest area comparable to the size of women's breasts. Consequently, he gave Shen the diagnosis of "female pseudo-hermaphroditism," and, as a treatment, he surgically constructed "an artificial vagina" (人工造膣, rengong zaozhi). The Shenbao report claimed that Shen "turned into a woman swiftly" and showed a photograph of Shen's genital area after the sex-change operation. Therefore, Shen's medicalized sex transformation provided Mao an important precedent for interpreting Yao Jinping as just another case of female pseudo-hermaphroditism.[115]

Another physician, named Wang Guning (王顧甯), provided a slightly different diagnosis for Yao's condition. Wang graduated with a doctorate from the Faculty of Medicine at Katholieke Universiteit Leuven, and he was a former surgeon at the Beijing Railway Hospital and worked as a neurologist at the Royal Manchester Hospital in England. Wang claimed to have expertise in sexual pathology and interpreted Yao's incident as one among the many female-to-male sex transformation cases that sat at the forefront of global medical research. Instead of explaining her condition with the popular notion of pseudo-hermaphroditism, Wang introduced a more sophisticated-sounding medical term, "the masculinization of the adrenal gland" (副腎化男體, fushen huananti), to further impress the public with the depth of his professional knowledge. According to Wang, doctors around the world had yet to agree on a consistent set of symptoms for this biological disorder, but it had been identified mainly among women living in the temperate regions. Mature women who had this disorder would undergo bodily changes that made them look like men. However, these changes were typically due to long-term effects of hormonal imbalance. Therefore, Wang contended, Yao's attribution of her sex change to a lightning strike must be a false illusion as it could not be the actual reason for her transformation. Following other medical experts, Wang concluded that the final word on Yao's sex change could only be reached after "a licensed practitioner has carefully scrutinized her body."[116]

Enthusiastic about the potential breakthroughs that the Yao case may bring to the medical field, all of these experts were careful not to arrive at a conclusive diagnosis too hastily prior to the physical exam. On the day that Yao was transported to the National Shanghai Medical College, its director, Yan Fuqing, met with General Li Du and was surprised to learn that her sex had changed abruptly overnight. According to Yan, the female-to-male transformations resulting from female pseudo-hermaphroditism were typically gradual. In order to figure out what was really going on behind Yao's self-proclaimed sex change, Yan promised to assign the best practitioners in the hospital to this case, including the chair of the gynecology department, Dr. Wang Yihui (王逸慧), and the chair of the urology department, Dr. Gao Rimei (高日枚).[117]

Word that Yao's story was merely a hoax soon shocked the public. On March 21 the *Shenbao* coverage of Yao's clinical examination was introduced with the headline "Yao Jinping Is Completely Female." The *Shenbao* journalists had confirmed this startling finding with General Li over telephone on the evening of March 20. According to Li, because Yan Fuqing highly valued the groundbreaking prospect of Yao's case, he assigned six of his best doctors (two Westerners and four Chinese) to conduct a thorough examination of Yao. At nine o'clock in the morning of March 20, they tried to persuade Yao to take off her clothes so that the medical team could examine her body closely, but she persistently refused to cooperate. Eventually, the team had to rely on anesthesia to bring Yao to sleep, and, upon close investigation, the doctors realized that Yao's body was genuinely female without a slightest hint of genital transformation. This disappointing discovery was confirmed by eleven o'clock in the morning. Pressed by Li immediately afterward, Yao explained that in self-presenting as a man, her sole intent was to join the army in Xinjiang and reconnect with her father. This was not possible for a woman. Li quickly forgave Yao and promised her family an annual support of 300 yuan plus the cost required to send her back to school in Tianjin.[118] According to *Xinwenbao*, the deceiving erected male organ detected (indirectly) by Yao's grandmother and Dr. Mao Wenjie was nothing more than a bundle of cloth wrapped in a rod-like fashion.[119]

While news articles put Yao in the spotlight, the Shanghai medical team came under fire as well. Some of their peers felt uneasy about the practice of forcing Yao into an unconscious state. For instance, in the immediate aftermath of Yao's forced examination, *Medicine Review* (醫藥評論, *Yiyao pinglun*) published an article focusing on its legal and ethical implications. The author, likely a medical researcher or practitioner, wrote under the pseudonym Han Gong (憨公) and raised three main objections to the ways in which the Yao case had been handled by the Shanghai medical team. First, human hermaphroditism had been known to the medical profession for centuries, so even if Yao's condition was confirmed as such, the author asked, "is this finding really that exceptional and unprecedented to be able to advance medical science in significant ways?" Second, the author expressed discomfort at the way the team induced Yao to sleep under the instructions of General Li Du rather than respecting the consent of Yao. The author poignantly asked, "Does General Li Du adequately represent Yao's interest? What if behind Yao's refusal to being examined rests a life-threatening reason?" It seemed especially unacceptable to the author that "the hospital went out of their way to sacrifice [Yao's] free will." Finally, the doctors should have adhered to an ethical standard that protected rather than breached the confidentiality of a patient's medical history, but the speed with which the Shanghai team exposed their findings to the public might have made a significant impact on Yao. Who would claim the responsibility, for instance, "if Yao became so emotionally entrenched that she decided to commit suicide?"[120] Despite the accelerating interest in the Yao story, some medical professionals found it only right to proceed with caution, value the ethical implications of their work, and resist conforming to public hype.

The Lure of Yao

For the most part, the public viewed Yao Jinping neither as a freak nor as someone embodying the negative connotations conventionally attached to the *renyao* figure. Instead, when confronted with this highly sensationalized case of sex transformation, Chinese readers reacted in a surprisingly sympathetic tone. In light of the level of publicity that it received, the Yao story provoked interest in nearly every corner of urban Chinese culture in the spring of 1935. In contrast to Yao's bogus transformation, some magazines claimed, stories of genuine sex change prevailed in foreign countries such as Japan, Egypt, and the United States.[121] Above all, the majority of observers endorsed the value of science that was promoted in the sexological literature. Some commentators stressed the importance of gathering sufficient scientific evidence before jumping to any hasty conclusions about Yao's bodily change; others, following the leading voice of doctors, assumed that her sex change was already real and argued the other way around: Yao's experience was valuable for unlocking the secret of nature and thus pushing science forward. More often, though, Yao's ostensible transformation was perceived with a growing sentimentalism that framed her behavior and motivations in extraordinarily positive (usually filial) terms. In the mid-1930s, the press coverage of Yao Jinping generated a "public passion" on an unprecedented scale toward the issue of sex change.[122]

On March 18, the day after Yao Jinping's name made headlines in China, a commentary that appeared in *Xinwenbao* attempted to offset the sudden peak of public interest and anxiety surrounding the Yao story. The writer, Du He (獨鶴), began by pointing out the prevalence of both female-to-male and male-to-female transformations in the Chinese historical record. The popular tendency to dismiss these cases as outright impossible, according to Du He, should be corrected. In fact, around this time another case of female-to-male transformation was widely reported outside China. The message was clear: this coincidence of "Sino-Western reflection" (中西對照, *zhongxi duizhao*), in the author's words, suggested that Yao's experience was not exceptional. Locating the cause of Yao's transformation in congenital physiological defects, Du He argued against seeing it as an irregular or surprising event. Here the view articulated by Chinese sexologists (such as Liu Piji) in the 1920s had filtered down to the popular level: modern science could throw light on puzzles of life previously less well understood.

Interestingly, Du He insisted that Yao was already a man regardless of his physiology. Apart from the fact that Yao was consistently referred to in the masculine pronoun, "he" (他) instead of "she" (她), the entire discussion proceeded on the assumption that Yao had undoubtedly turned into a man. To Du He, Yao embodied a masculine gender worth praising rather than being disparaged about:

> Yao Jinping missed his father deeply when he was a girl. He cried day and night. Now that he has become a man, he promised himself to find and reconnect with his father. It is evident that he is not only filial but also masculine-hearted by nature. People like him and those who are associated with him should be applauded and granted extra love and care. With positive support, he can turn into a "good man" (好男兒, haonan'er). His physiological changes should not be the focus of discussion, which would render him as a biological oddity.

By placing an equal, if not greater, emphasis on gender embodiment, the author differed from the sexological elites who upheld science as the only answer to all aspects of sexual life. Notwithstanding his reinforcement of gender stereotypes, Du He wanted to convey a larger point regarding the societal treatment of people who changed their sex: that their social status should not be stigmatized by scientific narratives of abnormality.[123]

Others perceived the relationship between science and Yao's unconventional changeover in a less antagonistic way. According to an article that appeared in the Shanghai tabloid newspaper *Crystal* (晶報, *Jingbao*) on March 21, "Research on Female-to-Male Transformation" (女轉男身之研究, *nüzhuan nanshen zhi yanjiu*), the significance of Yao's experience and the value of scientific research should be more adequately understood in reciprocal terms. The author, Fang Fei (芳菲), opened with the observation that there had been plenty of historical documentations of sex change in China, "but, without reliable evidence, they are not trustworthy." Fang claimed that Yao's transformation, on the other hand, provided a rare and important opportunity for the scientific assessment of similar phenomena. Even "the pierced ears and the bound feet" in Yao's case "do not constitute solid evidence because they are the result of human labor [人為, *renwei*]." In contrast, such natural changes in Yao's physiology as genital transformation, the flattening of breasts, and the development of an Adam's apple and, according to Fang, "the most surprising observation that all of these were induced by a lightning strike" warranted "further investigation by researchers."[124]

By and large, Fang's discussion endorsed the outlook of scientism bolstered by May Fourth sexologists. In her view, Yao's case presented researchers and medical doctors a golden opportunity to study the nature of sex transformation based on hard evidence and, by extension, to advance the status of the Chinese scientific community. Because actual human sex change was "such a rare event in life" (人生難得之遭逢, *rensheng nande zhi zaofeng*), Fang encouraged experts in medicine and biology to not let this opportunity slip. Similar to Du He, Fang's assumption here, before knowing the eventual outcome of Yao's story, was that Yao had already become male. But unlike Du He, Fang did not take science as a powerful force of cultural authority that necessarily pathologized and marginalized the social status of people like Yao Jinping. Fang argued instead that precisely due to its rarity, Yao's unique experience should actually make her proud after "abandoning any feeling of shame and offering [herself] as a candidate for scientific research."[125]

At the peak of Yao's publicity, some tabloid writers followed the leads of earlier sexologists and brought to public discourse similar bodily conditions, such as hermaphroditism and eunuchism. In a *Crystal* article entitled "Reminded of A'nidu Because of Yao Jinping," the author, Xiao Ying (小英), recalled a lady named A'nidu from the Shanghai Courtesan House after learning about Yao's story. A'nidu, who passed away a few decades before the Yao incident, turned out to be the foster parent of the famous Shanghainese courtesan Wendi Laoba (文第老八). According to Xiao Ying, A'nidu's body was masculinized in ways similar to Yao's transformation: she "had a *yang* presence but a *yin* face," and "she wore women's clothes to emulate a *ci* [female] appearance, yet her large physique resembled a man." When A'nidu was still alive, many assumed that she was an "underdeveloped man" (發育未全之男子, *fayu weiquan zhi nanzi*). Xiao Ying regretted that A'nidu's body was not subjected to postmortem examination. For Xiao Ying, the difference between a man and a woman (男女之別, *nannü zhibie*) could not be determined solely based on genital appearance: the internal structures of the reproductive system mattered, too. Writing in a language similar to Chai Fuyuan's notorious *ABC of Sexology* (especially his discussion of "incomplete male growth"), Xiao Ying applied some of his sexological ideas about natural reproductive anomalies to the case of A'nidu.[126]

With respect to Yao, Xiao Ying's main point was that any claim put forth about her sex change could only be inconclusive prior to a thorough medical exam. "Although Yao Jinping is publicly known to have transformed from *yin* to *yang*," Xiao Ying carefully asserted, "her lower body parts have not been properly investigated. The claims that her grandfather, Yao Qingpu, made about her penile development have not been verified. Most doctors judge the case to be the 'ci becoming xiong [雌孵雄, cifuxiong]' type, but it seems to be too early to draw this conclusion." Unlike Du He and Fang Fei, Xiao Ying did not take for granted Yao Jinping's sex transformation. Xiao Ying brought up A'nidu precisely to underscore the importance of a careful physical checkup, especially in order to achieve a reliable assessment of Yao's anatomical status. "It would be most welcome," Xiao Ying wrote, if "the determination of Yao Jinping as either ci or xiong by doctors" could "be reported in various newspapers and print venues." In arguing that the result of Yao's sex determination should be publicized rather than emphasizing her marginal and stigmatized status, Xiao's intention, similar to the previous two writers, was to promote the value of science in an age of social and political uncertainties.[127] Another article embraced this scientific faith by criticizing the media's emphasis on a "lightning strike" and a "strange dream" as ungrounded explanations for Yao's transformation.[128] This echoed some of the criticisms leveled by medical experts about superstitious claims.

Xiao Ying also avowed that the whole publicity on Yao Jinping brought back memories of eunuchs. Four days after the article on A'nidu appeared, she contributed another piece to *Crystal* called "Reminded of Eunuchs Because of Yao Jinping's Female Body." As reflected in its opening sentence, news of Yao's unchanged female sex was widely reported by this point: "The female status of the self-proclaimed female-to-male Yao Jinping was confirmed after being medically examined by a group of seven doctors—Chinese and Western—at the National Shanghai Medical College." But Xiao Ying did not offer a straightforward rendition of what happened. "The doctors discovered that her fake male genital appearance was made possible by a phallic-shaped bundle of cloth and not the result of an actual transformation of the reproductive organ. This bears striking similarities to the castration surgeries operated on eunuchs for the immediate effect of dismemberment [若昔之太監淨身脫然而落也, ruoxi zhi taijian jingshen tuoran erluo ye]. [Upon uncovering the truth behind Yao's sex change,] Professor Yan Fuqing and his medical team must have enjoyed a good laugh."[129]

Yao Jinping reminded Xiao Ying of castration in both realistic and metaphoric ways. Realistically, Yao's female body was laid bare in front of medical experts like a eunuch's body lacking a penis. In Xiao Ying's metaphoric formulation, the doctors' discovery of Yao's true sex became a performative restaging of castration itself—the public enactment of a medical procedure that "removed" Yao's highly publicized male identity. Whether Yao Jinping was reminiscent of the hermaphroditic body (via A'nidu) or the allegorical experience of castration (via eunuchs), her intention to become a false male never abhorred Xiao Ying. Xiao Ying merely approached Yao's sex change from the angle of rendering medical science as the cradle of truth.

After the public exposure of Yao's disguise, or perhaps because of it, the tabloid press continued to identify physicians as the most authoritative to answer questions related to reproductive defects. Shortly after Xiao Yin associated Yao's sex change with hermaphrodites and eunuchs, another *Crystal* writer reported on the opinion of an eminent gynecologist named Yu Songyun (余忪筠). As the national spotlight on Yao was just beginning to recede, the subject of her sex change came up in a conversation the writer had with Yu, who established the Gynecological Clinic of Zhongde Hospital (中德醫院平民產科醫院, *Zhongde yiyuan pinmin chanke yiyuan*) in Shanghai. Yu suggested that even if Yao's case did turn out to be a real "*ci* becoming *xiong*" transformation, a national sensation would still be an overreaction given that she would not be the first in China anyway.

Five years earlier, in 1930, Yu had delivered a child born with the genital appearance of both sexes. The medical team diagnosed the child as more biologically female than male, so Yu distinguished her from Yao Jinping and categorized her instead as an example of "*xiong* becoming *ci*" (雄孵雌, *xiongfuci*). The parents refused to listen to the doctors, who tried to dissuade them from viewing the baby's rare birth defects with disgust. The baby was eventually transported to the Jiangping Children's Home (江平育嬰堂, *Jiangping yuyingtang*) to be raised there. The hospital kept her file, which included her photographs, her date of birth, and the names of her parents. Based on his experience as a practicing gynecologist in Shanghai, Yu Songyun also encountered births with "an external fleshy bulge in the shape of grapes" (產肉葡萄一束者, *chanrou putao yishuzhe*) and "internal organs formed on the body's exterior" (產五臟六府在外之兒者, *chanwuzang liufu zaiwai zhi erzhe*).[130] Similar to Xiao Yin, the writer of this article did not view Yao Jinping as a freak of nature. His goal in reporting Yu's clinical experience converged with the aim of the other tabloid writers and contemporary sexologists: to deepen a middlebrow print culture that promoted a vision of modernity grounded in the pursuit of accurate scientific knowledge. In this context, clarifying the epistemic ambiguity surrounding sex change became a spirit embodied by all participants of this new cultural production.

Meanwhile, many critics argued that China's feudalism and patriarchy were responsible for Yao's fabrication. A contributor to *Reading Life* (讀書 生活, *Dushu shenghuo*) anticipated that after the revelation of Yao's hoax, some people would grab onto "such an amusing story" to hark back to and defend "old morale," taking away the lesson that "extraordinary things can happen to filial [people]." This author vehemently criticized those who ignored medical knowledge about hermaphroditism and bought into unscientific explanations of Yao's transformation.[131] Another writer underscored the "innovative" nature of Yao's incident, as it happened at a juncture when science carried an insurmountable degree of cultural authority. According to this author, Yao's "creation" was highly original because it precisely called into question long-standing "feudalistic beliefs" that may lead some to surmise that "God could turn a woman into a man simply to reward her for extreme filialness."[132] Moreover, since Ruan Lingyu (阮玲 玉), one of the most iconic female celebrities of the time, committed suicide in the same month that Yao's story hit the headlines, many observers inferred a close connection between these two tantalizing events.[133] One writer thus claimed, "as we all know, Ruan Lingyu's suicide is the consequence of immense patriarchal pressure. Isn't it obvious that the root of Yao Jinping's desire to turn into a man lies a similar problem?"[134] Even Lu Xun criticized the media sensationalism poured onto Luan's suicide and Yao's transition. In his preface to Xiao Jun's *Village in August* (1935), Lu Xun hinted at the urgency of pulling the Chinese nation out of a dark and feudalistic past through enlightenment knowledge.[135] One writer even went so far as to claim that the success of the feminist movement would eventually "wipe out the practice of cross-dressing" because, as Yao's story demonstrated, "the only way for women to enter the public sphere is to dress up as men."[136]

These critics had put their finger on a familiar trope in Chinese culture. The "feudalistic beliefs" that they decried had long justified a woman's desire to change gender as a sign of filial expression. For example, the positive image of Hua Mulan, a woman warrior who joined the military on behalf of her father, had circulated in popular culture since the Northern Wei dynasty (386–535). In the late Ming, stories of female-to-male transformation "were subject to less questioning, scrutiny and skepticism" than transformations in the opposite direction because, to quote Charlotte Furth, "gender transgression often merely serves the accepted social hierarchies by a controlled display of their inversion."[137] In the late Qing period, a Shanghai lithograph recounted a tale in which a filial daughter cooked her own flesh to feed her sick father. However, being a girl and the only child, she prayed to transition into a man so that she could continue the family bloodline. One day she had a strange dream in which an old lady with white hair gave her a sugarcane and two oranges. Upon waking up, she developed male genitals in her lower body. She eventually got married and had two children (figure 4.5).[138] The close association of female-to-male transformation with a daughter's filialness provided Yao a powerful cultural resource for rationalizing her lie. It was precisely this loaded connotation that modernizing thinkers wished to overturn in their effort to fortify the conviction to scientism.

And the public attention soon shifted to the role of other cultural agents. Some critics held reporters, journalists, and popular writers responsible for the public disappointment about Yao's deceiving sex change. According to a *Crystal* article that appeared on March 23, "At first, the news of [Yao's] transformation into a man due to a lightning strike came from Tianjin," and "reporters and journalists from all major presses visited Yao, interviewed her family, gathered together narratives of her past, and vigorously spread the word about the incident." "Because they demonstrated a conspicuous lack of common knowledge," the author, Xin Sheng (辛生), contended, "newspaper reporters and journalists must take at least half of the responsibility for [Yao's] fraud." According to Xin Sheng, "Current scientific knowledge posits that the reversal of male and female physiology overnight is impossible. From the start, the author and his friends have firmly believed that the sudden national spotlight on Yao would only extend nonsense and superstitious attitudes toward the supernatural and the strange [荒誕神怪之不良觀念也, *huangdan shenguai zhi buniang guannian ye*]. Now that the truth is unveiled . . . it is truly a joke." For Xin Sheng, news such as Yao Jinping's sex-change story attempted to lure Chinese readers with shocking accounts of unusual phenomenon, rare biological problems, and astonishing medical solutions. He took them as unhealthy press coverage that contributed nothing positive and "deeply hope[d] that press editors do not publish any more circumstantial writings of this sort without the support of solid evidence."[139] Rather than blaming Yao for her self-fashioned

Figure 4.5 Late Qing lithograph "A Filial Daughter Becomes a Man" (1893).
Source: "Xiaonü huanan," 1893.

sex change, some educated observers compared the role of journalism to that of modern scientific knowledge as important vanguards of a more reliable civil society.

In the wake of the Yao story, creative writers, too, began to articulate their own vision of cultural modernity.[140] Apart from serious tabloid commentaries, the publicity showered on Yao Jinping inspired a few poems and song lyrics that appeared in both mainstream and tabloid presses.[141] The tone of much of these creative pieces tended to cast Yao as a filial subject and, like the above article, attribute the growing disappointment with the outcome of Yao's medical examination to a shallow grasp of scientific principles. One poem lamented the absence of comparable public interest and historical record on male-to-female transformation.[142] This poem, simply titled "Female Becoming Male" (女化男身, *Nühua nanshen*), expressed the author's affection toward another man and his wish to reincarnate as a woman, revealing a subtle appreciation of how his desire differed from homosexuality.[143]

In these tabloid accounts, not a single author passed a moralist judgment on human sex change. None of the commentators cast Yao Jinping in a negative light, and, before she was medically examined at the National Shanghai Medical College, some observers even described her bodily state as a rare and unique biological condition that could potentially provide scientific researchers and medical doctors a multitude of research possibilities. All of them invoked medical knowledge to *naturalize* birth defects and human anomalies. But more importantly, Yao Jinping's story played a decisive role in turning the mass circulation press into a platform for expressing a normative ethos of scientism. This gradually transformed "sex change" into a more general category of human experience not confined to congenital bodily defects. Despite its shocking outcome, Yao's story invited a wide range of reactions that looked beyond the single medical explanation of pseudo-hermaphroditism. As the commitment to the power and authority of science deepened, the idea that even non-intersexed individuals could change sex gradually took shape in Chinese popular culture by the 1940s. The next section recounts an episode of this crystallization process through a close reading of the science fiction short story called "Sex Change" (1940), arguably the first transsexual autobiographical narrative in Chinese history.

Gu Junzheng's "Sex Change"

In 1940 Gu Junzhen's science fiction short story "Sex Change" (性變, *Xing-bian*) was serialized in the magazine *Scientific Interest* (科學趣味, *Kexue quwei*). A popular science writer, novelist, and translator, Gu became an editor in the translation department of the Commercial Press in 1923. He then relocated to the Kaiming Bookstore (開明書店), which was established in 1926, and became one of its chief editors in 1928. His interest in popular science literature began in the early 1930s and led him to cofound the magazine *Scientific Interest* in 1939. His three other more well-known science fiction short stories—"The London Plague" (倫敦奇疫, *Lundun qiyi*), "Below the North Pole" (在北極底下, *Zai Beiji dixia*), and "A Dream of Peace" (和平的夢, *Heping de meng*)—also appeared in *Scientific Interest* in the 1940s, and they dealt mostly with the theme of wartime turbulence and chaos and a disturbed world order. In questioning the category that seemed most fundamental and fixed of all, "Sex Change" stood out for diverging from the predominant emphasis of the science fiction genre on war and anti-imperialist nationalism at the peak of the Second Sino-Japanese War (1937–1945).[144]

The narrator begins by recalling a homicide case that occurred roughly eight to nine years prior. Also known as "the case of a mad murderer" (瘋子殺人案, *fengzi sharen'an*), the incident involved the abrupt disappearance of the famed biologist Dr. Ni Weili (倪維禮) and his daughter, Ni Jingxian (倪靜嫻). Equally mysterious was an old woman who was found dead along with an unconscious teenage boy in Dr. Ni's research laboratory, both of whose identity had since remained unknown. On the same day, Ni Jing-xian's fiancée, Shen Dagang (沈大綱), showed up in a nearby police station and confessed that he was responsible for the crime. The case seemed all the more puzzling because Shen's motivation was unclear. His subsequent suicide added another layer of mystery to the case. According to the forensic report, Shen's death was caused by self-poisoning one to two hours before he turned himself in, suggesting that his motivation for committing suicide probably had nothing to do with guilt.

The narrator then refers to the entry titled "The Case of Shen Dagang's Surrender" in a book called *Mystery Cases of the Twentieth Century* written by the supposedly authoritative criminal psychologist Huang Huiming (黃慧明), who is of course, like Dr. Ni, a fictional character. In deciphering

Shen's motivation for killing the Ni family and, eventually, himself, Huang eliminates the possibility that it stemmed from conflicts over money or relationship (because Shen's salary was quite high at that point, and he remained deeply invested in marrying Ni Jingxian). Huang raises two related question. First, who are the old lady found dead and the teenage boy found unconscious in Dr. Ni's laboratory? Their identities are still unknown, and the boy suddenly disappeared one day from the hospital where he had been taken for treatment. Second, what happened to the bodies of Dr. Ni and his daughter? If the Ni family was indeed killed by Shen, as revealed by himself, what did he do with their bodies? Most popular accounts simply explained the incident away by suggesting that Shen Dagang had gone mad. But Huang considers this too simplistic of an explanation and concludes instead that, without the necessary clues and sufficient facts that can shed new light on the above two questions, "The Case of Shen's Surrender" must remain one of the greatest mysteries of the century.

In citing the perspective of an authoritative criminal psychologist, the narrator of "Sex Change" seems to hint at the possible limitations of modern science. However, to begin with Shen Dagang's crime, he also sets it up as an enigma for which the story of "Sex Change" itself can offer a crucial solution. The narrator thus writes: "But Mr. Huang, you are wrong. The answer to the true mystery you described can be found here [in the following pages]."[145] As such, the structural underpinning of "Sex Change" can be viewed in a question–and–answer format, with an introductory "question" section that delineates the parameters of a homicide mystery and the rest of the narrative bringing forward the "answer" that supposedly holds the key to resolving it. Here, metaphorically, the medical possibility of "sex change" and its desirability are mediated through the genre of science fiction, as the story of "Sex Change" testifies the value of medical science by playing the role of scientific discovery itself that promises to provide answers to a commonly misunderstood problem—in this case, Shen Dagang's motivation for committing the crime and the fate of Dr. Ni's family.

After describing "The Case of Shen's Surrender," the narrator immediately brings the reader back to where it all started: Shen Dagang's return from the city where he has been working and his decision to pay Dr. Ni a visit on a sunny day in late spring. On his way to Dr. Ni's research laboratory, Shen reflects on his career development and the growth of his love for Ni Jingxian over the last two years. Previously, Dr. Ni had refused his daughter's request to marry Shen on multiple occasions, explaining that Shen's career instability constituted a major obstacle. Now with a stable income, Shen is excited about the prospect of proposing to Ni Jingxian again even though they have not been in touch for over two years. But upon Shen's arrival in his office, Dr. Ni immediately focuses their conversation on his most recent research breakthrough, leaving Shen almost no opportunity to bring up the marriage proposal.

Dr. Ni's ability to convey the latest scientific theories and research on sex designates one of the most unique features of the story: its accurate recounting of modern scientific knowledge. The main source that Gu Junzheng relies on in developing Dr. Ni's extensive overview of the scientific study of sex seems to be the writings of the renowned life scientist Zhu Xi, whose work we encountered in chapter 2. Gu begins this part of the narrative by citing Zhu's *Humans from Eggs and Eggs from Humans* (1939) and ends with a reference to his *Scientific Perspectives on Life, Aging, Illness, and Death* (1936).[146] In addition to drawing from the work of the best-known authority on reproductive biology in twentieth-century China, the story touches on the embryological theory of sexual development to underscore the point that, in Dr. Ni's words, "all new embryos display a common feature: they are sexless. They all have the potential to develop male and female characteristics."[147] Dr. Ni also discusses the chromosomal theory of sex determination, explaining that, whereas in women the sex chromosomes are the two X chromosomes, men have one X chromosome and one Y chromosome. But he continues, "although it might seem that sex is naturally determined at the moment of conception, something that happens randomly and cannot be altered by will, all of this is not set in stone."[148]

Evidently, in his discussion Dr. Ni begins to move toward a definition of sex as something malleable. After noting his dissatisfaction with the genetic theory of sex determination, Dr. Ni describes biological sex using the metaphor of a "balance" (天平, *tianping*), something that, when tipped one way or the other, would result in the predominant expression of maleness or femaleness. Here Dr. Ni points out European scientists' recent discovery of parasitic castration, a natural phenomenon in which anthropodan animals such as bees or crabs would switch their sex after their gonads have been attacked by parasites.[149] Speaking of parasitic castration "makes the old professor even more excited," leading him to make the following remark: "Consequently, I think the sex of human beings is not predetermined. If we know the criteria of sex determination, we would be able to change people's sex."[150] To add credibility to his comment, Dr. Ni brings up Eugen Steinach's classic experiments that induced male-to-female (雄化雌, *xionghuaci*) and female-to-male (雌化雄, *cihuaxiong*) transformations in rats. And to make all of these ideas about sex change sound even more plausible and convincing, Dr. Ni finally introduces Shen Dagang to the idea of "sex hormones," the internal secretions that play a decisive role in sexual maturation. Like the Chinese sexologists and tabloid writers discussed earlier, Dr. Ni seizes this opportunity to use the example of "eunuchs of the Qing dynasty" to highlight the significance of sex glands: as a result of not having a functional gonad, these castrated individuals "remain beardless even at an old age, and their physical appearance resembles neither a man nor a woman."[151] These passages demonstrate that the scientific theory of universal bisexuality has now been absorbed into and rearticulated in the cultural domain of popular fictional literature, and Chinese indigenous examples of reproductive anomalies such as eunuchs continue to operate as a cross-cultural epistemological anchor for crystallizing foreign ideas about sex and sex transformation.

Moreover, the careful application—and not just the nominal referencing or presentation—of modern scientific knowledge could be said to be a staple of an early wave of literary production that simultaneously pushes for a greater degree of flexibility and creativity in the science fiction genre.[152] In "Sex Change," this is best exemplified by the biomedical breakthrough for which Dr. Ni prides himself throughout his conversation with Shen Dagang. According to Dr. Ni, the "experimental product" of his latest breakthrough is a white potion that "can turn a woman into a man both biologically and psychologically in four days after injection into the bloodstream."[153] As the reader would soon discover, Dr. Ni has belabored the various scientific theories of sex and introduced this recent invention of his to Shen only because he has used it to change the sex of his daughter, Ni Jingxian, thus making it impossible for Shen to propose to, let alone marry, his only child. The example of Dr. Ni's sex-change potion reflects a tremendous degree of informed creativity on the part of the author, Gu Junzheng, who has not only appropriated and accurately presented Western scientific ideas about sex but also built from them and deliberately proposed a new method of human sex transformation beyond the existing scope of medical technology. This is best captured in Dr. Ni's own words before he shows Shen the actual potion:

> Scientists have now confirmed that secondary sex characteristics in humans are determined entirely by the secretions of the sex glands, so these sex characteristics can be easily modified with the surgical techniques of castration, transplantation, or [hormonal] injections. However, there is still no procedure that can change an individual's primary sex characteristics. In other words, although scientists can make a woman look like a man and a man look like a woman, they are still unable to turn a woman into a man and a man into a woman completely. But allow me to inform you now that, after many years of research and experimentation, I have found a way to alter sex characteristics on the primary level.[154]

Although the potion is a fictional entity, its material possibility and functional comprehensibility is circumscribed by the existing biomedical lexicon of sex. Whereas bodily modification techniques such as castration, tissue transplantation, and the administration of synthetic hormones constitute a crucial source of imagination, the author's presentation of the potion as the sole technological innovation that can transform one's true sex achieves a level of literary production and originality that exceeds any existing epistemological configuration of medical science. This thus marks a radical departure from the science fiction novels written before the Republican period.

As a story about a topic as ahead of its time as sex alteration, the plot of "Sex Change" ironically embraces and reflects broader cultural claims about the relationship of science to gender. Dr. Ni's rationale for creating the potion, for instance, is undergirded by a prevailing discrimination in Chinese culture that values sons over daughters. After being told that his intended bride-to-be had turned into a boy, Shen Dagang presses Dr. Ni for an explanation. Posing "an implicit sign of victory," Dr. Ni responds:

> You think I would back off and just let you take [my daughter] away from me? You fool! You have no idea how much I love her. For years I have focused on my research day and night for the simple reason that I wanted to turn her into a son! You fool! Do you think I would

let some stupid kid to propose to her just because he selfishly thinks that he loves her and to use her to threaten me? This is something that I would never allow to happen, because she is my child. If she is a boy, I would not have to worry about anyone proposing to her. If I have a son, I can make him pursue my unfinished work. His accomplishment can open a fresh chapter for the Chinese scientific community. How wonderful and valuable would that be?![155]

The white potion gives Dr. Ni a son by transforming his daughter, a female character, into a masculine subject, a supposed sign of scientific progress. Dr. Ni's explanation implies that what women want and long for plays no role in the determination of their fate. Instead, it is only the men—the father and the potential husband—that participate in the manipulation, a power play of sorts, of women's lives. Dr. Ni's words make it evident that whether his daughter actually wants a sex change is insignificant. What is at issue here is his own desire for his daughter's sex change (which fulfills his ambition to contribute to the progress of science and China), mirroring Shen Dagang's subsequent desire for Ni Jingxian to undergo a second sex change (so that she can be turned back into a girl). In other words, the male voice and opinion dominate the entire structural dynamics of the relationship between Dr. Ni, his daughter, and Shen Dagang, relegating the female voice, not only here but throughout the narrative, to the background and even a status of nonexistence. Medical technology, the plot seems to imply, helps men to perpetuate the value of their sexed existence.

It can be said that the author is making an implicit criticism of ideas about gender and the body in traditional Chinese culture. Or, more specifically put, the story of "Sex Change" can be interpreted as formulating an indirect critique that plays off on the gender dynamics of a society in which such corporeal practices as footbinding thrived. Both footbinding and Ni Jingxian's sex change involve the transformation (if not "mutilation") of the female body, but mainly for the explicit pursuit of male pleasure, desire, and ambition.[156] By narrating the story about Ni Jingxian's change of sex through the power struggle between Dr. Ni and Shen Dagan, the author similarly reveals the underlying patriarchal biases, unfair assumptions, and male selfishness of such gendered customs as footbinding.

However, throughout the narrative of "Sex Change," the reader is never exposed to the voice of Ni Jingxian, such as regarding how she feels about her predetermined fate to change sex and its consequent effects on her life.

Her only spoken dialogue in the story appears immediately after Shen Dagang comes face to face with the masculinized version of her: "Ah, Dagang . . . this wasn't my intention. I thought my father has told you that already."[157] These meager words reveal the author's intentional effort to make room for the expression of female agency in the text only through the masculinist discourse—the voice of Dr. Ni and his reasons for changing his daughter's sex. And perhaps what distinguishes Ni Jingxian's sex change from footbinding is again the role of medical scientific invention. In the story, the sex-change potion symbolizes scientific progress, and what it can do symbolizes male success and accomplishment. Even as the narrator of "Sex Change" is revealed in the end to be (the post–sex change) Ni Jingxian her/himself, this exposure only further suggests that the act of uncovering "truth" (in this case, the truth behind the homicide mystery introduced at the beginning of the story) can be done and articulated only by a masculine subject (for the narrator is really no longer the female Ni Jingxian but a married physiologist and father of two children).

The story's perpetuation of patriarchal values is also exemplified by its overall message that science remains a masculine endeavor. Pressured by Shen to turn Ni Jingxian back into a girl, Dr. Ni comes up with another potion for which he needs an experimental subject. Running out of patience, Shen immediately injects the new potion into Dr. Ni's body, exclaiming "you are the most convenient experimental subject, old fool!"[158] Contrary to the positive tone associated with Dr. Ni's success in changing her daughter's sex, his own transformation leads to a disastrous final episode, for which the author gives the subtitle "A Tragedy." Unfortunately, after Dr. Ni becomes an old woman, she is no longer capable of creating the magical sex-change elixir again. After Shen has repeatedly begged the old woman to remake the potion that can potentially bring back the female Ni Jingxian, "The old woman adamantly stares at Shen and frowns. She finds his request distasteful and says nothing. She is no longer a professor passionate about the progress of science. She has completely forgotten about science, as if she has never learned a thing about it."[159]

This passage conveys the author's explicit association of science with men (or the masculine gender), implying that the pursuit of science remains outside the scope of women's sphere. Correspondingly, an underlying message of the story implies that male-to-female transformation is less preferable than female-to-male transformation, which again reinforces a central component of Chinese society that puts fathers and sons instead of mother and daughters at the center of kinship relations. Reminiscent of how most tabloid writers approached Yao Jinping's intention to alter the public appearance of her sex, the depiction of Ni Jingxian's female-to-male transformation in "Sex Change" is layered with various positive signs of scientific progress and gendered modes of ambition. On the other hand, Dr. Ni's sex change resulted in the shattering of hope (specifically, Shen Dagang's hope). And the truth behind the entire incident would only be recovered and uncovered years later through, once again, the voice of a masculine subject who was previously female.

"Will Not Surrender Until I Become a Woman"

At the start of the twentieth century, a few scattered accounts of human sex change surfaced in news pictorials, popular magazines, and medical journals. They featured both male-to-female and female-to-male transformations and represented a geographical diversity: stories came from Liaoning, Shandong, Hunan, Henan, and beyond. Although these extraordinary tales explored complicated social issues (such as marriage), involved graphic descriptions of bodily transformation, and even delved into the unusual prospect of reproduction, they rarely resorted to medical understandings of hermaphroditism to justify why certain individuals turned into the opposite sex.[160] Astonishing and odd, these stories built on earlier rumors of women becoming men and men morphing into women, and they cohered neatly with the binary construction of sex upheld by scientists and doctors since the late nineteenth century.[161] But in the late 1920s and 1930s, and especially after Yao Jinping's celebrity, a new meaning of sex as hormonal, scalar, and malleable opened up different ways of not just thinking about but also *explaining* human sex transformation.

This chapter has traced an evolving discourse of "sex change" in the mass circulation press from the 1920s to the 1940s. Relevant scientific ideas including the theory of constitutional bisexuality first articulated in early twentieth-century sexology were filtered through media sensationalism and publicity on sex change, climaxing in Yao's media blitz, and finally diffused and absorbed into the popular imagination, as exemplified by fictional works like Gu Junzheng's "Sex Change." As new biomedical interpretations of reproductive anomalies reached a wider public, the concept of "sex change" gradually moved away from a specialized term circulating primarily in the scientific literature and became a more general category of experience with which individuals hoping to alter their bodily sex could come to associate.

But the idea of transforming physical sex alone was insufficient. The dissemination of information about surgical intervention made it a convincing reality. Toward the midcentury, Chinese readers were exposed to testimonies of people seeking *and* receiving sex-reassignment operations in Europe and the United States. The most famous examples of these, and widely reported across the world, were European women athletes who became men. Zhang Ruogu (張若谷), a renowned author and the founding editor of *Greater Shanghainese* (大上海人, *Da Shanghairen*), discussed some of these examples in an article in *Sex Science*. In 1935 Zdeňka Koubková, a Czechoslovakian track athlete who won two medals at the 1934 Women's World Game, "decided to abandon her world record" and "changed her name to [Zdeněk] Koubek" after surgery. In summer 1936 the British javelin, discus, and shot-put champion Mary Weston changed her name to Mark and divulged that "I decided to consult a medical specialist two months ago. After two operations and six weeks of recovery, I have become a man." As Weston put it, the Charing Cross Hospital in London had seen at least "twenty or so similar cases" by the time of her treatment.[162]

Chinese readers soon learned more about Weston's surgeon, the South African–born Dr. Lennox Ross Broster, who attracted considerable press attention to London for his "sex-change" operations in the 1930s.[163] His work on endocrine diseases, especially what would later be known as adrenal congenital hyperplasia, formed the centerpiece of an article titled "Marvelous Stories of Sex Change" (性變奇談, *Xingbian qitan*) published in *West Wind* in 1937 (later serialized in *Desert Pictorial* [沙漠畫報, *Shamo huabao*] in 1940). According to the author, Li Xinyong (李心永), humans with intersexed conditions ("*yin-yang* persons") have existed throughout history since the Greco-Roman world. However, the last few decades saw significant breakthroughs in this area of research. In narrating the experience of notable intersexed individuals in France, Poland, Czechoslovakia, Britain, and the United States, Li repeatedly pitched genital surgery as the fulcrum of their life histories. The article broadcasted in detail Broster's infamous adrenalectomy procedure (the removal of the adrenal glands), including how he came to this idea, the kind of patients who went to him for help, the steps involved in the treatment process, and the success with which he "turned women with mustaches into beauties and effeminate men into machos" (使有鬍鬚的女人變成美女，使柔弱如女子的男性變成雄糾糾的大丈夫, *shi youhuxu de nüren biancheng meinü, shi rouruo ru nüzi de nanxing biancheng xiongjiujiu de dazhangfu*). Though it was obvious that Broster was not working alone in the field of intersexuality research (for example, the works of Oscar Riddle at the Carnegie Institution of Science, Francis Crew at Edinburgh, and medical endocrinologists at Johns Hopkins were mentioned in passing), Li singled out his accomplishments and praised him as "a magician at the Charing Cross Hospital." Yet the overall tone of the article tended to stress the "corrective" function of Broster's surgeries, implying that they ultimately helped to "restore normal sexual development" (恢復性能的正常發展, *huifu xingneng de zhengchang fazhan*).[164]

In the first half of the twentieth century, an affluent discourse of mutability, on top of the visual and carnal aspects explored in the previous two chapters, turned the concept of sex into an intellectual and cultural mainstay in China. Whereas biologists highlighted sex dimorphism in visual illustrations and sexologists collected scientific data about people's sexual desire, experts drawn to the endocrine sciences argued that everyone could easily convert to either sex. This completes our mapping of the three coordinates of this modern "epistemic nexus." By the late 1940s, the press had dampened an immutable view of sex and begun to foreground stories of sex-reassignment operations on a more regular basis. In 1947, for example, *Victory* (勝利, *Shengli*) announced a twenty-four-year-old woman in Sweden, who "has recently turned into a man after receiving a surgery. . . . The hospital refuses to comment on this case or disclose the name of the patient."[165] Another article in *New Woman Monthly* (新婦女月刊, *Xinfunü yuekan*) focused on two sisters in Rome, Italy, who "changed into the opposite sex after their parents agreed to surgical intervention." As a result, both of them "gradually grew into a man with robust physique" (逐成為健壯男兒, *zhucheng wei jianzhuang nan'er*) and "got married to their neighbor's two daughters."[166] In the immediate postwar years, these reports hinted at a global possibility for transforming bodily sex through technology.

In this context, *West Wind* received a startling letter from a male reader in Beiping named Feng Mingfang (馮明方) in 1948.[167] In writing the letter, entitled "Will Not Surrender Until I Become a Woman" (不變女子誓不休, *bubian nüzi shi buxiu*), Feng's intention was twofold: to request an adequate answer for "a problem that has troubled me for a long time" and to make his concerns public so that "others with a similar condition can benefit from reading about it." As Feng put it, his father "treated me like a daughter before the age of seven" and "dressed me up like a girl."[168] In high school, he often "looked into the mirror and admired my own pale and soft skin, round and smooth hip, and curly body . . . oftentimes forgetting that I am actually a man." He constantly asked himself "isn't it a pity for such a beauty to be trapped inside a male body?" Recently, he "has a pressing urge with growing intensity . . . to be castrated and grow larger breasts in order to become a true woman [變成一個道地的女子, *biancheng yige daodi de nüzi*]." The "more I try to suppress this longing, the stronger and more persistent it becomes," so he begged the editors to "save me, hurry!" In their reply, the *West Wind* editorial team cited the recent findings of Alfred Kinsey in the United States and recommended psychoanalytic consultation, but they also expressed "great sympathy toward [his] pain."[169]

Written in an era of immense political unrest and without the existence of an organized underground network for gender variant people in China, the publication of Feng's quest for sex change indicates a broader desire for reaching out to people with a similar experience. Whereas a binary division of sex provided an earlier generation of feminist activists a powerful language, the burgeoning writings of scientists, doctors, educators, researchers, reformers, popular authors, critics, and readers colluded in renouncing its tenacity by the midcentury. With a mercurial definition of sex, alongside a growing awareness of surgical possibility, Chinese people soon "discovered" their first transsexual in postwar Taiwan. But this version of transsexuality emerged at a historical juncture when Chinese geopolitics began to assume new configurations under the ambit of global Cold War Asia. This new chapter in the mutually generative history of China's geobody and the body corporeal is where we turn next.

CHAPTER V

Transsexual Taiwan

An Episode of Transnational Spectacle

On August 14, 1953, the *United Daily News* (聯合報, *Lianhebao*) announced the striking discovery of an intersexed soldier, Xie Jianshun (謝尖順), in Tainan, Taiwan. The headline read "A Hermaphrodite Discovered in Tainan: Sex to Be Determined After Surgery."[1] By August 21 the press had adopted a radically different rhetoric, now trumpeting that "Christine Will Not Be America's Exclusive: Soldier Destined to Become a Lady."[2] Considered by many as the "first" Chinese transsexual, Xie was frequently dubbed as the "Chinese Christine" (中國克麗斯汀, *Zhongguo Kelisiding*). This was an allusion to the American ex-G.I. transsexual star Christine Jorgensen, who received her sex reassignment surgery in Denmark two years prior and became a worldwide household name immediately afterward due to her personality and glamorous looks. The analogy reflected the growing influence of American culture on the Republic of China at the peak of the Cold War.[3] Within a week, the characterization of Xie in the Taiwanese press changed from an average citizen whose ambiguous sex provoked uncertainty and national anxiety to a transsexual icon whose fate contributed to the global staging of Taiwan on a par with the United States. Centering on the making of Xie's celebrity, this chapter argues that the publicity surrounding her transition worked as a pivotal fulcrum in shifting common understandings of transsexuality, the role of medical science, and their evolving relation to the popular press in mid-twentieth-century Sinophone culture.[4]

The feminization of the Chinese Christine became a national story in Taiwan at a critical juncture in the making of Cold War East Asia. Predating the Jinmen shelling crisis of August 1958, this episode of sex transformation commanded public attention at the conclusion of the Korean War (1950–1953) and the nascent "liberating Taiwan" campaign (1954–1955) on the mainland. The Chinese Communist Party's (CCP) involvement in the Korean and First Indochina Wars further consolidated the U.S.–Guomindang (GMD) alliance and amplified the long-standing CCP-GMD tensions while achieving what historian Chen Jian identifies as Mao's aspiration for continuing the momentum of his communist revolution at home and abroad.[5] Between 1951 and 1965, Taiwan received steady annual support of $100 million from the United States. American interest in stabilizing Taiwan's military system and economic growth was folded into the U.S. AID Health Program drafted and distributed in 1954. In the next two decades, American guidance and recommendations would gradually replace the Japanese model and play a central role in shaping the long-term public health policy and priorities in Taiwan.[6] Meanwhile, the CCP and the U.S.-backed GMD engaged in repeated confrontations across the Taiwan Strait in the 1950s, thereby making this area one of the main "hot spots" of the Cold War. It was within this historical context of postcolonial East Asian modernity—providing the conditions for such mimetic political formations as the Two Koreas and the Two Chinas—that the mass circulation press introduced the story of Xie Jianshun to readers in Taiwan. This chapter offers a preliminary glimpse of where the parallel contours of culture and geopolitics converged in early Cold War Taiwan.

Dripping with national and trans-Pacific significance, Xie's experience made "*bianxingren*" (變性人, transsexual) a household term in the 1950s.[7] She served as a focal point for numerous news stories that broached the topics of changing sex, human intersexuality, and other atypical conditions of the body.[8] People who wrote about her debated whether she qualified as a woman, whether medical technology could transform sex, and whether the two Christines were more similar or different. These questions led to the persistent comparisons of Taiwan with the United States. But Xie never presented herself as a duplicate of Jorgensen. As Xie knew, her story highlighted issues that pervaded post–World War II Sinophone society: the censorship of public culture by the state, the unique social status of men serving in the armed forces, the limit of individualism, the promise and pitfalls of science, the relationship between military virility and national sovereignty, the normative behaviors of men and women, and the boundaries of acceptable sexual expression. Her story attracted the press, but the public's avid interest in sex and its plasticity prompted reporters to dig deep. As the press coverage escalated, new names and strange medical conditions grabbed the attention of journalists and their readers. The kind of public musings about sex change that saturated Chinese culture earlier in the century now took center stage in Republican Taiwan.

This chapter adds greater historical depth to this episode of *trans-culturation* by cross-examining other accounts of unusual bodily condition that the press brought to light in response to the Xie story. In reconstructing the broader context in which *bianxingren* acquired widespread currency, I focus on examples of "trans" formation that made explicit reference to Xie, especially during moments when Xie and her doctors intentionally withheld information from the public. By trans formations, I refer to concrete examples of "transing"—a phenomenon proposed by Susan Stryker, Paisley Currah, and Lisa Jean Moore that "takes place within, as well as across or between, gendered spaces" and "assembles gender into contingent structures of association with other attributes of bodily being."[9] The examples aside from Xie's life history tell stories of gender transgression, defects of the reproductive system, uncommon problems related to pregnancy, the marriages of individuals with cross-gender identification, transsexual childbirth, human intersexuality, and sex metamorphosis itself, emanating from both domestic contexts and abroad. Although these stories came to light within an overall narrative of Xie's transition, they provide crucial evidence for the growing frequency of sex-change-related discussions in Chinese-speaking communities in the immediate post–World War II era.

The excavation of a series of trans formations in postwar Taiwan maps gender and sexual marginality onto the region's global (in)significance in and beyond the 1950s.[10] After weaving together Xie's narrative with other stories of gendered corporeal variance, this chapter concludes with a historiographical framework in which these diverse examples of transing could be adequately appreciated. Within the field of Sinophone postcolonial studies, these stories have broader historical import, bringing new analytic angles, new chronologies, and new conceptual vocabularies. The Sinophone world, in this context, refers to Sinitic-language communities and cultures situated outside of China or on the margins of China and Chineseness. By contesting the epistemic status of the West as the ultimate arbiter in queer historiography, the history of trans formations in Sinophone Taiwan offers an axial approach to provincializing China, Asia, and "the rest." The queerness of the trans archive delineated in this chapter is underpinned not only by the very examples of trans subjectivities retrieved and documented but also by the enabling effect of their Sinophonicity to de-universalize existing historiographical hegemonies, whether defined in the conventions of writing the twentieth-century "Chinese" or "Taiwanese" past.[11] This queer emergence of transsexuality in postwar Taiwan concludes our study, therefore, by turning our attention to the social origins of a new historical epoch.

Discovering Xie

When the story of Xie first came to public spotlight, the press sutured a direct reference to Jorgensen: "After the international frenzy surrounding the news of Miss Christine, the American ex-G.I. who turned into a lady after surgery, a *yin-yang* person [hermaphrodite] has been discovered at the 518 Hospital in Tainan." This opening statement reflects a transitional moment whereby sex-change surgery developed from a common clinical management of intersexuality into a new basis of sexual subjectivity. Early on, the *United Daily News* released an article suggesting that Xie had in fact been fully aware of his feminine traits since childhood but had kept it a secret until its recent "revelation" under the close attention of doctors in Tainan. Born in Chaozhou, Guangdong, on January 24, 1918, the thirty-six-year-old Xie had joined the army when he was sixteen, lost his father at the age of seventeen, and lost his mother at eighteen. He came to Taiwan with the Nationalist army in 1949. "At the age of twenty," the article explained, "his breasts developed like a girl, but he had hidden this secret as a member of the military force rather successfully. It was finally discovered on the 6th of this month, upon his visit to the Tainan 518 Hospital for a physical examination due to recurrent abdominal pains and cramps, by the chair of external medicine Dr. Lin. He has been staying at the hospital since the 7th [of August]."[12]

The initial national excitement focused on deciphering Xie's sex, sexuality, and gender. In their first impression, the public was given the opportunity to imagine Xie's sexually ambiguous body with extensive descriptions. According to *Taiwan Shin Sheng Daily News* (台灣新生報,

Taiwan xinshengbao), "the abnormal bodily features of the *yin-yang* person include the following: protruding and sagging breasts, pale and smooth skin, soft hands, manly legs, squeaky and soft voice, a testicle inside the left lower abdomen but not the right, a closed and blocked reproductive organ, no [male] urinary tract, a urethra opening between the labia, a small symbolic phallic organ, and the capacity to urinate in the standing posture."[13] Another article stated that Xie's "head appears to be normal, mental health is slightly below average, facial features are feminine, personality is shy, other bodily parts and dietary habits are normal."[14] Dr. Lin Chengyi (林承一), a graduate of the Tokyo Zhaohe Hospital and the external medicine department of the Jingjing Medical School, diagnosed the case carefully and conjectured that Xie needed at least three operations. As stated in the *China Daily News* (中華日報, *Zhonghua ribao*) report, the first operation was scheduled to take place on August 20 and would involve the following three major steps: exploratory laparotomy (the opening of the abdominal cavity) to detect the presence of ovarian tissues; labia dissection to examine the vaginal interior, determine the length of the vagina, and confirm the presence (or absence) of the hymen; and finally, "if ovaries and vagina are found inside the womb, removing the penis can turn Xie into a woman; otherwise he becomes a man."[15] From its premise that Xie intentionally concealed his femininity, to its elaborate description of Xie's convoluted biology, and to its presentation of the criteria involved in Xie's sex determination, the press operated as a cultural vehicle through which medical biases toward Xie's body were expressed liberally. Through and through, Xie was *assumed* to be a biological woman trapped inside a male body, whose feminine-like features gradually revealed themselves under the fingertips of medical experts and in the eyes of the public. This assumption led one observer to complain about the lack of sympathy and humanistic concerns in this wave of flagrant newspaper accounts.[16]

On the day following the public discovery of Xie, the media called attention to a radical departure of his experience from the familiar story of the American Christine. Whereas the American transsexual celebrity had a deep-seated desire to be transformed into a woman, the Republican Chinese soldier had an unshakable longing to remain a heterosexual man. A *Shin Sheng Daily* headline declared, "*Yin-Yang* Person Uncovers a Personal Past . . . Hoping to Remain a Man." The article quoted Xie for resenting to become "Christine No. 2."[17] Another article noted that "the *yin-yang* person Xie Jianshun is still in love with his lover of more than two decades—the rifle" and that he "personally desires to become a perfectly healthy man."[18] Most tellingly, the piece divulged Xie's heterosexual past, including his romantic affairs with women, and graphic descriptions of his previous sexual encounters. The media narrative reminded the reader of, rather than downplayed, Xie's physical defects: "At the age of seven, Xie fell sick. At the time, his penis was tied to his labia, but given his living situation in the countryside, going to a doctor for surgical intervention was not immediately feasible. His mother therefore simply tore them apart by hand. From that point on, he urinated from both secretion openings."[19]

Reporters disclosed that Xie's "unpleasant experience with his physical abnormality" started at the age of twelve. That year, his grandmother introduced him to a girl, with whom he was arranged to marry. Although he was just a child, his fondness for the girl grew by day. One day, when no adult was home, he initiated an intercourse with the girl but ultimately failed because of his "biological defect." They ended up getting around the problem "by using their hands." Since then, Xie "acquired the habit of masturbation without the ability to produce sperm, being in a state of more physiological pain."[20] After joining the army, he fell in love with another girl. Her father even accepted their marriage proposal. This seemingly positive news, however, upset Xie. Given his "physiological shortcomings," Xie wanted to avoid leading the girl into an unhappy union. At the time he still did not have the courage to come clean about his reproductive problems. He therefore ran away from the girl and the relationship, a decision some writers interpreted as "a comedy of marriage escape" (逃婚喜劇, *taohun xiju*).[21] Despite this tragic turn of events, a twenty-nine-year-old man from Anhui announced his eagerness to look after Xie for the rest of his life.[22]

The most significant message that the above biographical synopsis seems to convey is squarely concerned with his forthcoming sex determination. Will Xie turn into a man or a woman? What does *he* want? A *United Daily News* article ended with a confident note: "He firmly hopes to remain male, to be able to return to the army and pick up the rifle again" to "defeat mainland China and eliminate the communists."[23] *Taiwan Min Sheng Daily News* (台灣民聲日報, *Taiwan minsheng ribao*) corroborated that Xie "prefers to talk about military activities" and "is not interested in domestic life or matters of concern to women."[24] Indeed, the newspapers mentioned more than once that Xie "experiences 'sexual' desire when interacting with women, but none toward men."[25] Construed as a respectable citizen of the

Republican state, Xie was heterosexualized and masculinized as a national subject fulfilling his duty, even as he faced the possibility of being stripped of his manliness within a week. At least for a brief moment, Xie was able to vocalize his desire of *not* wanting to change his sex through the mainstream press. And it was the first time that readers heard his voice. The remark, "If my biology does not allow me to remain a man but forces me to become a woman, what else can I do?" marked the first utterance of his opinion in the press.[26] On the second day of his media exposure, readers started to sympathize with Xie and considered him, unlike the American Christine, a rather normal, however unfortunate, heterosexual man.[27]

If doctors and reporters hastened to purport a clear picture of Xie's hidden sex and sexuality, they tried to detect his gender orientation in a more cautious fashion. As soon as the 518 Hospital scheduled Xie's first "sex-change surgery" (變性手術, *bianxing shoushu*), the relevant experts proposed a plan to determine Xie's gender self-awareness. They sent a group of female nurses to mingle with Xie five days before the operation. Given Xie's long-time career involvement in the military, "the hospital considers his previous social interactions with men insufficient basis for determining how Xie feels deep down inside as man- or woman-like. In preparing for Xie's sex reassignment surgery, a number of 'attractive' nurses were asked to keep Xie company and chat with him on 15 August."[28] Through Xie's interaction with these nurses, it was hoped that "a better understanding of his/her inner sense of self as a man or woman could be reached by drawing on the clues from his emotions and facial expressions, which should reflect his inner sense of self."[29] It is worth noting that neither the medical profession nor the popular press locked him into a particular gender category at this juncture. Despite their assumptions about Xie's biological hidden (female) sex, doctors at the 518 Hospital actually believed that they had adopted a more careful and objective approach to unearth his psychological gender. And despite its covert announcement of his heterosexuality, the press refrained from reaching any conclusion yet about Xie's gendered sense of self.[30]

The First Operation

The first turning point in the framing of the Xie Jianshun story in both medical and popular discourses came with his first operation. Again, the press collaborated with Xie's physicians closely and kept the public informed about their progress. On August 20, the day of Xie's first operation, the *United Daily News* published a detailed description of the surgical protocols scheduled for three o'clock that afternoon: "The operation scheduled for today involves an exploratory laparotomy, followed by a careful examination of his lower cavity to detect the presence of uterus and ovary. If Xie's reproductive anatomy resembles that of a typical female, a second operation will follow suit as soon as Xie recovers from this one." "In the second operation," the description continued, "the presently sealed vaginal opening will be cut open, and the vaginal interior will be examined for symptoms of abnormality. If the results of both operations confirm that Xie has a female reproductive system, the final step involves the removal of the symbolic male genital organ on the labia minora, converting him into a pure female [純女性, *chunnüxing*]. Otherwise, Xie will be turned into a pure male [純男性, *chunnanxing*]."[31]

By bringing the reader's eyes "inward" toward Xie's internal anatomical configurations, the communique repeated the epistemological claims of the medical operation intended for his sex determination. Step by step, the newspaper article, presumably relying on the information provided by Dr. Lin's team, told its reader the surgical procedures and criteria for the establishment of Xie's female sex. Yet no symmetrical explanation was given for establishing a male identity for Xie. The narrative only concluded with the brief remark, "Otherwise, Xie will be turned into a pure male." One wonders what would happen if Xie's interior anatomy was found to be drastically different from the normal female sex. What were the doctors planning to do then with his "sealed vaginal opening?" If Xie could be transformed into a "pure female" by simply cutting off his "symbolic male genital organ," what would turning him into a "pure male" entail? Would that also involve the removal of something? Or would that require the addition of something else? Even if female gonads were found inside his reproductive system and the second operation followed suit, what happens next if his vaginal interior showed signs of anatomical abnormality? On what grounds would the doctors evaluate the resemblance of his vagina to that of an average woman at this stage? To what degree could his vagina deviate from the internal structure of a "normal" vagina before it is considered too "abnormal?" The media coverage answered none of these questions. Under the pretense of keeping its readers informed, the press actually imposed more assumptions (and raised more questions) about Xie's "real" sex. *Shin Sheng Daily* asserted that it was "easier to turn Xie into a woman than into a man," and *China Daily News* learned that Xie had been so anxious that he cried numerous times about the potential makeover.[32] By the day of his first operation, the medical and popular discourses congruently prepared the lay public for a sensational outcome—an unprecedented sex-change episode in Chinese culture. Xie's sex was arguably already "determined" and "transformed" before the actual surgery itself. This reciprocated the ambiguity surrounding the purpose of his first operation: Was its goal the determination or transformation of his sex?

The following day, the Taiwanese public confronted a lengthy summary of Xie's surgery in the news billed "Soldier Destined to Become a Lady." This echoed the headline of the New York *Daily News* front-page article that announced Jorgenson's sex reassignment in December 1952, "Ex-GI Becomes Blonde Beauty." The *United Daily News* piece included a more telling subtitle: "The *Yin-Yang* Person's Interior Parts Revealed Yesterday after Surgery: The Presence of Uterus and Ovaries Confirmed." The *Shin Sheng Daily* headline read "Yin-Yang Person's Yin Stronger than Yang," while *China Daily News* concluded that Xie's bodily biology was "thoroughly female."[33] From this point on, Xie was frequently dubbed as the "Chinese Christine." Whereas in the first week of publicity reporters had used either the masculine pronoun "he" (他, *ta*) or both the masculine and the feminine pronouns, they thoroughly changed to the feminine pronoun "she" (她, *ta*) in referring to Xie in all subsequent writings.

In discussing Xie's operation with the public, Dr. Lin's asserted that "Xie Jianshun should be converted into a woman in light of his physiological condition" and that this procedure would have "a 90 percent success rate."[34] The news report described Xie's first surgery with remarkable detail:

Xie's operation began at 3:40 P.M. yesterday. Dr. Lin Chengyi led a team of physicians, including Le Shaoqing and Wang Zifan, and nurses, including Jin Ming. Because this is the first clinical treatment of an intersexed patient in Taiwan, Dr. Lin permitted out-of-town doctors and news reporters to observe the surgical proceeding in the operating room with a mask on. After anesthesia, Dr. Lin cut open the lower abdominal area at 3:50 and examined its interior parts. The operation ended successfully at 4:29, with a total duration of 39 minutes. It also marked a decisive moment for determining the sex of the *yin-yang* person Xie Jianshun.[35]

This excerpt thus familiarized the reader with the clinical proceeding of Xie's surgery, thereby reinforcing Xie's status as an object of medical gaze even after the surgery itself. Ultimately, this careful textual restaging of Xie's medical operation translated its *clinical* standing into a glamorized *cultural* phenomenon in postwar Taiwan.

Xie's growing iconicity as a specimen of cultural dissection also hinged on the detailed public exposure of the surgical findings. According to the press release:

> After a thirty-minute inspection of the [lower] abdominal region, the *yin-yang* person is confirmed female given the presence of ovarian tissues. The uterus is 6 cm long and 3.5 cm wide, which is similar to the uterus size of an unpenetrated virgin (含苞未放處女, *hanbao wei-fang chunü*), but slightly unhealthy. Not only are the two ovaries normal, the existence of Fallopian tubes is also confirmed. Upon physical inspection prior to the surgery, no testicle can be detected on the lower right abdominal region and only an incomplete testicle can be found on the left. Because Xie Jianshun once had chronic appendicitis, her appendix is removed during this operation. The five viscera are identified as complete and normal. Based on the above results, have [the doctors] decided to perform a [sex-change] surgery on Xie Jianshun? The answer is with 90 percent certainty.
>
> According to what her physician in charge, Dr. Lin, told the reporters following the operation, the [sex] transformation surgery will take place in two weeks after Xie Jianshun has recovered from this exploratory laparotomy. The procedure for converting [him] into female begins with the cutting open of the presently closed *labia majora* and *labia minora* (將閉塞之大小陰唇切開, *jiang bisai zhi daxiao yinchun qiekai*). After that, a close inspection of [her] vagina will be necessary to see if it is healthy and normal. Anyone with a uterus has a vagina. After both the *labia majora* and *labia minora* have been split open and the symbolic phallic organ has been removed from the latter, [Xie]'s transformation into a pure woman will be complete.[36]

Based on these descriptions alone, the reader was able to join Dr. Lin's medical team and examine Xie's physical body carefully, not unlike what happened on the previous day at the 518 Hospital. This narrative even defined the parameters around the anticipation of this unprecedented medicalized sex change in Chinese culture. Although one type of interrogation was conducted in the "private" (closed) space of the operation room and the other was carried out in the "public" (open) domain of printed publications, medical science and the popular press ultimately converged as mutually reinforcing sites for the anatomization of Xie's sex transformation. One policeman rushed to confess his excitement. He publicly declared his admiration for Xie and his interest in dating her after the operation.[37]

As the outcome of Xie's first operation attracted growing publicity, the press further aligned itself with the medical profession by keeping Xie in a public "closet." This "closet" was characterized in a way different from what gay and lesbian scholars have typically conceived to be the staple features of queer lives in the past: hidden, secretive, and "masked."[38] Instead of concealing one's sexual orientation in public, Xie's closet allowed the public to hide his transsexuality from himself. Following the surgery, to quote the exact words in the *United Daily News*, " 'Miss' Xie Jianshun opened her eyes and looked at her visitors with a slightly painful expression. But she seems to be in a good psychological state. While not a single word has slipped out of her mouth, and although she has not consulted the doctors about the outcome of her surgery, she is at present oblivious of her fate—that she is destined to become a lady."[39] When a snapshot of the surgical proceedings and a photo of Xie became available for the first time in public on August 22, the news of future medical plans to change his sex (including female hormonal therapy) still remained unknown to Xie (figure 5.1).[40] Xie was finally "brought out of the closet" nine days after the exploratory laparotomy, which many deemed a success.[41] On the afternoon of August 29, Dr. Lin debriefed Xie, and, being the last person to know about his fate, Xie agreed to cooperate in all subsequent medical procedures that would bring about a full sex reassignment.[42] Prior to that, by maintaining his sex-change operation as a secret from Xie himself, both the doctors and the press generated a public "closet" that delineated a cultural division between the desire of the transsexual individual and the desire of others. Only in this case, however ironically, Xie, the transsexual, had once expressed strong *reluctance* to change his sex.[43]

Why did the medical team not inform Xie of its decision immediately following the operation? As Dr. Lin explained it, his colleagues learned from the nurses that Xie expressed great anxiety about being converted into a woman after having lived as a man for thirty-six years. Given his

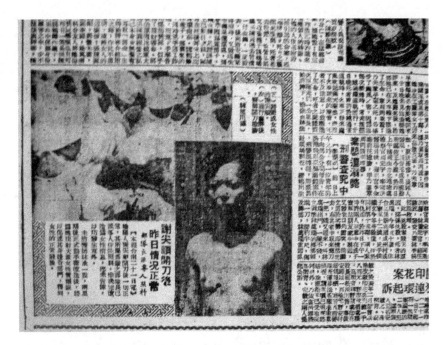

Figure 5.1 Xie Jianshun after the first operation (*United Daily News*, August 22, 1953). *Source: Lianhebao* 1953m.

steadfast desire to remain biologically male, Dr. Lin's team was afraid that, if Xie found out about their decision to convert his sex so abruptly, he would take his own life, which was implicated in his earlier conversations with the nurses.[44] Although the doctors attempted to uncover Xie's gender orientation (by sending a group of "attractive" women nurses to socialize with him) just a few days before the first operation, the surgical outcome—reinforced by the sensationalist tone of the press—nonetheless suggested that, for them, biology trumped psychology.[45] Despite the fact that Xie's condition was really a case of human intersexuality, the doctors insisted over and again that they were surgically transforming his sex.

From the beginning, the exploratory laparotomy operation lacked a clear objective. Although the doctors announced their attempt to determine Xie's sex based on the configurations of his anatomy, they repeatedly proposed a series of surgeries to be performed on Xie's body and called these "sex-change" operations. After the exploratory laparotomy, bolstered by the breathtaking accounts that stormed the newspapers nationwide, they successfully maintained a "public closet" that prevented Xie from intervening their plan to reassign his sex. Xie's refusal to be transformed into a woman shifted from public knowledge to an open secret. The doctors continued to push for an opposite surgical outcome, and, as the journalistic sensationalism surrounding his medical condition accumulated, they behaved as vanguards of medical science in the Republic of China by hinting at their ability to alter Xie's sex just like the doctors abroad. In the shadow of Christine Jorgensen, the construction of Xie Jianshun's (trans) sexual identity was driven less by his self-determination—his eventual signature on the surgical consent form notwithstanding—and more by the cultural authority of the surgeons involved and the broader impact of the mass circulation press.

The Chinese Christine

Nine months after the New York *Daily News* announced the sex-change surgery of Jorgensen, readers in postwar Taiwan were told that they, too, had their own "Chinese Christine." A newspaper article titled "The Chinese Christine" provides a poignant cross-cultural comparison of the two transsexual icons.[46] The writer, Guan Ming (管明), began by describing Jorgensen's situation in the United States, noting the substantial measure of fame and wealth that her sex reassignment had brought her. Guan also rightly noted how the Jorgensen story became harder to "sell" when news of her incomplete female anatomy went public. (Jorgensen did not undergo vaginoplasty until 1954, and, prior to that, many physicians considered Jorgenson's sex change unsuccessful.) Indeed, after Jorgenson returned from Denmark, American journalists soon questioned her surgically transformed sex. *Time* declared, "Jorgenson was no girl at all, only an altered male," and *Newsweek* followed suit.[47]

In contrast, Guan observed, "Our 'Chinese Christine,' Xie Jianshun, has turned into a 100 percent biological woman, overtaking the 'incomplete female' Christine Jorgensen." Unlike the American celebrity, Xie was inclined to continue living as a man, "let alone earning money [with a dazzling transsexual embodiment]." Guan added that Xie was even "afraid of losing his privilege to stay [in the military] after sex reassignment."[48] The stark contrast in their reactions to changing sex was also observed by a *China Daily News* writer.[49] Based on these differences, Guan concluded, Jorgensen's transformation generated an international sensation in part because of her "opportunistic inclinations" and the "widespread curiosity in society"; Xie's sex alteration, in contrast, transpired as a legitimate surgical solution for a congenital bodily defect. "But no adequate social resources were yet available for people like Xie," wrote Guan.[50] At the time of expressing his views, Guan of course could not anticipate the kind of spiritual and financial support that Xie would later receive from various military units in southern Taiwan on a sporadic basis.[51] More problematically, Guan had mistaken Xie's first exploratory laparotomy operation for a full sex-transformation surgery. He also overlooked the convention among experts in the Western medical profession, in the years before Jorgensen, to declare sex-change surgeries as an acceptable treatment for intersexed conditions.[52]

Nonetheless, Guan's comparison of the two transsexual icons nicely illustrates how sexualized bodies circulating in the early Cold War–era public milieu represented an ambivalent platform on which claims about national similarities (e.g., between the United Sates and the Republic of China) could simultaneously infuse broader claims about cultural (and even civilizational) divergence between "China" and the West.[53] On the one hand, by systematically referring to Xie as the "Chinese Christine," Taiwanese journalists and commentators interpreted her medical condition and Jorgensen's transsexual experience as more similar than different. On the other hand, they brought Xie's intention to remain biologically male to full public disclosure and at one point even suggested the possibility that Xie may be a "true" hermaphrodite and Jorgensen only a "pseudo" one.[54] For Guan in particular, whereas the global reputation of Jorgensen's transsexuality could be attributed to the social norms of "opportunistic" thinking and curiosity in the West, Xie's publicity in postwar Taiwan reflected the ethical responsibility of Chinese doctors who aimed to provide the best care for exceptional medical conditions. In either case, the popular press portrayed Xie's condition and her sex-change surgery as a rare and important event in medical science, thereby modeling such advancement in postwar Taiwan after the latest surgical breakthrough in Western biomedicine. In this way, the story of Xie Jianshun helped to situate Taiwan on the same global horizon as the United States.[55]

Despite the prevailing tendency to compare the two transsexual icons, Xie reacted to her unforeseen stardom in a manner exceedingly different from that of the glamorous American Christine. Whereas Jorgensen enjoyed her international fame, collaborated with various media agents to help shape it, and took other deliberate measures to promote it, Xie did not seize the press coverage of her genital surgery as an opportunity to boost her own reputation. To Xie, the popular rendition of her body as a valuable medical specimen and a concrete ground for U.S.-Taiwan idiosyncratic comparison was less important than her desire to be treated properly and resume a normal and healthy life. Little did Xie realize that the significance of her celebrity came not only from the direct comparisons with Jorgensen but also from the underlying similarities between the evolving perceptions of transsexuals in the popular imagination (due to her publicity) and the subsequent flood of other stories in Taiwan. Both the Christine analogy and the surfacing of other similar sex-change stories in Taiwan were, in many ways, inflected by the global reach of the Jorgensen narrative. As the nominal label of "Chinese Christine" suggests, "the power behind the culture of U.S. imperialism comes from its ability to insert itself into a geocolonial space as the imaginary figure of modernity, and as such, the natural object of identification from which the local people are to learn."[56]

The Second Operation

When the Republican government officials took a more serious interest in her case, Xie resisted their top-down decisions. Xie's second operation was initially scheduled to take place within two weeks after the first, but the only news that reached the 518 Hospital four weeks after the exploratory laparotomy was a state-issued order to transport her to Taipei. The reporters wrote, "In order to ensure Xie's safety, and in the hope that a second operation will be carried out smoothly, it has been decided that she will be relocated to Taipei. After being evaluated and operated upon by a group of notable doctors in a reputable hospital, [Xie's sex change] will mark a great moment in history."[57] Xie refused, however. She immediately wrote to government bureaucrats to express a firm preference for staying in Tainan and being operated upon again there.[58]

To her dismay, Xie paid a price for challenging the authorities. They neglected her and delayed her operation for at least three weeks following her request. The press reappeared as a viable venue for voicing her dissent. On October 17, Xie disclosed her deep frustration with her last menstrual experience, which occurred roughly a month prior. "Given her vaginal blockage, wastes could only be discharged from a small [genital] opening, leading to extreme abdominal pains during her period," an article with the title "The Pain of Miss Xie Jianshun" explained. Since another menstrual cycle was right around the corner, she urged Dr. Lin, again, to perform a second operation as promptly as possible. But Dr. Lin despairingly conceded that he must receive a formal response from the central government before he could initiate a second surgery. All he could do at this point, as one might have expected, was to reforward Xie's request to the higher officials and wait.[59] At the end of the month, Xie's former officer, Fu Chun (傅純), paid her a visit, bringing her three hundred dollars to help her get through during this difficult period.[60]

By late November the prolonged waiting and the accumulated unanswered requests forced Xie to agree bitterly to relocate to Taipei. The newspapers announced the fifth of the following month as the date of her arrival and the Taipei No. 1 General Hospital (台北第一總醫院, *Taibei diyi zongyiyuan*) as her second home. A medical authority from the Taipei hospital anticipated their takeover of Xie's case: "In light of Xie's biology, there is no leap of faith in how successful the second operation will proceed to complete Xie's transition. The only thing that remains to be determined is whether Xie is a pseudo or true hermaphrodite [偽性或真性半陰陽, *weixing huo zhenxing banyinyang*]. This can be accomplished by taking a sample from one of Xie's incomplete testes [一顆不完全的睪丸, *yike buwanquan de gaowan*] and determine whether it could produce semen."[61] The doctor reinforced the popular perception of Xie's condition as an extraordinary phenomenon of nature by labeling it "truly rare in the world's medical history."[62]

In early December the *United Daily News* announced "Chinese Christine Coming to Taipei Today for Treatment," and many gathered around the Taipei main station that day expecting to greet the transsexual celebrity in person.[63] Despite the great measure of patience and enthusiasm with which her Taipei fans waited, their hopes ended up in despair. Xie's anticipated relocation on December 5 failed to materialize, which disappointed those who were eager to witness the legendary transsexual icon. Journalists reported that "Xie's Taipei trip has been cancelled or postponed due to unknown reasons" and offered no estimation of her new arrival date.[64] On December 9, *Min Sheng Daily* quoted Xie as declaring, "I have made up my mind not to relocate to Taipei" and "since my body belongs to me, I should have total control over it."[65] To the public's dismay, it would be at least six more weeks before Xie quietly showed up at the No. 1 General Hospital in Taipei.

The media had heretofore functioned as a key buffer among the medical professionals, Xie Jianshun, and the Taiwanese public. The national dailies in particular served as the primarily means through which readers could learn Xie's thoughts and gauge her feelings. Those who followed her story closely relied chiefly on the press for the ins and outs of her treatment. Recall that doctors even allowed news reporters to witness the first exploratory laparotomy operation and, afterward, to disclose publicly their decision to turn Xie into a woman before telling Xie herself. Similarly, Xie considered the press as the most immediate (and perhaps reliable) way to publicize her desire to remain biologically male before the operation and her unwillingness to leave Tainan afterward. Almost without the slightest degree of hesitation, both Xie and her physicians readily collaborated with journalists to escalate the initial scoop of media reporting into a nationwide frenzy.

Although the reporters continued to clamor, the coverage took a dip near the end of 1953. In 1954 only three articles in the *United Daily News*, one in *Min Sheng Daily*, one in *Central Daily News* (中央日報, *Zhongyang ribao*), and none in either *China Daily News* or *Shin Sheng Daily* followed up on Xie's situation. After the cancellation of her December trip, the first update on Xie's condition came in as late as mid-February 1954. It was only by that point that her reticent move to Taipei on January 16 was revealed to the public. The name of her new surgeon in charge at the No. 1 General Hospital was Jiang Xizheng (姜希錚). Yet, despite the surprising news that Xie was now in Taipei, the closest impression one could gain from reading this article was a description of her hospital room: "Xie Jianshun's room features simple decorations, with one bed, a tea table, a long table, and a chair. There is a window at the end of the room, but the curtains are almost always closed in order to prevent strangers from taking a peek at [her] secrets."[66] What these words reflected was not only the physical distance between Xie and any curious visitors; these words also captured the metaphorical distance between Xie and the readers who found it increasingly difficult to gather information about her based on the newspaper reports alone. Even as the *United Daily News* indicated that Xie was now taking hormones so that she was closer to becoming "the second Christine," it failed to identify the source of that information and the degree of its reliability.[67]

The long silence in the press coverage might suggest that the public's interest in Xie's story had begun to wane. However, the next *United Daily News* article, which appeared in mid-March, indicated otherwise and put forth a more plausible explanation: "The hospital has been especially secretive about the exact location of her room so as to avoid unsolicited visits from intrusive strangers. Meanwhile, perhaps as a result of her male-to-female transformation, Xie Jianshun has become increasingly shy in front of strangers, so she has asked the hospital staff not to disclose any further information about her treatment to the public while she is hospitalized. Deeply concerned with her psychological well-being, the doctors agreed as a matter of course."[68]

Also painting a hyperfeminine image of Xie, both *Min Sheng Daily* and *Central Daily News* confirmed that the hospital was "working closely with Xie to protect her privacy."[69] In other words, the dip in the press coverage had less to do with the public's declining interest in Xie than with a mutual agreement between Xie and her attending physicians to refrain from speaking to media representatives. This constituted the second turning point in the evolving relationship between the medical profession and the coverage of Xie in the mass media. The popular press no longer played the role of a friendly intermediary between the public, the doctors, and Xie herself. To both Xie and her medical staff, the publicity showered on them after the first operation seemed to have impeded rather than helped their plans. Xie, in particular, may have considered the authorities' indifference toward her request to stay in Tainan to be a result of nationwide media coverage, thereby holding her prolonged waiting against the reporters. Apart from a brief comment about how Xie displayed "more obvious feminine characteristics" post-hormonal injections, the March *Lianhebao* article included no new information on her situation.[70]

As the voice of the newspaper accounts became increasingly speculative, and as the mediating role of the press gradually receded to the background, available information about Xie's second operation proved to be less certain and more difficult to ascertain. The tension between the reporters and those who tried to protect Xie from them peaked around late June, when the *United Daily News* reported on Xie's story for the third and final time in 1954. The article opened with a sentence that mentioned only in passing Xie's "more 'determinant' operation performed recently at the No. 1 General Hospital." Framed as such, Xie's "second" operation was barely publicized, and even if readers interpreted this statement to mean that Xie had undergone a second operation, the doctors withstood the temptation to give an update on it. When the reporters consulted Xie's medical team on June 24, they were met with a persistent reluctance to respond to any questions and to permit visitation rights for nonmedical personnel. A staff at the No. 1 General Hospital was even quoted for saying, "We are not sure if Xie Jianshun is still staying with us."[71]

In contrast to the sensationalist tone and mundane details that dominated the discussion of the first surgery, the way that the media covered the second operation was less fact oriented and more congested with suppositions. The major newspapers glossed over any information that would support the claim that Xie had become more feminized after relocation. Despite the best intentions of the hospital staff to distance the media people from Xie,

> a journalist has conducted an investigation inside the hospital and found signs that suggest that Xie Jianshun has become more lady-like and that she is undergoing an accelerated metamorphosis. . . . Despite the high surveillance under which Xie Jianshun is monitored, her face can still be seen sometimes. According to an individual who claims to have seen Xie Jianshun in person lately, it is difficult to discern whether Xie Jianshun has transformed into a woman completely. Nonetheless, based on what he saw, Xie has grown her hair longer, and her face has become paler and smoother. The general impression one would gain from looking at Xie now is that Xie Jianshun has transformed into a woman gradually over time [謝尖順已日漸頃向於女性型, *Xie Jianshun yi rijian qingxiang yu nüxingxing*].[72]

Not only did this account fail to mention what the second operation accomplished, it only surmised the outcome based on some unknown source. Unlike the step-by-step recounting of the surgical protocols involved in the first operation, the doctors' strategy for pursuing Xie's bodily transformation in the immediate future remained opaque.

Sensationalism Beyond Xie

As doctors, the authorities, and Xie became more self-conscious about what they said in public, the press met increasing obstacles in sensationalizing new narratives about the alleged "first" Chinese transsexual. After the exploratory laparotomy, reporters lacked direct access to information about Xie's medical care, so they began to look for other tantalizing stories of gender transgressive behavior or bodily ailment. Between late 1953 and late

1954, the popularity of Xie's transsexual narrative instigated the appearance of other similar accounts of unusual body morphology—though sometimes deviating from the actual transformation of sex—in the Taiwanese press. During the pivotal moments when the media attention on Xie took a back seat, these stories of physical trans anomaly came to light in the shadow of Christine's glamour and thereby played an important role in sustaining popular interest in sex change in Cold War Taiwan.

The media coverage of the Xie story enabled some readers to consider the possibility of experimenting with their own gender appearance. For instance, toward the end of September 1953 the *United Daily News* published an article, "A Teenage Boy Dressed Up as a Modern Woman," which included a photograph of the transgender individual in question (figure 5.2a and 5.2b). The nineteen-year-old cross-dresser, Lü Jinde (呂金德), was said

Figures 5.2a and 5.2b Lü Jinde in male appearance, on the left, and female attire, on the right (*United Daily News*, September 25, 1953).
Source: *Lianhebao* 1953b.

to "appear beautifully," had "a puffy hairstyle"; wore "a Western-style white blouse that showed parts of her breast, a blue skirt, a white sling-back, and a padded bra on her chest"; and carried "a large black purse" on a Thursday evening in Taipei.[73] This "human prodigy" (人妖, renyao) was found with "foundation powder, powder blush, lipstick, hand mirror, and a number of photos of other men and women" in her purse, and her face was described as "covered with a thick layer of powder" and with "a heavy lipstick application." She also "penciled her eyebrows so that they look much longer." "All of these," the writer claimed, "were aimed to turn her-self into a modern woman."[74] Indeed, this may have been the first instance in which the term "renyao" was explicitly associated with the intentionality of transvestism in postwar Taiwan. The only other exception was the infa-mous trial of Zeng Qiuhuang (曾秋皇) in the early 1950s, where Zeng was found guilty of committing a series of fraudulent offenses. However, the association of Zeng with the label "renyao" centered not only on his "neither-man-nor-woman" (不男不女, bunan bunü) identity and his ambig-uous social role of having been married to both men and women but also on the crimes he had committed, which rendered him defiant of the proper legal expressions of human behavior.[75]

Lü, who used to work as a hairdresser, was identified by one of her for-mer clients living in the Wanhua (萬華) District. This client followed Lü around briefly before turning to the police, explaining that Lü "walked in a funny way that was neither masculine nor feminine." After being arrested, Lü told the police that "because I enjoy posing as a lady [做個小姐, zuoge xiaojie], starting roughly two months ago, I have been wandering around the street in female attire [男扮女裝, nanban nüzhuang] after sunset on a daily basis." Lü confessed that cross-dressing enabled him to "align with his psy-chological interest" (合乎他的心意, hehu tade xinyi) and that, sexually, he was "attracted to women." "Although Lü emulated a modern lady quite successfully," the writer of this article insisted, "his feminine attire still fails to conceal his masculine characteristics, which are easily recognizable in the eye of any beholder."[76] One exceptional observer considered Lü's cross-dressing behavior unproblematic, pointing out the counterexample of the increasing number of women who had begun imitating the roles of men in society.[77] Most commentators reacted conservatively, though, claiming to have witnessed "an immoral, confusing, and gender ambiguous persona that provokes disgust" (不倫不類非男非女的樣子, 叫人看了要嘔吐, bulun bulei feinan feinü de yangzi, jiaoren kanle yao outu).[78]

Another story of gender transgression falls more appropriately in the category of what historians of gender and sexuality in America and Europe have called "passing."[79] The twenty-three-year-old Ding Bengde (丁甫德) dressed up as a man and was arrested for having abducted another young woman named Xu Yueduan (許月端). Xu's mother turned Ding in to the police after the two girls reappeared in Xu's hometown, Huwei District (虎尾鎮), and accused Ding of seducing and abducting Xu. Ding explained that she came to Huwei with the sole purpose of meeting a friend. She had to be able to earn a living to support herself and her family, so she decided to cross-dress as a man in public. This "passing" would lower her chances of being mistreated by her coworkers and other men. Labeled by the press as a "male impersonating freak" (女扮男裝怪客, nüban nanzhuang guaike), Ding denied the accusation made by Xu's mother. Similar to the coverage of Lü Jinde, the press coverage fascinated its readers with engrossing details about Ding's masculine appearance: "The cross-dressing freak wore a long-sleeve white shirt, a pair of white pants, no shoes, a sleek hairstyle, and natural body gestures, making it difficult for people to discern his/her true sex [使人見之難別雌雄, shiren jianzhi nanbie cixiong]." The reporters, moreover, hinted at a "deeper meaning" to this case, which the police were still in the process of figuring out. Perhaps by "deeper meaning" they had in mind the possibility, however remote, of a lesbian relationship between Ding and Xu. But neither the concept of homosexuality nor the word "lesbianism" appeared in the textual description of this incident.[80]

Apart from explicit gender transgressive behaviors, other astonishing accounts of bodily irregularity made their way into the press. In writing about these stories, the reporters always began by referring to Xie Jianshun's experience as a departing point for framing these rare disorders of the reproductive system. For example, a gynecologist came across a young woman with two uteruses in Tainan, where news of Xie's sex transformation originated. This coincidence led the reporter to declare, "While the date for the second gender reassignment surgery of Xie Jianshun, the Christine of Free China, remains unsettled, another case in which a surgery was pursued to treat biological anomalies was uncovered in Tainan."[81] The woman was pregnant and near the end of her third trimester when she showed up at the Provincial Tainan Hospital (省立台南醫院, Shengli Tainan yiyuan) for treatment, and it was clear from the start that this case bore very little resemblance to Xie's transsexuality.

Upon discovering two uteruses inside her womb, Dr. Huang Jiede (黃皆得) decided that, for this woman's delivery, he would first perform a caesarian section, followed by a tubectomy (tubal ligation). The purpose of the tubectomy, according to Huang, was to prevent "gestation in both uteruses, which may lead to undesirable side effects in the future." Reporters pressed Huang for further clarification on the safeness and necessity of the C-section procedure. Huang explained that normal vaginal birth would be difficult in this case "because [the patient] has two uteruses." He stood by his decision "to deliver the baby with a C-section, which is the safest option." Interestingly, in contrast to the tremendous degree of publicity accorded to Xie Jianshun, journalists complied with the medical team's instruction to withhold the personal information, including the full name, of this particular patient. What is certain, though, is that the media exposure of this bi-uteral condition hinged on its potential for forthright comparison to the Xie story, given that both shared a certain feature of "rareness [to be investigated by] the medical community."[82]

In November 1953 the press discovered another individual with uncommon pregnancy problems. Only this time the patient was a man. Born in 1934, the farmer Liao (廖) had experienced persistent cramps and abdominal discomfort over the past two decades. The pain had become more pronounced over time, especially in recent years, reaching an intolerable state that forced Liao to seek medical assistance with the company of his family. Although this was not the first time that Liao consulted doctors about his condition, it was the first instance that he received surgical (and possibly terminal) treatment for it. Dr. Yang Kunyan (楊坤焰), the president of the Jichangtang Hospital (吉昌堂醫院, *Jichangtang yiyuan*), situated on Zhongzheng Road in the Luodong District of Yilan County (宜蘭縣羅東鎮中正路, *Yilanxian Luodongzheng Zhongzhenglu*), operated on Liao on November 7. News of this male pregnancy was circulated at least on two levels: the local district level and the county level. On the local district level, the report explained that "because [Liao's male body] does not allow for natural delivery, Dr. Yang could remove [the head of the fetus] only surgically."[83] The county-level coverage of Liao's condition was more detailed: "Dr. Yang found a growth in Liao's abdomen and excised the pink fleshy bulge that weighed four hectograms [四公兩, *sigongliang*]. The doctors were unable to determine the causes of this tumor even after careful research and investigation. After removing it from Liao's body, they found a head [with some hair], a pair of eyes, a nose, and a mouth on the fleshy growth. The only

missing parts [which would otherwise make this growth resemble a fetus]
are the arms and legs."[84]

In the shadow of Xie Jianshun's transition, the question of Liao's sexual
identity was high on the reporters' radar. The district-level reporter wrote,
"Everyone is curious about where Liao's baby comes from and whether he
will be transformed into a man or a woman. In Dr. Yang's perspective,
Liao is indisputably male [道地的男人, *daodi de nanren*]. Therefore, after
recovering from the delivery and the laparotomy incision surgery, Liao will
be able to leave the hospital and enjoy the rest of his life like a normal man."
Similarly, the county-level coverage disclosed Dr. Yang's confirmation that
Liao "was neither a woman nor a hermaphrodite."[85] The venturing into
Liao's sexual identity led to greater clarification of his physical ailment.
"The growth," Dr. Yang speculated, "may have been the result of twin
conception during his mother's pregnancy and that one of the fetuses
formed prematurely and remained in his body."[86] The district-level coverage introduced Liao's male pregnancy with the opening sentence: "The
ex-soldier Xie Jianshun, now a lady, has become a household name in Taiwan, being the focus of the most popular current event of the year."[87] The
county-level coverage stressed the value of the Liao case by noting the
strong interest that numerous medical experts had expressed toward it:
"This rare event has taken the county by storm. The medical profession
places great emphasis on this case, believing that it bears a tremendous value
for medical research."[88] Although neither the woman with two uteruses
nor the pregnant man expressed medical symptoms related to sex change
per se, the Xie Jianshun story provided an immediate optic for coming to
terms with these problems. The papers claimed that, like Xie's transsexuality, these were extraordinary biological phenomena with the potential
of contributing to the advancement of biomedical research. On their end,
in both cases, the doctors justified surgical intervention for these "unnatural" bodily defects.

In the midst of Xie Jianshun's relocation to Taipei in December 1953,
the press recounted the story of another transsexual: Gonggu Bao (宫古
保), a foreign criminal who at times disguised herself as a man but more
often appeared as a woman, and who had lived in various parts of Asia at
different points of her life. Born in Siberia in 1902, Gonggu entered the
world as Gonggu Baozi (宫古保子). Her father was Chinese, and her mother
was half Koryak and half Japanese. After her mother died due to malnutri-
tion during the Russo-Japanese War (1904–1905), her father married

another Japanese woman and moved to Tokyo. At the age of seven, Gonggu Baozi discovered that her facial and other physical appearances began to exhibit "masculine traits." Doctors performed plastic surgeries on her (how intrusive these surgeries were in terms of direct genital alteration is unclear from the newspaper account), but she still appeared "neither womanly nor manlike" (不像女的, 也不像男的, *buxiang nüde, ye buxiang nande*). Given the situation, her father and stepmother decided to change her name to Gonggu Bao, believing that by adopting this new, more masculine name, she was destined to pass as a normal man.[89]

Unfortunately, at the age of thirteen, Bao began to menstruate. This horrified her, as someone who had been assigned a male identity for half of her life up to that point. She began to alienate herself. She never played with other kids at school. Her parents, hugely disappointed at the situation, decided to send her away to live with her grandmother. Since the age of fifteen, so the newspapers claimed, Gonggu had committed at least thirty-eight crimes all over the world, including in Shanghai, Hong Kong, Singapore, Japan, and even Alaska and Canada. But more importantly, what Gonggu Bao's story confirmed was that Xie Jianshun's sex change was neither exceptional nor the first in Asia. Although their life trajectories proceeded in vastly different social, cultural, political, and historical contexts, Gonggu and Xie followed the same legacy of bodily transformation through medical intervention.[90] Moreover, the renewed interest in Gonggu Bao implied that it would be too simplistic to consider her, like Christine Jorgensen, merely as a historical precedent to Xie's popularity; rather, it was precisely the ways in which the popular press served as a central vehicle for disseminating the possible idea of sex change that enabled the stories of Gonggu, Jorgensen, and Xie to command public interest as interwoven and interrelated in Cold War Sinophone culture.

In addition to Gonggu, journalists in the same month uncovered two more domestic stories of human intersexuality. In both cases, the newspaper accounts referred to Xie's experience as a window into these anomalous medical discoveries. The first concerned a thirty-five-year-old man, Mr. Zheng (鄭), whose ambiguous genital anatomy came to the attention of doctors responsible for screening new military recruits at Yuanli District (苑裏鎮). After a long and careful consideration, the military physician ultimately agreed on the label of "the middle sex" (中性, *zhongxing*) for designating Zheng's gender status.[91] The second story concerned a nineteen-year-old woman, Lin Luanying (林鸞英), who was a frequent client of a tofu shop in Yeliu Village (野柳村) of Taipei County owned by the widow Li Axiang (李阿向). Building on two years of customer relation, Lin became very intimate with Li's eldest son, Hu Canlin (胡燦林), and with parental consent, Lin and Hu decided to get married on December 8 of the lunar calendar. As the wedding day drew near, however, Lin began to panic. She believed that something was terribly wrong with her body, so she consulted a doctor at the Yilan Hospital (宜蘭醫院, *Yilan yiyuan*) and "tried to fix her problem." The media framed her visit in voluntary terms, describing her as "a *yin-yang* person like Xie Jianshun," who also went to the doctors for a checkup after experiencing physical discomfort. "The major difference [between them]," though, "was that Lin was soon turning into a bride." After performing an investigative operation on Lin (presumably similar to Xie's first exploratory laparotomy), the doctors were surprised by the incomplete development of her genitalia, with the external absence of labia majora and labia minora and the internal absence of a uterus. The doctors were "astounded by what they saw, but claimed to lack the technical expertise to help her" (醫師見而興嘆, 乏術開闢桃源, *yishi jian'er xingtan, fashu kaipi taoyuan*). Lin's condition, they suggested, proved to be more complicated than the simple determination of gonadal tissues that had made sex reassignment in Xie's case possible.[92]

As the cast of characters mounted, newspapers published more sensational stories. The most heartbreaking and tragic of these was probably the story of Wang Lao (汪老), a fifty-seven-year-old intersex person who committed suicide in March 1954 due to chronic loneliness and depression. The media interpreted her biological condition as "identical to Xie Jianshun" with the exception that her intersexuality had never been properly attended to by doctors.[93] The most optimistic and encouraging story was probably that of the five-year-old Du Yizheng (杜異征). While the result of Xie's transition was still up in the air, surgeons in Taichung (台中) claimed to have successfully converted this boy into a girl, giving this child a normal life and the public an additional boost of confidence in Taiwan's medical practice. As the press framed it, this case represented a landmark achievement in the Taiwanese medical profession and enabled parents to have a stronger faith in the way doctors approached clinical cases of intersex children.[94]

But the story of the transsexual Liu Min (劉敏) stood out as the most puzzling and intriguing of all (figure 5.3). On December 10, 1954, the *United Daily News* billed the story as one about the transformation of "a

Figure 5.3 Liu Min (*United Daily News*, December 13, 1954). Source: *Lianhebao* 1954b.

fair lady into a heroic warrior."[95] Three days later another article opened with the enigma itself: "For a woman who had given birth and then transforms into a man within a decade is an event that reasonably arouses suspicion on all fronts."[96] Born in Beiping as Liu Fangting (劉芳亭), Liu Min came to Taiwan with the Nationalist army as Xie Jianshun had done in the late 1940s. She had recently acquired visible male physical traits owing to medical complications, but what amazed everyone was the fact that she also had a daughter, Xiaozhen (小真), with her husband. Is it possible for a transsexual to give birth? With the unknown fate of Xie Jianshun (including the effect of her sex-change surgeries on her eventual ability to conceive) still lurking in the media background, the combination of Liu's past pregnancy with her recent sex metamorphosis seemed all the more bizarre, pertinent, and worth pursuing. For over a week, Liu's life history prompted the speculations and opinions of people from all walks of life, and the initial coverage soon escalated into a nationwide obsession.[97]

It turned out that Liu had never delivered a baby. Xiaozhen was her half-sister and, accordingly, adopted child. In 1938, after marrying her cousin, Liu felt regular distress in her abdominal region, not unlike Xie's early conditions. (By that point, Liu's biological father had already left her and her mother for over a decade.) Her relatives considered these cramps to be signs of actual pregnancy. Upon learning this, her mother immediately disclosed her own recent pregnancy to Liu (without stating who the father was). But her mother's economic and health situation at that point made it unfeasible to raise a second child. Her mother therefore pleaded Liu to raise her stepsister as her own child in the pretense of casting this whole situation as the outcome of her ostensible pregnancy. With her husband's agreement, Liu accepted her mother's request and promised to never reveal this secret to anyone. Meanwhile, over the years, Liu had her uterus surgically removed in Beijing, which led to startling changes in her genital area, including "the closing up of her vagina" (陰道逐漸閉塞, *yindao zhujian bisai*) and "the formation of a phallus organ on top of her labia" (大小陰唇之上便開始長出男性生殖器, *daxiao yinchun zhishang bian kaishi zhangchu nanxing shengzhiqi*). According to Liu, the reporter to whom she shared this secret was only the fourth person to know about it.[98]

By the end of 1954 reporters had lost almost all contact with Xie Jianshun and her medical team. Xie's case gradually moved from current events to yesterday's news, but as other stories of unusual bodily problems arose and resurfaced, the media reminded the public that manhood, womanhood, and their boundaries were neither as obvious nor as impermeable as they once had seemed. From Lü Jinde's transvestism to the lady with two uteruses, and from Liao's male pregnancy to Lin Luanying's intersexed condition, the earlier publicity on Xie provided cogent leverage for both the journalists and health care professionals to relate other nominal stories of bodily irregularity to the idea of "transsexuality." Although not all of them were directly or necessarily about sex change per se, these stories enabled some readers to take seriously the possibility of sex/gender transgression. With an elevated awareness of the malleability of gender, they began to learn what the label *"bianxingren"* meant and appreciated the immediate role of medical intervention in the reversing of one's sex. Through the press coverage, stories of intersexuality and sex transition recast earlier questions about human identity in a new light. The alleged authority of doctors to unlock the secret of sexual identity, in particular, became more firmly planted in the popular imagination.

Transformation Complete

The story of Liu Min finally pushed medical experts to come clean about Xie's situation. After the news of Liu's fake pregnancy broke, readers channeled their curiosity back to Xie. In January 1955 a newspaper article with the headline "Xie Jianshun's Male-to-Female Transformation Nearly Complete: The Rumor of Surgical Failure Proved to Be False" shattered any doubts about the stunted progress in Xie's transition. After the first operation, given the way that Xie's doctors had intentionally refrained from leaking any word to the press, the public was left with an unclear impression of what was going on inside the hospital specifically and how Xie was doing more generally. Rumor soon had it that the doctors' long silence meant Xie's procedures were ultimately unsuccessful. According to the article, the cause of this rumor "can be traced to an incident reported last month in Tainan of a *yin-yang* person. The general public's memory of Xie was refreshed by this story of the *yin-yang* person in Tainan, and as a result of this reminder, the public began to revisit the question of whether Xie had been successfully transformed into a woman." In an attempt to dispel any doubts, doctors from the No. 1 General Hospital were quoted for confirming that "the rumor is absolutely false." They clarified that "Xie Jianshun's sex transformation has in fact proceeded rather smoothly and is reaching its final stages." Xie, the doctors promised, "is living a perfectly healthy life." But when the reporters requested to speak to Xie in person, they were turned away and were told by the hospital staff that this kind of request "could only be fulfilled with a permit from the state authorities."[99] Interested readers would acquire a similarly opaque impression from reading the coverage in *Central Daily News*.[100]

The initial upsurge of the renewed interest in Xie survived only briefly. It would take another eight months—after the doctors had performed Xie's "third" and "final" operation—before her name would make headlines again.[101] On August 31, 1955, the *United Daily News* carried an extended front-page article with the headline, "A New Chapter in the Nation's Medical History: The Success of Xie Jianshun's Sex-Change Surgery."[102] On the following day, *Central Daily News* teased the public by announcing that "The Details of Xie Jianshun's Sex-Change Operations Will Be Publicized Shortly."[103] According to Xie's physician in charge, "Contrary to a number of fabricated claims, Xie Jianshun's final operation took place very smoothly on the morning of 30 August. With respect to the protocols and results of this decisive surgery, the medical team promises to release all of the relevant information in a formal report in due course." The papers glossed over the aim of this operation with the succinct words "to unclog her fallopian tubes," the obstruction of which had caused her periodic discomfort for months.[104] Apparently, Xie felt dizzy after the operation but recovered by the next morning. The representatives from the No. 1 General Hospital explained that both Xie's own request and the uniqueness of her case constituted their main reasons for holding off on disclosing its clinical details. Since Xie had explicitly asked the medical staff to abstain from speaking to journalists and reporters, the doctors assumed the responsibility of protecting her privacy from media exposure. On the other hand, the doctors believed that her sex-change operations "promise to mark an important medical breakthrough in the country" (此一手術尚為我國醫學界之創舉, ciyi shoushu shangwei woguo yixuejie zhi chuangju), so they wanted to be extra careful in making any kind of public statement. Silence seemed to be the best demonstration of their precaution before the final verdict.[105]

On the following day, newspapers declared "the success of Xie Jianshen's sex-change surgery," pitching it as "a fact that can no longer be shaken." Although the staff at the No. 1 General Hospital pledged to disclose the surgical specifics in the near future, readers in Taiwan had already learned a great deal on the day following the operation. Xie's popularity first skyrocketed two years ago, in August 1953, when doctors, scientists, the press, and the lay public "discovered" her. Despite the detailed coverage of her first operation, or because of it, Xie and the people in her immediate circle became much quieter in their dealing with reporters. As its media coverage began to thin out in 1954, the Xie story grew more and more mysterious while other stories of uncommon body morphology abounded in the press. Even the timing and completion of her second operation were never thoroughly announced until this point. The pertinent accounts now clarified that, in the months following her first operation, Xie not only resisted relocating to Taipei but ardently opposed to changing her sex. The second operation eventually took place in April 1954, and it involved "the removal of the two symbolic male gonads." After the second operation, Xie "began to develop stronger female sexual characteristics" (體內女性生理性能轉強, tinei nüxing shengli xingneng zhuanqiang), which included the enlargement of her breasts and the onset of regular menstruation. Because Xie's reproductive system lacked a full vaginal canal, her periodic menses caused regular discomfort when excreted with urine through the urethra. As she "started to learn how it feels to live like a woman" (開始嘗到做女人的滋味, kaishi changdao zuo nüreng de ziwei), these physiological reorientations made her more reluctant to identify as female. After wrestling with the idea of relocating to Taipei, she struggled with and eventually failed to convince her surgeons not to transform her sex.[106]

Amid a world of uncertainties brought about by World War II and its immediate aftermath, the media used the metaphor of the Cold War to depict Xie's relationship with the doctors. If the estimated timing of the second operation were true, sixteen months had elapsed before Xie entered her recent surgery. To quote the exact words used to frame this extended period of time, "The Cold War [冷戰, lengzhan] between Xie Jianshun and the hospital lasted until 5 April of this year." What was frozen during this period was not only Xie's reaction to the decisions made by her physicians in charge but also the overall fate of her medical treatment (or sex transformation). Distinguishing her ambition from the intent of her doctors, Xie requested a second relocation to a different hospital, but her request was ultimately denied. What "ended this Cold War," according to the newspapers, was a letter that she wrote to the president, Chiang Kai-Shek, in which she expressed her disdain toward how the doctors handled her case and the absence of adequate nutrition provided by the hospital.[107]

In response to the letter, the Ministry of National Defense sent two representatives to the No. 1 General Hospital to resolve the tension between Xie and her medical attendants. Xie's complaint about how she was mistreated at the hospital, they found out, was a misleading "expression of her wrong set of mind" (內心理不正常發出的牢騷, neixinli buzhengchang fachu de laosao). They told her that the recurrent cramps that she experienced were due to the menstrual periods, which typified the bodily experience of female reproductive biology. In order to alleviate this somatic (and not psychological) discomfort, the doctors needed to construct a functional vaginal canal for her. Ultimately, the two National Defense representatives succeeded in persuading Xie to accept doctors' advice and complete her sex transformation with one final surgery. The *United Daily News* speculated that "perhaps it is due to her prejudice against the hospital staff, or perhaps it is due to her loyalty to the military, she agreed to a third operation after contemplating for only ten minutes or so."[108] The year-long "Cold War" thus concluded with the direct intervention of not the medical experts but state authorities. Whereas, according to historian Elaine May, the contemporaneous structural norms of American families helped to offset the nation's domestic and foreign political insecurities, Cold War's metaphoric power, as evident in the example of Xie's transsexuality, was diffused in the public discussion of sexually malleable bodies in the context of postwar Taiwan, situated on the fringes of China and Chineseness.[109]

Before doctors released their official report of Xie's case, details of the third operation and its influence on Xie were already openly discussed by those in her immediate circle. For example, the new surgeon in charge, Zhang Xianlin (張先林), uninhibitedly commented on the nature of Xie's latest operation. Whereas most peopled considered this operation the most critical and fate determining, Zhang regarded it merely as "a simple reconstructive surgery" (簡單的矯形手術而已, *jiandan de jiaoxing shoushu eryi*). Because Xie's reproductive system was already confirmed female, according to Zhang, the operation involved the enhancement of her female biology by "removing her symbolic phallic organ" (把她那象徵性的陰莖予以割除, *bata na xiangzhengxing de yinjing yuyi gechu*) and, more importantly, "the construction of an artificial menstrual canal" (開闢出一條人工的排經道, *kaipichu yitiao renggong de paijingdao*), which would allow her to release menses normally. The operation, which Zhang considered to be a breeze, began at eight o'clock in the morning and ended at ten after nine.[110] To assess the efficacy of the operation, the doctors vowed to administer an X-ray examination in two weeks.[111] When the *United Daily News* in Taiwan and the *Kung Sheung Daily News* (工商日報, *Gongshang ribao*) in Hong Kong published half-nude photos of the "post-op" Xie on September 8, representatives from the No. 1 General Hospital quickly dismissed them as a sham.[112] As a sign of their interest in looking after Xie's psychological well-being, within three weeks after the operation, the Ministry of National Defense awarded Xie one thousand New Taiwan dollars to help her defray the cost of purchasing new feminine attires.[113] This generous sum offered Xie greater freedom in constructing a social image—and a new sense of self—that aligned neatly with her new biological sex (figures 5.4 and 5.5).[114]

As doctors sought to clarify what happened during and after Xie's third operation, newspapers continued to report on other astonishing stories of sex change. In early May 1955, for instance, the case of a soldier with a medical condition similar to Xie's was reported in Chiayi County (嘉義縣, *Jiayixian*). This twenty-eight-year-old "gender ambiguous soldier" (性別可疑的軍人, *xingbie keyi de junren*), Xu Zhenjie (徐振傑), was born in Henan. The individuals who first raised an eyebrow on his gender identity were those from within his troop. According to them, Xu was always reserved and quiet, and what especially made others suspicious was that he never showered with other men and always left his clothes on whenever he joined them. Initially, the doctors who examined his body had only a

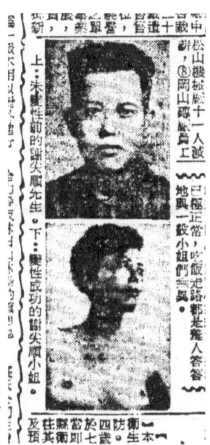

Figure 5.4 The success of Xie Jianshun's transformation (*Central Daily News*, September 9, 1955). *Source: Zhongyang ribao* 1955c.

Figure 5.5 Xie's new public image after transition (*China Times*, October 10, 1956). *Source: Zhongguo shibao* 1956a.

vague sense of the structural differences between his genitals and that of other male soldiers, but they could not reach a consensus on his "true" sex. After this incident came to light, the gynecologists and nurses at the Chiayi Hospital recalled that during his previous visit for a checkup, Xu complained about his own gender confusion and unfortunate fate.[115]

The story of Xu echoed elements of the earlier anxiety and fascination with Liu Min, whose fake pregnancy stimulated renewed public interest in Xie. What grabbed everyone's attention, again, was the intriguing relationship between transsexuals and childbirth. At one point Xu tried to convince his family that his reproductive organ was "more feminine than masculine"; in fact, he suspected, he "may be one hundred percent female." If that were the case, had he ever menstruated or become pregnant? "Faced with these questions," Xu only "kept silent and turned away shyly." It fact, when Xu first joined the army near the Dianmian (滇緬) borders, he self-identified as female. Having eventually enlisted in a men's troop, however, Xu became close friends with other men in the army. His relationship with one of them became especially intimate, and after revealing his congenital condition to the person, Xu had his child. After delivering the baby, however, Xu considered his own gender identity even more perplexing and distressful. By the time of his military discharge, "one could hardly tell the difference between Xu's mannerism and physical appearance from other men's."[116] Even as Xu tried to dissociate himself from a masculine past, the press homed in on a sturdy, masculine image. Although the question of whether Xu had actually experienced menstruation and childbirth (and what happened to the child if he did) remained up in the air, the press seemed to be more interested in using these questions as a foil against which to juxtapose his current embodiment of a heteronormative masculinity.[117]

In the United States, after the initial stories on Christine Jorgensen dwindled, reporters produced a flood of sensational copy on sex-change operations in newspapers, periodicals, and magazines. Much like the way the coverage in Taiwan centered on Xie Jianshun, each new story confirmed that Jorgensen was not alone and that a number of other people similarly desired to alter their bodily sex. The stories came from all over the world, but those from Britain and the United States attracted the most attention from the American press.[118] In the mid-1950s, these stories began to make their way across the Pacific and reached the Chinese-speaking audience. One of these stories in particular, that of Tamara Rees and James Courtland, appeared in Taiwan's *United Daily News* in July 1955.[119] By reading the brief coverage in the *United Daily News*, Chinese readers learned not only of the names, ages, and occupations of the couple—the thirty-one-year-old paratrooper Rees (李絲) and the thirty-year-old businessman Courtland (卡德倫)—but also the details of Rees's transsexual experience in a piecemeal fashion. The article clearly indicated the time and location of Rees's sex-change operation—in Holland in January 1954—thereby hinting at a much broader and global dimension to sex reassignment surgeries beyond Taiwan and the United States. Of course, what the Chinese coverage did not include were the minor details of Rees's transition. For instance, born in 1924, Rees had already begun taking hormones and passing as a woman in Los Angeles before she traveled to Amsterdam for her genital surgery. After she married Courtland in July 1955, one magazine even called the wedding "history's first transvestite marriage."[120] And when the psychoanalyst Ralph Greenson (1911–1979) later published an article on Rees in the *International Journal of Psycho-Analysis* in 1964, he interpreted her gender confusion as a case of "homosexuality similar to that of neurotic adults."[121]

In contrast to the American stories, the majority of news of sex change in Taiwan emerged from the medical screening of new recruits at military units. In September 1955, a twenty-five-year-old young man by the name of Wu Kunqi (吳坤祈) was identified with a "dual-sexed genital organ" (兩性器官, *liangxing qiguan*) by the doctors at Zhongshantang (中山堂) in the Gangshan District of Kaohsiung County (高雄縣岡山鎮, *Gangshanzheng Gaoxiungxian*). Wu's medical screening revealed "a tiny hole below his penis" with "a size penetrable by a finger"; his penis "lacked a urethral opening," and his "urine came out of the tiny hole" rather than the penis. When asked by the doctors, Wu admitted that he often ejaculated from the tiny hole as well.[122] In the same month, the father of the twenty-one-year-old Zeng Qingji (曾清吉) arrived at Madou (麻豆), also in southern Taiwan, and requested the doctor responsible for screening new recruits to exempt Zeng from conscription because of his congenital sexual disorder. After a careful examination of Zeng's body, Dr. Wang Baikun (王百焜) found both a penis and a vaginal opening in his genital area. Like Wu's, Zeng's penis did not have a urethral opening, but there was a tiny hole surrounded by a pair of labia underneath the penis. Unlike Wu, whose body could produce semen, Zeng discharged regular small-quantity menses. According to Dr. Wang's diagnosis, then, Zeng was a "pseudohermaphrodite" (假性陰陽人, *jiaxing yinyangren*), and given his predominant female biological constitution, he could be easily transformed into a woman through the surgical removal of his male genital organ.[123] Similar to the experiences of Xie Jianshun and Xu Zhenjie, all of these later accounts of sex change embraced a principal narrative of hiding one's ambiguous sexual identity. Both Wu and Zeng expressed great disappointment when their intersexuality was uncovered by the medical professionals. Most importantly, in these stories, doctors always construed surgical sex transformation as the most desirable solution after bursting these extremely personal secrets wide open in public.

On October 28, 1955, the *United Daily News* carried a front-page story that finally proclaimed "The Completion and Success of Xie Jianshun's Sex-Change Operation."[124] The story continued on page 3, which contained a full-length official report on Xie's medical treatment released by the No. 1 Army Hospital. *China Times* (中國時報, *Zhongguo shibao*) similarly presented Xie as someone who "has transitioned from a masculine man into a beauty" (從此鐵漢成佳麗, *congci tiehan cheng jiali*).[125] The two-page *Central Daily News* coverage celebrated the "successful and perfect" (圓滿成功, *yuanman chenggong*) transformation of Xie.[126] The excitement was also broadcasted in *China Daily News*, *Shin Sheng Daily*, and *Public Opinions* (公論報, *Gonglunbao*).[127]

The official report revealed numerous aspects of the Xie story that overthrew earlier speculations. Of these revelations, the most surprising was probably the fact that Xie's most recent operation was actually her fourth and not her third operation. Recall that Xie's second operation received little publicity in the previous year. By June 1954, from reading the scattered newspaper accounts, interested readers could gain a vague impression that doctors in Taipei had performed a second surgery on her, but its date, nature, and purpose lacked transparency. According to this official report, however, Xie's second operation, which was also an exploratory laparotomy but with the additional step of removing parts of her male gonadal tissues, had taken place on April 10, 1954. Based on the samples extracted from her body during this operation, the doctors confirmed Xie's status as a true hermaphrodite, meaning that she had both ovarian and testicular tissues in her gonads. The doctors also clarified that by that point, her "testicular tissues were already deteriorating and unable to generate sperm," but her "ovarian tissues were still functional and able to produce eggs." In light of a stronger presence of female sexual characteristics, the medical team performed a third operation on August 26, 1954. After the surgery, Xie's penis was replaced by an artificial vaginal opening. All this happened more than a year prior. Xie's most recent and fourth genital surgery, which took place on August 30, 1955, was simply a vaginoplasty. Now with "a normal woman's vaginal interior" (陰道內腔與正常女性一樣, *yindao neiqiang yu zhengchang nüxing yiyang*), Xie Jianshun's "transformation from a soldier into a lady is now indisputable."[128] Brought to light by the report, Xie's personal triumph encapsulated the postwar fears and hopes about the possibilities of medical science.[129]

Also on October 28, the second page of the *United Daily News* included the sixteenth installment of "The Story of Miss Xie Jianshun," a biography of Xie that had been serialized daily since October 13. The concluding installment appeared on November 18, which meant that for over a month, Taiwanese readers were exposed to Xie's life story with familiar moments and surprising details.[130] This extended exposure seemed to reflect the fact that the Xie story continued to sell even two years after the initial frenzy. No less significant, again, was the similarity in the presentational strategies of the Taiwanese and American presses. The stylistic objective of "The Story of Miss Xie Jianshun" closely resembled that of the series "The Story of My Life," which appeared in *American Weekly* three days after Christine Jorgensen returned to the United States from Denmark. Jorgensen's series was billed as "the story all America has been waiting for," which would have been an equally appropriate description for the Xie installments with a nominal substitution of the word "Taiwan" for "America."[131]

But the two series bore significant differences as well. Whereas the first-person confessional format of the American version gave Jorgensen a chance to convey story in her own voice, the third-person observational tone of the Taiwanese version allowed the writer, Yi Yi (憶猗), to narrate Xie's experience with a unique voice that was at once authoritative and absorbing. This mode of narration, of course, built on the earlier public image of Xie, who had been constantly portrayed as a nationally and transnationally significant figure but never for reasons acknowledged by herself. Although Jorgensen's full-length personal memoir was eventually published in 1967 and its film adaptation released in 1970, by that point Xie had lost all contact with the press and faded from the public sphere.[132] The final media blitz surrounding the Xie story occurred in the late 1950s, during which it was reported that Soong Mei-ling (宋美齡, 1897–2003), Chiang Kai-shek's wife, and a number of celebrities had visited Xie in

Taipei and that Xie had begun working at the Ta Tung Relief Institute for Women and Children (大同婦孺教養院, *Tatong furu jiaoyangyuan*) under the new name Xie Shun (謝順) after *nine*, not four, surgeries.[133] Ever since the birth of "the Chinese Christine," the comparison of Xie to Jorgensen had intrigued, satisfied, and resonated with observers time and again, but never without limits.

Queer Sinophone (Re)Production

In their initial diagnoses of Xie, doctors frequently spoke of a hidden female sex. In contrast, the press provided a cultural space for Xie to articulate a past heterosexual romantic life and the desire of *not* wanting to change his sex in a masculinist voice. Early on, both medical and popular discourses adhered to a neutral position in discussing his psychological gender. Both discourses were fundamentally reoriented by the time of his first operation. The pre-op coverage of his first surgery only foreshadowed a highly sensational outcome—the characterization of Xie as the "Chinese Christine," the first transsexual in Chinese society. By elevating Xie's iconic status as both the object of medical gaze and the specimen of (trans)cultural dissection, medical and popular discourses foreclosed any space of epistemic ambiguity concerning Xie's "innate" sex, gender, and sexuality. Many believed that Xie was destined to become a woman. Or, more aptly put, he became nothing but a transsexual star following the footsteps of the American Christine. In the next two years, the press covered fewer and fewer stories on Xie and began to report more widely on other surprising accounts of unusual bodily conditions. After her fourth surgery in May 1955 Xie's popularity as the first transsexual in Chinese culture, on top of these other pathological "symptoms" of postcolonial modernity, helped establish the global significance of Taiwan vis-à-vis the neocolonial hegemony of the United States. The persistent comparisons of Taiwan with America, through the comparisons of two versions of transsexuality, became an important arena for articulating a sense of Sinophone difference from Anglophone culture as well as of Republican Taiwanese nationalism.

It is interesting to note that, in the context of the 1950s, the Chinese term "*bianxingren*" carried almost none of the psychopathological connotations that distinguished its English counterpart, "transsexual." This probably reflected the relatively late involvement of psychiatric experts in dealing with patients diagnosed with *bianxing yuzheng* (變性慾症, transsexualism) in Taiwan.[134] It was not until 1981 that the national spotlight on the male-to-female transsexual Jiang Peizhen (江佩珍) opened that new chapter in the history of transsexuality in Taiwan. According to Jiang's psychiatrist and past superintendent of the Tsyr-Huey Mental Hospital in Kaohsiung County, Dr. Jung-Kwang Wen (文榮光), the story of Jiang Peizhen made a huge impact on enhancing the public awareness of transsexualism in Taiwan in the early 1980s. Her case pushed doctors, especially psychiatrists, to come to terms with patients requesting sex reassignment or showing symptoms of gender identity disorder. Physicians also began to consult the Harry Benjamin Standards of Care that had been adopted by American medical and psychological experts since 1979.[135] Personal testimonies of transsexuals attested to the breadth, significance, and cultural reach of the Jiang story. Miss Lai (賴), a former male-to-female patient of Wen, noted how the possibility of sex reassignment surgeries was brought to her attention only by the media coverage of Jiang.[136] In the 1980s Xie Jianshun and her surgeons had disappeared altogether from the public sphere, and this seemed to confirm that one era had ended. For the new generation of transsexuals and doctors like Miss Lai and Wen, the heroine from the 1980s onward was Jiang.[137]

Nevertheless, the saga of Xie Jianshun, and that of other media reports of "transing" that followed in her wake, attest to the emergence of transsexuality as a form of modern sexual embodiment in Chinese-speaking society. As one *United Daily News* front-page article asserted, "to reconstruct a thirty-something year-old man into female is unprecedented in the history of clinical medicine [in Taiwan]."[138] Xie's story, in particular, became a lightning rod for many post–World War II anxieties about gender and sexuality, and it called dramatic attention to issues that would later drive the feminist and gay and lesbian movements in the decades ahead.[139] In a different way, these stories of trans formation bring to light a genealogy that exceeds, even subverts, familiar historicizations of Taiwan's postcoloniality. They illustrate the ways in which the Chinese community in Taiwan inherited a Western biomedical epistemology of sex from not only the Japanese colonial regime (a conventional reading of Taiwan's colonial past) but also, more importantly, the sophisticated scientific globalism that characterized the Republican period on the mainland.[140] This genealogy from Republican-era scientific modernity to postwar Taiwanese transsexuality, connected via the Sinitic language but also made possible culturally by the migration of over one million people from the mainland in the late 1940s, underscores the ways in which the Nationalist government regained sovereignty in Taiwan beyond a monolithic framing of Japanese postcolonialism.[141] Parallel to British colonial Hong Kong, Taiwan experienced the highly institutionalized establishment of Western biomedical infrastructure under Japanese occupation.[142] In the 1950s, when Mao "nationalized" Chinese medicine in continental China, both Taiwan and Hong Kong represented the most advanced regions in modern Western medicine situated on the geo-margins of the sinosphere.[143] Adding to its catalytic role in the transmission of Western biomedical knowledge and practice, British colonialism was instrumental for establishing Hong Kong as a more permissive cultural space when other parts of mainland China were strictly governed by a socialist state.[144] These historical factors thus allowed for the immense media publicity showered on Xie Jianshun and on sex change more broadly. Together, the rapid technology transfer of Western biomedicine and the availability of a fairly open social and cultural milieu enabled the Sinophone articulation of transsexuality to emerge first and foremost across the postcolonial East Asian Pacific Rim.

The examples of gender transformation unearthed here must be identified with the broader horizon of Sinophone production, by which I mean a broadening of queer Sinophonicity to refer to a mode of cultural engendering coalescing around the multiple peripheries of dominant geopolitical and social formations. The queer historicity of the transvestites, the bi-uteral woman, the pregnant man, intersexed persons, and other transing characters who came to light in the shadow of Xie rested on epistemological-historical pillars that came from outside the geopolitical China proper, including the legacies of Japanese postcolonialism, American neoimperialism, the recontextualization of the Republican state's scientific globalism, and Taiwan's cultural and economic affiliations with other subregions of Cold War East Asia, such as Hong Kong and Japan. Between the end of the Korean War in the mid-1950s and the reopening of the Chinese mainland in the late 1970s, Japan, Okinawa, South Korea, and Taiwan became U.S. protectorates. "One of the lasting legacies of this period," according to Kuan-Hsing Chen, "is the installation of the anticommunism-pro-Americanism structure in the capitalist zone of East Asia, whose overwhelming consequences are still with us today."[145] Inherent in the concept of the Sinophone lies a more calculated awareness of the implicit role played by Communist China in the stabilization of this (post–)Cold War structure in transnational East Asia. Sociologist Marshall Johnson and anthropologist Fred Yen Liang Chiu's theory of subimperialism is useful here because the various examples of trans formations explored in this chapter "are not the unfolding of master imperial or orientalist logics. Rather, they exist through agencies whose contingent patterns always admit the possibility of otherwise."[146] None of the queer subjects and embodiments that emerged in 1950s Taiwan can be sufficiently appreciated according to the historical logic of Republican China, Communist China, imperial Japan, or modern America alone, but their significance must be squarely situated at their discrepant and diffused intersections.

Considering Xie's celebrity and influence as a Sinophone (re)production of transsexuality is also instructive in four other regards. First, the Sinophone approach pushes postcolonial studies beyond its overwhelming preoccupation with "the West." Postcolonial critics have problematized the West either by deconstructing any variant of its essentialist invocation or by provincializing the centripetal forces of its greatest imperial regimes, such as Europe and America. Naoki Sakai's essay "Modernity and Its Critique: The Problem of Universalism and Particularism" (1988) and Dipesh Chakrabarty's "Provincializing Europe: Postcoloniality and the Critique of History" (1992) are perhaps the most representative studies of each approach, respectively.[147] At other times, critics have attempted to recuperate nativist examples from the histories of third-world nations. Certain modern concepts often understood as imposed from the outside and sustained by the colonial system, they argue, were actually already internal to the indigenous civilization. The work of Ashis Nandy is exemplary in this regard.[148] But the West appears to be analytically deployed as a universalized imaginary Other in all three strategies. By perpetually being treated as the theoretical heart in historical narration and cultural criticism, the West continues to function as "an opposing entity, a system of reference, an object from which to learn, a point of measurement, a goal to catch up with, an intimate enemy, and sometimes an alibi for serious discussion and action."[149]

On the contrary, viewing trans formations in postwar Taiwan as historical *events* of Sinophone production repositions our compass—and redraws our map—by recentering the non-West, Asia, and China more specifically. In his provocative book, *Asia as Method*, Kuan-Hsing Chen invites postcolonial scholars to "deimperialize" their own mode of investigation by moving beyond the fixation of "the West" as a sole historical-theoretical caliber of civilizational, national, imperial, colonial, and Cold War predicaments.[150] In his words,

> In Asia, the deimperialization question cannot be limited to a reexamination of the impacts of Western imperialism invasion, Japanese colonial violence, and U.S. neoimperialist expansion, but must also include the oppressive practices of the Chinese empire. Since the status of China has shifted from an empire to a big country, how should China position itself now? In what new ways can it interact with neighboring countries? Questions like these can be productively answered only through deimperialized self-questioning, and that type of reflexive work has yet to be undertaken.[151]

My narration of the history of Chinese transsexuality, centering on the cases of Xie Jianshun and others, can be viewed as an attempt at doing this kind of reflexive work. While the dispersed circuits of knowledge that saturated the Chinese Christine's glamour question any straightforward conclusion about the Chineseness or Americanness of Xie's transsexuality, the other contemporaneous stories of trans corporeality represent a highly contingent and conditional response to the nascent genealogy of sex change emerging out of regional currents and global tensions. This chapter in 1950s Taiwanese history refocuses our attention from the "influence" of Western concepts and ideas to the inter- and intra-Asian regional conditions of subjectivity formation—from denaturalizing the West to provincializing China, Asia, and the Rest.

Second, by provincializing China, the Sinophone framework enables us to see and think beyond the conventions of China studies. In the spirit of marking out "a space in which unspoken stories and histories may be told, and to recognize and map the historically constituted cultural and political effects of the cold war," I have aimed in this chapter to raise a series of interrelated questions that challenge the various categorical assumptions that continue to haunt a "China-centered perspective."[152] Was Xie Jianshun's transsexuality "Chinese" or "American" in nature? And "transsexuality" in whose sense of the term? Was it a foreign import, an expression (and thus internalization) of Western imperialism, or a long-standing indigenous practice in a new light? How can we take the Republican state's administrative relocation in the late 1940s seriously? Is it possible to speak of a "Republican Chinese modernity" that problematizes the familiar socialist narrative of twentieth-century Chinese history? Which China was alluded to by the Chineseness of the label "Chinese Christine?" In the yet-to-appear discourse of Taiwanese nativism, did the Republican regime exemplify settler colonialism, migration, immigration, or diaspora? To better comprehend the historical context, we might also ask, "Is the [GMD] regime a government in exile (which would mean that it resides abroad), a regime from another province, a defeated regime, or simply a cold-war regime?"[153] Evidently, the complexity of the history far exceeds the common terms used to describe the historical characteristics of postwar Taiwan. To call the GMD a regime from the outside or a colonial government only partially accounts for its proto-Chineseness or extra-Chineseness, and precisely because of the lack of a precedent and analogous situations, it is all the more difficult to historicize, with neat categorical imperatives or ways of periodization, the social backdrop against which and the epistemic conditions under which the first Chinese transsexual became a comprehensible concept. So the queerness of Sinophone Taiwan, as evinced through trans-archiving, calls into question not only the conventions of China studies; it invariably casts light on a symmetrically nested agenda to decenter the normative orientation of Taiwan studies as well. Like China, Taiwan has never been a *straight*-forward geobody, a political container of sorts, carrying evolving historical cultures that merely reflect a series of colonial governmentality displacements.

Third, understood as "a way of looking at the world," the epistemological rendition of the Sinophone as "an interruptive worldview" not only breaks down the China-versus-the-West binary; it also specifies the most powerful type, nature, and feature of transnationalism whose interest-articulation must lie beyond the hegemonic constructions of the nation-state. According to Françoise Lionnet and Shu-mei Shih, the transnational "can be less scripted and more scattered" and "is not bound by the binary of the local and the global and can occur in national, local, or global spaces across different and multiple spatialities and temporalities."[154] If "China" and "Chineseness" have indeed evolved over the course of the history of sex change from castration's demise to the growing influence of Western biopolitics, then the changes over time we witness in this history have less to do with the "coming out" of transsexuals than with the shifting trans-nationalism of queer Chinese cultures: from the growing global hegemony of Western conceptions of lifehood and sexuality in *major* transnational

China to the rhizomic interactions of geopolitical forces, historical conditions, and cross-cultural contours in *minor* transnational China. In other words, the peripheral ontologies of Sinophone queerness demand carefully executed place-based analyses while never losing sight of the ever-shifting parameters of the norms and centers of any given regional space. The transnationalism and interregionalism of the trajectory from Republican China to postwar Taiwan make it evident that any hegemonic understanding of "China" and "Taiwan" as sovereign nation-states will always fall short in capturing the genealogical grounding of those queer livelihoods, maneuvers, and experiences encapsulated in the two categories' politically contested relationality.

Although I have used mid-twentieth-century Taiwan as the exemplary context of queer Sinophone (re)production, its implications obviously extend beyond Taiwan and the early Cold War period. By bringing the theoretical category of the Sinophone to bear on the non-identitarian history of trans formations narrated here, my aim is to bring together, historically, the reciprocal rigor of queer and Sinophone theoretical critiques, thematizing the coproduction of gender heteronormativity and the hegemonic (Chinese) nation-state as they are articulated through one another. Together, the queerness of Sinophone perspectives and the anti-Sinocentric logic of queering settle on unsettling the overlapping recognitions of Xie Jianshun's transsexuality as a Chinese copy of a Western original, a Sinophone production of a Chinese original, a straight mimesis of a male-to-female transgendered body, a queer reproduction of an American blond beauty, and so forth. A social history of trans formations in Sinophone Taiwan that exceeds a conventional Japanese postcolonial paradigm comprises the broad spectrum of these potential straightforward convergences and postnormative divergences. The resulting historiographical task challenges a homogenous postcolonial interpretation of twentieth-century Taiwan that figures in either Chinese imperial hegemony or Japanese colonialism (or American neocolonialism for that matter) as its exclusive preoccupation. The intraregional emphasis on these intertwined historical legacies, therefore, accounts for a more sophisticatedly layered "postcolonial Taiwan," one that complements but complicates the model developed by the literary critic Fang-Ming Chen, yet always insisting on the multiplicity of its possible limits and meaningful points of entry.[155] The history of contemporary Taiwan therefore invites multiple interpretative strategies and approaches to account for its "colonial" (read: global) past—a historicism that decenters rather than recenters the hegemony of formal imperial giants such as China, Japan, the United States, and so forth.

This brings us to the last, yet perhaps the most important, contribution of the Sinophone methodology: to appreciate the formation of Sinophone modernity that began to distinguish itself from and gradually replaced an older apparatus of colonial modernity in the course of twentieth-century Chinese history. The year 1989 is a pivotal turning point for reflecting on the historical development of late twentieth-century Chinese and Sinophone cultures. The People's Republic of China (PRC) government's military action to suppress the Tiananmen Square protests of 1989 has been widely condemned by the international community. Taking place two years after the lifting of martial law in Taiwan, the incident has been construed as a direct reflection of the sharp divergence in democratic characteristics of various Chinese-speaking communities (e.g., across the Taiwan Strait). The latest rendition of this perceived divergence is none other than the 2014 Umbrella movement to challenge the PRC's suppression of electoral democracy in Hong Kong.[156] If the Cold War structure of East Asian capitalist zones had in fact remained intact by as late as the turn of the twenty-first century, it would still be heuristically useful to periodize contemporary Chinese history along this temporal axis.[157] In this legacy of the Cold War, and despite its termination, American culture, in both its elite and popular forms, continued to operate as one of the defining forces shaping Taiwanese culture even after Richard Nixon's normalization of American diplomatic relations with Communist China (completed in 1979) at the expense of ties with Taiwan.[158] It is for this reason that the Taiwanese lesbian, gay, bisexual, and transgender movement took shape in the way it did, as it mirrored the development of these subcultures in the United States.

In the post-1987 era, the Taiwanese social and cultural space soon became home to a vibrant group of queer authors, scholars, activists, and other public figures who passionately emulate North American gay and lesbian identity politics and queer theoretical discourse.[159] Apart from social movement and academic theorization, gay men and lesbian women in Taipei in particular have constructed an urban geography of their own with unique subcultural tempos, rhythms, and patterns. As Jens Damm has observed, "Taipei is the only city—probably not only in Taiwan but the whole of East Asia—where a huge open space, the Red House district, has been successfully developed into an area where gays and lesbians have openly created their own urban infrastructure, with bars, restaurants, shops and information exchange opportunities."[160] Since the 1990s, cultural flows between the PRC, Taiwan, and Hong Kong have steadily accelerated. Critics now tend to trace the roots of queer political activism in mainland China in the early twenty-first century to the initial influx of Western queer theory (酷兒理論, *ku'er lilun*) and the rise of the gay and lesbian movement (同志運動, *tongzhi yundong*) in Taiwan and Hong Kong in the 1980s and 1990s.[161] In terms of lexical circulation, the Chinese vernacular translations of "gay" (同志, *tongzhi*) and "lesbian" (拉拉, *lala*) acquired political valence and enjoyed wide currency first in Hong Kong and Taiwan, respectively, and were then imported back into mainland Chinese culture. Similarly, the first gay pride parade in Chinese-speaking communities took place in Taiwan in 2003, followed by Hong Kong in 2008 and Shanghai in 2009. Many gay and lesbian activists in Taiwan and Hong Kong today often claim that they have relatively little to learn from the mainlanders and that the trajectory of activism-strategy appropriation would flow in one direction only (rather than reciprocal in nature), that is, from Sinophone communities to the PRC.[162]

The queer Sinophone framework underscores the ways in which the particular polities mediating the transmission of foreign/Western knowledge to China (such as Japan in the early Republican period as often viewed through the lens of colonial modernity), at least in the areas of gender and sexuality, have been gradually replaced by Sinophone communities by the end of the twentieth century. Taken together, what the cases of gender transgression recollected in this chapter reveal is a much earlier moment of historical displacement, in the immediate postwar era, when the sociocultural articulation of nonnormative genders and sexualities was rerouted through—and thus re-rooted in—Sinitic-language communities and cultures on the periphery of Chineseness.[163] The transition from colonial to Sinophone modernity around the mid-century, therefore, is something that we are only beginning to appreciate.

Conclusion

China Trans Formed

This book began with the story of eunuchs in late imperial Peking, and it ends with the public awareness of transsexuals in Sinophone Taiwan. By looking at a seemingly marginal phenomenon—the transformation of "sex"—throughout, I aim to put the transformation of "China" at the center of historical inquiry. This recurring motif fore-grounds the differences between eunuchs and transsexuals less as a natural mutation over time than as the culmination of historical contingencies. Yet some might argue that these two groups of historical actors are more simi-lar than different. Certainly, the idea of eunuchism implies achieving some kind of systematic surgical procedures on the body; so does transsexuality. For eunuchs, their genital alteration in particular was a cornerstone of their new social, cultural, and political identity, and so is this the case for trans-sexuals. Eunuchs were oftentimes looked down upon as social outcasts, yet at times glorified by others as martyrs of their day; so are transsexuals in the history of the medical and legal battles they have fought. Often seen as either a rare specimen or culturally inferior, eunuchs' existence and treat-ment (including their privilege, power, and function) in Chinese society had been a conspicuous subject of debate, especially among a supporting cast of cultural elites, and similar views can be held accountable for the experience of transsexuals today. However, to collapse these similarities under the nominal designation of eunuchs as "premodern transsexuals" elides the nuances and complexities of the process whereby sex became a meaningful category of experience in twentieth-century China.

As this book shows, the modern formulation of *xing* qua sex rested on the rise of new structures of knowledge that enveloped an "epistemic nexus"—a new regime of conceptualization—around the relationship between the visual realm, the subjectivity of desire, and the malleability of the body. In the second half of the nineteenth century, missionary doctors such as Benjamin Hobson introduced Western anatomical concepts of the human reproductive body. As we saw in chapter 2, these new concepts were accompanied by visual illustrations that featured a dissection-based "anatomical realism," a new aesthetic convention that formed a distance between the viewer and the image not apparent in earlier Chinese medical illustrations. By reorienting the burden of proof away from the system of theoretical correspondence and into the realm of anatomical visual comprehension, this new realistic distance translated Western anatomical images of the body into a more "scientifically objective" image of Western anatomy itself. Illustrations of the Western anatomical body were endorsed and reproduced by Chinese natural scientists in the early twentieth century. These cultural elites also circulated images that highlighted the morphological differences between male and female organisms, the similarity between the human body and mechanical objects, and the distinctions between the two sexes on the subcellular register. Biologists like Zhu Xi used the allegorical figure of the hermaphrodite to anchor their discussion about the scientific basis of sex and to underscore its discursive visual context.

The influence of these techniques of visualization persisted into the second half of the twentieth century, but already in the aftermath of the New Culture movement, sex acquired a new epistemic dimension. As chapter 3 shows, May Fourth iconoclastic intellectuals such as Zhang Jingsheng and Pan Guangdan modeled their work after European sexologists by collecting "data" on the sexual lives of Chinese people and by translating foreign sexological classics on the psychology of sexual variations. Of all the sexological vocabularies, the concept of homosexuality received the most traction among Chinese sex scientists in the 1920s and 1930s. Among other activities, they strategically promoted eugenics, argued about the proper credentials of sex educators, formed professional organizations to consolidate the disciplinary boundary of sexological science, and debated vociferously on the causes and prevention of homosexual relations. Sex, in their formulations, was no longer something to be observed in nature, but it was something to be desired, catalogued, and deciphered. In moving from a biological understanding of sex to the psychological realm of sexuality, the effort of Republican-era sexologists produced an epistemological break in the conceptualization of sexual desire: from a culturalistic to a nationalistic style of argumentation that made homosexuality a nodal point of referencing human difference and social identity. This generated a conceptual titration of gender variation from same-sex sexuality, crystalizing the two phenomena's independent epistemic status in the twentieth century.

Apart from being the object of observation and the subject of desire, the modern concept of sex acquired its comprehensibility through a third epistemological coordinate: a malleable essence of the human body. Whereas the development of the new epistemic structures around the visuality and carnality of sex relied on biological and psychological models, the elasticity of sex emerged from a new glandular model that quantified sex in chemical terms. Drawing on the findings of European animal sex-reversal experiments, Chinese sexologists introduced a theory that construed everyone as inherently bisexual. As demonstrated in chapter 4, this bisexual theory emphasized the innate transmutability of sex and fitted nicely with the biochemical definition of sex as a variable of endocrine secretions, according to which everyone has both male and female sex hormones. Chinese writers grabbed onto the nascent idea of sex hormones and the theory of constitutional bisexuality to shed new light on existing ideas about the effects of castration, hermaphroditism, and gender translocations. In the 1930s and 1940s this growing awareness of sex mutability leveraged the ascending media publicity on human sex transformation and filtered elite scientific ideas about sex into vernacular culture.

By the early 1950s the visibility, carnality, and malleability of sex made it possible for some individuals to be identified with the label "transsexual" (*bianxingren*) in Chinese-speaking communities. Prior to *bianxingren*, writers typically used "*ci* becoming *xiong*" to describe female-to-male transformations and "*xiong* becoming *ci*" to refer to male-to-female transformations in the natural world. They reserved the parallel terms "*nühuanan*" (woman-to-man) and "*nanhuanü*" (man-to-woman) for human sex changes. The circulation, popularity, and subtle interchangeability of these expressions around the time of the Yao Jinping incident revealed the growing authority of Western biomedicine in the 1930s and 1940s. But the word "*xing*" had yet to be fully integrated into these earlier formulations. The various scientific, social, and cultural developments eventually culminated in the 1950s, when the name of the first *bianxingren*, Xie Jianshun, hit the newspaper headlines in Taiwan.

The trajectory from eunuchs in late imperial Peking to transsexuals in Sinophone Taiwan tracks two coeval historical transformations of "the Chinese body." First is the transformation from a world in which surgically altered bodies did not correlate to specific medical notions of sexual deviance to a world in which similarly modified bodies are now assigned the visual, subjective, and transformative scientific connotations of sex—a trend in the biomedicalization of the human body that we can designate as the growing plasticity of sex. Second is the transformation from a historical context in which China's geopolitical borders were rapidly encroached by foreign imperial powers to a situation in which China's geopolitical frontiers, especially in regions like Taiwan and Hong Kong (replacing certain key agents of colonial modernity such as Japan), have begun to play an increasingly prominent role in mediating the transmission of foreign sexual knowledge and identity politics into mainland China—a trend in the reconfiguration of China's geobody that we can identify with the growing plasticity of Chineseness. Over the course of the twentieth century, whereas the definition of sex was gradually crystalized and its layers of complexity slowly unpacked by scientists, doctors, journalists, educators, tabloid writers, and other observers, the question of China's geocultural sovereignty over its bordering communities and the proper definition of Chineseness unfolded in an opposite, more opaque direction.

Ultimately, this hyperbolic geometry nested in the growing plasticities of sex and Chineseness sutures a genealogical relationship between the demise of eunuchism and the emergence of transsexuality. Both castration and sex-reassignment surgeries entail body modification processes, especially genital alteration, but eunuchs did not become women, and most transsexuals yearn for full sex transitions. Both eunuchs and transsexuals are often perceived as social pariahs, but some eunuchs wielded enormous political power in certain epochs of Chinese history and could thus be considered as living right next to the epicenter of the Chinese empire. On the contrary, transsexuals have always been viewed as an extreme minority in the human population—as individuals dwelling on the margins of society and who continue to pressure the boundaries of cultural norm, the consequences of medical pathologization, and the limits of the legal system.

Although the issue of self-volition looms large for subjects of both castration and sex-reassignment surgeries, the incentives for becoming a eunuch assume zero resemblance to a transsexual's deep-seated desire to become the opposite sex. In the course of the twentieth century, eunuchs and transsexuals converged in terms of the contested meanings of their bodily morphology, but they also diverged in significant ways in the evolving conditions of possibility for claims of scientific truth and the shifting structures of Chinese geopolitics.

At the dawn of the twentieth century, men and women in China began to understand their social differences in terms of modern scientific knowledge. The introduction of a Western biomedical epistemology of sex not only assigned eunuchs a "third sex" identity, but through that new identity it also eroded the very aura and possibility of their cultural existence. In the half-century before the Cold War, the reorientation of the visual, carnal, and malleable meanings of bodily sex provided the ground for the formation of a Chinese body (geo)politic that reverberated throughout the subsequent decades. Toward the end of the century, people were now able to observe, desire, and manipulate sex, and the excavation of this new imaginative space paved the way for the increasing visibility and political legitimacy of transsexuals. The genealogy from eunuchs to transsexuals embodies the very reasons why sex, as a product of history, still matters today.

Abbreviations

FNZZ	*Funü zazhi* 婦女雜誌 [Ladies journal]
JB	*Jingbao* 晶報 [Crystal]
LHB	*Lianhebao* 聯合報 [United daily news]
PGDWJ	Pan, Guangdan. 1994. *Pan Guangdan wenji* 潘光旦文集 [Collected works of Pan Guangdan]. 14 vols. Beijing: Peking University Press.
XKX	*Xing kexue* 性科學 [Sex science]
XWH	*Xin wenhua* 新文化 [New culture]
XZZ	*Xing zazhi* 性雜誌 [Sex magazine]
ZJSWJ	Zhang, Jingsheng. 1998. *Zhang Jingsheng wenji* 張競生文集 [Collected works of Zhang Jingsheng]. Ed. by Jiang Zhongxiao 江中孝. 2 vols. Guangzhou: Guangzhou chubanshe.
ZJWX	*Zhuanji wenxue* 傳記文學 [Biography literature]

Notes

Introduction

1. Although I use the phrase "gender-liminal" here, as I will argue later in chapter 1, the perception of eunuchs as the "third sex" was the product of an emergent nationalist discourse around the turn of the twentieth century. For a collection of rich ethnographic analyses of gender-liminal figures around the world, see Herdt 1993.
2. See my discussion of this term in chapters 1 and 4.
3. See, e.g., Robinson 2001; Shinian 2007; Tsai 1996; Wang 2004; Wang 2009.
4. Dale 2000; Hua 2014, 210–14; Rawski 2001, 162–66. Historians have begun to revise this interpretation. Norman Kutcher (2010), for example, identifies the degree of laxity with which Qing emperors managed eunuchs despite the official claim to the opposite.
5. It is arguable that even contemporary biological evidence for sex difference remains fraught and inconclusive. See, e.g., Jordan-Young 2010; Richardson 2013. My usage of the word "biology" in this book conforms to the definition offered by Michel Foucault (1973, 289), who repeatedly referred to the year 1800 as marking the emergence of biology as a formalized scientific discipline and its radical break from natural history in Europe. This convention differs from the one adopted by other sinologists who tend to apply selective definitional underpinnings of modern biology to a much earlier period in Chinese history. For an example of this somewhat anachronistic approach, see Furth 1988.

6. Lee 1996, 2005; Leung 2006; Yates 2005.

7. Furth 1999, 19–58.

8. Bray 1995, 1997; Furth 1986, 1987, 1999, 59–133.

9. Laqueur 1990; Schiebinger 1989, 189–213.

10. Furth 1986, 1999.

11. Zhang 2015.

12. On Li Shizhen and *Bencao gangmu*, see Elman 2005, 29–34; Métailé 2001; Nappi 2009.

13. Li Shizhen, *Bencao gangmu*, juan 52, as cited in Furth 1988, 5.

14. By putting the word "hermaphroditism" in quotation marks, I hope to underscore the concept's biomedical connotations and contend that, technically speaking, like "homosexuality," it did not exist in China before the twentieth century.

15. Wu 2010, 84–119; 2011.

16. Jia 1995; Rocha 2010.

17. For a study of how China turned "inside-out" in the late nineteenth century, see Meng 2006. On the introduction of Western anatomy and asylum practice, see, e.g., Asen 2016; Elman 2005, 283–421; Gulik 1973; Heinrich 2008; Luesink 2015; Ma 2014; Szto 2002, 2014.

18. Dikötter 1995.

19. In this study, I use "bioscience" as a broader concept than "biology" to refer to scientifically based knowledge or arguments pertaining to life promulgated by a wide spectrum of social actors, including but not limited to formally trained biologists and medical doctors.

20. Bailey 1990, 2007.

21. On the anti-footbinding movement, see Ko 2005, 9–68.

22. On the global image of the Chinese leper's crippled body in the early twentieth century, see Leung 2009, 132–76. On the concurrent rise of the various perceptions of a "weak" China, see Yang 2010.

23. On the history of bisexual theory in the biochemical sciences, see Clarke 1998; Fausto-Sterling 2000; Oudshoorn 1994; Sengoopta 2006.

24. On "plastic dichotomy," see Ha 2011.

25. See, e.g., Angelides 2001; Bullough 1994; Chiang 2008a, 2010; Hausman 1995; Irvine 1990; LeVay 1996; Logan 2013; Meyerowitz 2002; Minton 2002; Rosario 1997; Terry 1999.

26. Glosser 2003; Kuo 2012; Lee 2006.

27. Chiang 2014b; Dikötter 1995; Larson 2009; Leary 1994b; Lee 1999; Lee 2006, 186–217; Shih 2001; Tsu 2005; Zhang 1992.

28. Kang 2009; Martin 2010; Sang 2003.

29. Hershatter 1999.

30. On Jorgensen, see Meyerowitz 2002; Serlin 1995; Skidmore 2011.

31. See, e.g., Kuo 1998; Liu 2017; Wang 2017; Yang 2008.

32. Altman 2001; Bao 2011; Evans 1997; Farquhar 2002; Farrer 2002; Ho 2010; Jeffreys and Yu 2015; Kong 2011; Leung 2008; Liu 2015; Martin 2003; Rofel 2007; Yan 2003; Zhang 2015. My use of the term "transsexual" in this book follows the style of two of the most sophisticated monographs on the history of transsexuality to date: Meyerowitz 2002 and Najmabadi 2014. Though some scholars debate the problematic (even pathologizing) aspects of the term "transsexual," historians have kept this term in critical analysis to reflect greater historical accuracy and a more robust historicism. The term is also relatively more pertinent when examining its medical history.

33. For a similar argument, see Dikötter 1995. For a programmatic overview of Dikötter's thesis regarding the significance of the Republican period, see Dikötter 2008.

34. Dikötter 1995, 143.

35. This study joins recent revisionist efforts to reinterpret the significance of the Republican period through the lens of the history of science, especially the human sciences. See, e.g., Chiang 2001; Lam 2011; Schmalzer 2008; Tsu and Elman 2014.

36. See, e.g., Chiang 2008b. On the evolving discourse of same-sex desire, see chapter 3.

37. I use "genealogy/genealogical" in the way that Foucault has employed it in his historical analysis of the discursive formations of knowledge (*savoir*). See, e.g., Foucault 1977, 1980; Visker 1995.

38. The term "geobody" is first proposed by Thongchai Winichakul 1994. On the historical relationship between China's geobody and the human body as viewed through the lenses of war and military activities, see Hwang 2001, 2009.

39. On Sinophone articulations, see Shih 2007; and chapter 5 of this book.

40. In fact, the literary scholar Haiyan Lee (2006) has adopted a similar analytical framework in narrating the history of "love" in the late Qing and Republican periods.

41. Chiang 2015a.

42. Here is a quick sample of their titles: "婚姻進化新論"; "生殖器新書"; "男女交合新論"; "妊娠論"; "傳種改良問答"; "男女婚姻衛生學"; "處女衛生論"; "戒淫養身男女種子交合新論"; "日本小兒養育法"; "胎內教育"; "葆精大論"; "吾妻鏡"; "男女下體病要鑒"; "男女情交"; "產科學初步"; "全體通考." This list is taken from Zhang 2009, 131–32.

43. Dikötter 1995, 68; Farquhar 2002, 250–55; Liu, Karl, and Ko 2013, 10–13; Rocha 2010; Sang 2003, 99–126; Zhong 2000, 54.

44. *Xinhua zidian* 2004.

45. Legge 1875, 342.

46. Soothill and Hodous 1937, 258.

47. Cited in Rocha 2010, 607.

48. Rocha 2010, 608.

49. Jia 1995; Rocha 2010, 611–12.

50. By the 1920s, an equally popular translation is *xing kexue* (性科學), which highlights the scientific components of this domain of scholarly inquiry even more explicitly. See chapter 3 of this book.

51. Rocha 2010, 608–14.

52. Rocha 2010, 614.

53. Ochiai and Haga 1927, 2295.

54. Rocha 2010, 615 (emphasis original).

55. Liu 1995, 32.

56. Liu 1995, 33.

57. Meng 2006, 11.

58. Meng 2006, 58.

59. Meng 2006, 52.

60. Meng 2006, 20.

61. Meng 2006, 31–61.

62. Wang 2008, 126. See also Wang 1997.

63. Chang 1971; Dikötter 1992; Fogel and Zarrow 1997; Judge 1996; Rhoads 2000; Wong 1989. Of course, the debates on race and ethnicity have deeper historical roots in the Qing dynasty. See, e.g., Crossley 1999; Elliott 2001.

64. Austin 1975; Butler 1997.

65. Williams 1976. On historical epistemology, see Daston 1994; Davidson 2001; Hacking 2002, 2009; Rheinberger 2010a, 2010b.

66. Koselleck 2002.

67. Rheinberger 2010b, 2 (emphasis original). On the French tradition, see Bachelard (1938) 2002; Canguilhem (1943) 1991; Foucault 1994.

68. Rheinberger 2010b, 3.

69. Rheinberger 2010b, 3 (emphasis mine).

70. Rheinberger 2010b, 3.

1. China Castrated

1. Chen 1991b, 117.

2. Chen 1990a, 1990b, 1990c, 1990d, 1991a, 1991b. Previously published as "Nanxing kuxing taijiankao" (男性酷刑太監考) [An investigation of male castration and eunuchs], *Dacheng* (大成) 44 (1977).

3. Cai 2011; Zheng 1991.

4. For a social commentary that compares castration to footbinding in the early Republican period, see Zhi 1919.

5. By eunuchism, I am referring to the bodily state of castrated men, and I treat it like other forms of embodiment as a category of experience that needs to be historicized rather than foundational or uncontestable in nature. On the history of "experience," especially in relation to sexuality, see Foucault 1990; Halperin 2002; Scott 1991.

6. Chen 1991a, 126; Jay 1993, 460.

7. Other interchangeable words include 私白 (*sibai*), 淨身 (*jingshen*), 寺人 (*siren*), 宮人 (*gongren*), 腐人 (*furen*), 宦官 (*huanguan*), 宦寺 (*huansi*), 黃門 (*huangmen*), and 公公 (*gongong*). The more popular rendition, 太監 (*taijian*), became synonymous with 宦官 in the Ming.

8. Bo 1999.

9. Bo 2001; Robinson 2001; Shi 1988, 60–94, 115–55; Shinian 2007; Tsai 1996, 2002; Wang 2004.

10. For an amended assessment of the Qing period, see Kutcher 2010.

11. Jay 1993, 461.

12. Watson 1958.

13. Liang first published "*Zuguo dahanghaijia Zhenghe zhuan*" (祖國大航海家鄭和傳) [A biography of the admiral Zhenghe from the home country] in *Xinmin congbao* (新民叢報) 21 (1904) under the pseudonym "New Citizen of China" (中國之新民, *Zhongguo zhi xinmin*). The journal *Xinmin congbao* was founded by Liang in Yokohama, Japan, in November 1901. On Zheng's mediation of China and Southeast Asia, see the essays in Suryadinata 2005.

14. Notable exceptions include Dale 2010; Jay 1999.

15. Qing et al. 1994; Yu 1965.

16. See, e.g., Gu and Ge 1992, 316–54; Shi 1988, 8–12; Tang 1993, 5; Wang 1995, i; Xiao Yanqing (肖燕清) in Zhang 1996, 1901; Yan and Dong 1995, 3–6; Zou 1988, 306; Zhang 1996, 6. See also the negative depiction of eunuchs in Du 1996; Han (1967) 1991; Shinian 2007; Wang 2004; Wang 2009; Zhang 2004.

17. Taylor 2000, 38.

18. Historians today continue to have a difficult time in resisting the appeal of the trope of "emasculation," despite their critical positioning of their analyses of Chinese eunuchs. See, for example, Dale 2010.

19. World historians such as Scott Cook (1996) have designated the period between the mid-nineteenth century and the First World War "the age of high imperialism."

20. Chen 2008; and Wu 1999. On the Sick Man of Asia, see Heinrich 2008; and Rojas 2015. The "castrated civilization" trope, therefore, must be historically contextualized on a par with other relevant images of China in the early twentieth century, such as "Yellow Peril" and "the sleeping lion." See Tsu

2005, 88–96; Yang 2010. On how the Asian race became yellow, see Keevak 2011. On eunuchs as the "third sex," see, e.g., Mitamura 1970; Shi 1988; Wang 1995.

21. Eng 2001; Glosser 2003; Huang 2006; Kang 2009, 2010; Sommer 1997, 2000; Wu 2004; Wu and Stevenson 2010; Vitiello 2011; Zhang 2015; Zhong 2000.

22. Brownell and Wasserstrom 2002; Hinsch 2013; Louie 2002; Louie and Low 2003.

23. Halberstam 1998.

24. Ko 1997.

25. Zito 2006, 2007.

26. Ko 2005, 2.

27. The "archive" I am referring to here does not correspond to a physically existing archive. Rather, it refers to a repository of sources that I have collected that recount information about the castration operations performed in late Qing China. In bringing coherence to these sources, I join the work of Antoinette Burton (2003) and others to contest the assumed canonicity of a "panoptical" official archive, a site of knowledge that not only predetermines the standard by which disciplinary models are measured and gendered (or "dwelled") but also marginalizes those texts and sources that fall outside its purview. See also Bauer 2017.

28. Aisin Gioro 1964, 69.

29. Ma Deqing (馬德清) et al. in *Zhongguo renmin zhengzhi xieshang huiyi quanguo weiyuanhui wenshi ziliao yanjiu weiyuanhui*, 1982.

30. Faison 1996.

31. Ko 1997, 10.

32. Arondekar 2005, 12.

33. Arondekar 2009.

34. A possible addition here is the recollection of *gongnü* (宮女, palace maids), but this source genre is extremely rare. See one account documented in Hua 2014, 217–18.

35. On the anti-footbinding movement, see Ko 2005, 9–68. On the leper's crippled body in the era of Chinese national modernity, see Leung 2009, 132–76.

36. Hevia 2003, 124. See also Fan 2004; Forman 2013a; Liu 2004.

37. Hevia 2003, 123–55.

38. Morache 1869, 107.

39. Morache 1869, 133–34.

40. Jefferys and Maxwell 1910, 305–7.

41. Chen 2012, 47–85.

42. Morache 1869, 134.

43. Morache, 1869, 135.

44. Dale 2010, 38.

45. Takao Club, "George Carter Stent."

46. Stent 1877, 143.

47. Stent 1877, 143.

48. Ko 2005, 16.

49. See, e.g., Barbier 1996; Cheney 2006; Freitas 2009; Hatzaki 2009, 86–115; Howard 2014; James 1997; Lascaratos and Kostakopoulus 1997; McLaren 2007; Ringrose 2003, 2007; Scholz 2001; Tougher 2002.

50. Loshitzky and Meyuhas 1992, 31.

51. Loshitzky and Meyuhas 1992, 34.

52. Stent 1877, 170–71.

53. Jamieson 1877, 123.

54. "How the Chinese Make Eunuchs," 1877.

55. Andrews 2014, 96–105; Lei 2014, 21–44.

56. Tsai and Wu 1929, 480.

57. Wong and Wu 1936, 232–34.

58. Mitamura 1970, 28–35.

59. Humana 1973, 125–53.

60. Dale 2010, 42.

61. Cohen 1996, 9.

62. Dale 2000, 37; Liu Guojun (劉國軍) in Zhang 1996, 1690.

63. Wu and Gu 1990, 254.

64. Taylor 2000, 85–109.

65. Lascaratos and Kostakopoulus 1997.

66. Peschel and Peschel 1987.

67. Friedman 2001, 4.

68. Dong 1985, 21.

69. Dong 1985, 22 (see also p. 12).

70. Jia 1993, 22.

71. I thank Jean D. Wilson for providing an English translation of Wagenseil's article.

72. Wagenseil 1933, 416 (translation provided by Wilson).

73. Wu and Gu 1987.

74. Wu and Gu 1990, 251.

75. Wu and Gu 1990, 254–55.

76. Taylor 2000, 46–47.

77. Davidson 2001; Foucault 1990; Smith-Rosenberg 1985; Weeks 1981.

78. For the shifting medical and legal regulations of sexual violence in this period, see Freedman 1987.

79. Scott and Holmberg 2003, 502.

80. Freud 1924.

81. Lacan 1977.

82. Taylor 2000, 91.

83. It is interesting to note that Zhou (1935) used "castration" rather than "sterilization" to characterize these European operations.

84. Kevles 1985; Stern 2005.

85. Zhou 1935, 251–52.

86. Zhou 1934b, 267.

87. Yang 2003; Yan, Liu, and Yan 2009.

88. Ma and Li 2005; Wu and Wang 2003; Zhang 2011.

89. In comparing eunuchs across world historical cultures, historian Kathryn Ringrose (2007, 497) argues that "Chinese and Islamic cultures frowned on expressions of sexual pleasure of any kind between women of the court and their servants, making total ablation necessary."

90. Duhousset 1877.

91. Godard 1867, 154–55.

92. Duhousset 1896, 336.

93. Sandhaus 2011; Trevor-Roper 1976.

94. Fairbank 1977.

95. Bickers 2004.

96. Forman 2013b.

97. Backhouse 1943, 184–85.

98. Backhouse 1943, 185–87.

99. Kang 2009.

100. Stent 1877, 179.

101. Backhouse 1943, 197.

102. Backhouse 1943, 202.

103. Andrews 2014; Lei 2014; Wu 2013.

104. Korsakow 1898, 339. On Korsakow, see Korsakov 1904. I thank Jean D. Wilson for providing an English translation of Korsakow's article.

105. *New York Times* 1931.

106. Coltman 1891, 1901.

107. Coltman 1894, 28.

108. Coltman 1893, 329.

109. Coltman 1894, 28.

110. Coltman 1894, 28.

111. Coltman 1894, 29.

112. "Obituary: J. J. Matignon" 1900.

113. Matignon 1896, reprinted in Matignon 1899, 177–202.

114. Matignon 1896, 196.

115. Matignon 1896, 202.

116. Matignon 1899, 182.

117. Fraser 2011, 106.

118. Heinrich 2008, 76.

119. Heinrich 2008, 105.

120. On the evolving politics of "the Sick Man of Asia," apart from Heinrich 2008, see also Rojas 2015; Yang 2010.

121. Benedict 1996, 166; Delaporte 1986; Heinrich 2008, 15–37; Leung 2009, 132–76.

122. Derrida 1994.

123. Halberstam 2011, 113.

124. Wagenseil 1933, reprinted in Wilson and Roehrborn 1999, 4329; Jia 1993 (courtesy of Jia Yinhua).

125. Hunt 1999, 32.

126. Elman 1991; Ko 2005.

127. On the rise of Western anatomical knowledge in China, see, e.g., Elman 2005, 283–319 and 396–421; Heinrich 2008; Luesink 2015; Wu 2010; and chapter 2 of this book.

128. Jay 1993, 466.

129. Gu 1913.

130. Millant 1908, 234.

131. Millant 1908, 242.

132. Cohn 1987, 36–37.

133. Cohn 1987, 36–37.

134. Cohn 1987; Li 2008; Wang 1990; Wu et al. 2008.

135. Hobson (1851) 1857, section on 外腎 (waishen, "outer kidney"). The edition that I have relied on is the one available at the East Asian Library and the Gest Collection at Princeton University.

136. Chiang 2015b, 107–10; Valussi 2008, 72; Wu 2016.

137. Stent 1877, 170–71.

138. "Women weisheme qiaobuqi heshang he taijian," 1941, 69.

139. Die 1940; Hou 1942a; Liu 1936.

140. Stent 1877, 171, 181.

141. Ma Deqing et al. in *Zhongguo renmin zhengzhi xieshang huiyi quanguo weiyuanhui wenshi ziliao yanjiu weiyuanhui*, 1982, 224.

142. Hou 1942b.

143. Ma Deqing et al. in *Zhongguo renmin zhengzhi xieshang huiyi quanguo weiyuanhui wenshi ziliao yanjiu weiyuanhui*, 1982, 225.

144. Spiegel 2009, 5.

145. Mei 1997, 139.

146. Kutcher 2010, 467. On the policy of eunuch illiteracy, see Dale 2000, 27.

147. Kutcher 2010, 476.

148. On Zhang, see Zou 1988, 294; Yang Zengguang 1991, 14; Yan and Dong 1995, 108; Xiao Yanqing in Zhang 1996, 1903–7. On Ma, see Ma Deqing et al. in *Zhongguo renmin zhengzhi xieshang huiyi quanguo weiyuanhui wenshi ziliao yanjiu weiyuanhui*, 1982; Qiu 1990. On Sun, see Jia 1993, 24–26; Ling 2003, 17–21; Sun 1990a, 1990b.

149. Ma Deqing et al. in *Zhongguo renmin zhengzhi xieshang huiyi quanguo weiyuanhui wenshi ziliao yanjiu weiyuanhui*, 1982, 222–23.

150. Sun 1990a, 115.

151. Sun 1990a, 124–25. A more dramatized account of Sun's castration can be found in Jia 1993, 24–26.

152. More stories told by eunuchs of how they were treated by the royal family can be found in Zhu 1980a, 1980b.

153. Coltman 1894, 28.

154. Coltman 1894, 28.

155. Coltman 1894, 28.

156. Coltman 1894, 29.

157. Coltman 1893, 329.

158. Coltman 1893, 328–329.

159. Coltman 1894, 29.

160. See also Pujie's recollection in Pujia and Pujie 1984, 304–6.

161. Aisin Gioro 1964, 143; Pujia and Pujie 1984, 22.

162. Aisin Gioro 1964, 142.

163. Aisin Gioro 1964, 142.

164. Pujia and Pujie 1984, 28.

165. Pujia and Pujie 1984, 31–32.

166. "Qinghuangshi quzhu taijian," 1923, 18.

167. Pujia and Pujie 1984, 32.

168. Aisin Gioro 1964, 144.

169. Aisin Gioro 1964, 119.

170. Aisin Gioro 1964, 141–42.

171. Aisin Gioro 1964, 122.

172. Liu 1936, 69–70.

173. Scott 1991, 777.

174. Anderson 1990, 307–11.

175. Arondekar 2005, 26.

176. Cvetkovich 2003, 7.

177. Cvetkovich 2003, 7.

178. Ko 2005, 68.

2. Vital Visions

1. Elman 2004b; Rogaski 2004, 165–92.
2. I use the phrase "most of the world" in the way that the political historian Partha Chatterjee has formulated it in a different context. As Chatterjee (2004, 3) explains it, most of the popular politics of the governed "is conditioned by the functions and activities of modern governmental systems that have become part of the expected functions of governments everywhere." If we read the exposed castrated body as a mere example of bodily mutilation, then we are only rehearsing the discursive and functionalist governmentalization of the body everywhere, as "expected" to be found in most of the world. On a similar point, see also Chakrabarty 2000. For an insightful critique of the applicability of "political society," as formulated by Chatterjee, to postcolonial East Asia (and Taiwan more specifically), see Chen 2010, 224–45.
3. On "colonial modernity" in modern Asian history, see Barlow 1997, 2012; Kwon 2015; Loos 2006; Shin and Robinson 2001.
4. I focus on scientific claims because claims of a scientific nature fall within the realm of truth and falsehood. For most relevant examples of the historical analysis of the truth and falsehood of scientific knowledge, see Chiang 2015a; Daston 1994; Davidson 2001; Hacking 2002, 2009; Rheinberger 1997, 2010a. The idea of "geobody" comes from Winichakul 1994.
5. Meng 2006.
6. On asylum practice in late Qing China, which is beyond the scope of this chapter, see Chiang 2014b; Szto 2002. On the translation of Western-style medicine, see, e.g., Elman 2005; Gao 2009; Leung 2016; Rogaski 2004; Wright 2000; Yuan 2010.
7. I call these "techniques of visualization" rather than simply "modes of representation" because representation tends to assume a positivist ontological status for the object being represented. As we will see over the course of this chapter, the images should be comprehended as the product less of representation than visualization. They do not merely represent sex on different levels—of anatomical configuration, morphological appearance, and subcellular agents; rather, each of them involves a different kind of epistemological projection that constructs, rather than replicates, the object (sex) they claim to represent. "Technique," therefore, is invoked here strategically without a direct allusion to technological advancement. For a critique of the essentialist connotations of scientific representation, see Woolgar 1988.
8. Laqueur 1990. For critiques of Laqueur's thesis, see, e.g., Cadden 1993; Harvey 2004; King 2013; Park 2006; Stolberg 2003.

9. Furth 1999.

10. Furth 1986, 48. See also Furth 1987.

11. Bray 1997.

12. Wu 2010, 85–86.

13. Furth 1999, 15–53.

14. Wu 2010, 118–19.

15. On the role of Chinese partners in the making of this treatise, see Chan 2012.

16. Heinrich 2008, 113–47.

17. Hobson (1857) 1858, preface. The edition that I have relied on is the one available at the East Asian Library and the Gest Collection at Princeton University.

18. Heinrich 2008, 147.

19. Heinrich 2008, 14. For a parallel discussion of the "rhetoric of the real" in Renaissance anatomy, see Kemp 1996.

20. Elman 2006, 100–131.

21. Heinrich 2008, 113–47.

22. Hobson (1851) 1857, "Outer Kidney" (外腎經, *waishenjing*) section.

23. Hobson (1851) 1857, "Yang Essence" (陽精論, *yangjinglun*) section.

24. Hobson (1851) 1857, "Yin" (陰經, *yinjing*) section.

25. For textual descriptions, see Hobson (1857) 1858, "Scrotum" (腎囊証, *shennangzheng*) chapter, and "Scrotal Hernia" (腎囊水疝, *shennang shuishan*) section.

26. I thank Yi-Li Wu for clarifying the epistemic relationship between Hobson's anatomical terms and their role in earlier Chinese discourses of the body. On the history of *zigong*, see Wu 2010, 84–119.

27. Kuriyama 2002, 127–28 (emphasis original).

28. Anderson and Dietrich 2012.

29. For a survey of the visual representations of the body in China since the Song period, see Despeux 2005.

30. A more in-depth discussion of the distinction between "ideal" and "corrected ideal" anatomical representations can be found in Daston and Galison 1992, 88–93.

31. Fu 2008; Kuriyama 2002, 155–59; Miyasita 1967.

32. I wish to thank both Angela Leung for pointing this out to me and Pierce Salguero for confirming this through e-mail communications. On the *Manchu Anatomy*, see Asen 2009. Though beyond the scope of Asen's discussion, the *Manchu Anatomy* included plates of male and female reproductive organs.

33. Veith 2002, 98–100.

34. Furth 1999, 46.

35. Furth 1999, 45.

36. Unshuld 1986, 382; cf. the 39th difficult issue.

37. Unshuld 1986, 385.

38. Hobson (1851) 1857, "Inner Kidney" (內腎經, *neishenjing*) section.

39. Bhabha 1994, 177.

40. Quoted in Wu 2010, 104.

41. I thank Yi-Li Wu for pointing out the following three major differences between pre- and post-Hobson discussions of sexual difference in China: (1) that significance was newly accorded to genital anatomy as the definitional basis of sex; (2) that medicine emerged as a venue where the drawing of genitals became important (previously, these depictions can be found in erotic art, where the showing of male and female genitals was arguably de rigueur); and (3) that the depictions of internal genitals gradually shifted to the center of attention in sexual biology (e.g., erotic art had plenty of labia but obviously not uteri).

42. Elman 2006; Rogaski 2004.

43. Elman 2004a.

44. Elman 2005.

45. Dikötter 1992; Pusey 1998.

46. Nedostup 2009, 191–226.

47. On the historical significance of the debate, see, e.g., Adas 2004; Chou 1978; Chow 1960, 327–37; Elman 2006, 223–26; Furth 1970; Kwok 1965.

48. Horesh 2009; Yeh 2007.

49. Dikötter 1992, 1995, 1998; Jones 2011; Pusey 1983, 1998; Schneider 2003.

50. Chai (1928) 1932, 14.

51. Feng 1920, 1.

52. Wang 1926, 66.

53. On the popularization of the scientific theory of "sex antagonism" in Europe, see Sengoopta 2006.

54. Furth 1988; Zeitlin 1993, 98–131.

55. Feng 1920, 2.

56. Zhou (1927) 1928, 19–20.

57. Zhou 1931, 37.

58. Zhang 2009.

59. Daston and Galison 1992, 84–98.

60. See Fan 2004.

61. Daston and Galison 1992, 116.

62. Curtis 2012, 71.

63. Zhu (1939) 1946, (1940) 1948, (1941) 1948, (1945) 1948, 1946, (1946) 1948.

64. Schneider 2003, 1 (emphasis mine).

65. See Schneider 2003, 44–45, 132–33, 148–49. The only full-length biography of Zhou to date is Xie 1991. See also Jones 2011, 91–93; Xuefeng Wang 2006, 155–96.

66. On the "universal language" of science, especially in the context of scientific translations, see Elshakry 2008.
67. On Tan and Chen, see Jiang 2016; Schneider 2003.
68. Levine 1993, 84–85; Schneider 2003, 137; Zhu 1923.
69. Zhu (1939) 1946, ii.
70. See, e.g., Dikötter 1992; Jones 2011; Pusey 1983, 1998; Schwartz 1964.
71. Zhu (1939) 1946.
72. The second edition appeared in 1948, three years after the Second Sino-Japanese War. The main difference between the two editions is the additional materials Zhu included in the revised text. Under the influence of Western studies in endocrinology and biochemistry, Zhu adopted a chemical model of sex in this postwar material, which reflected the increasing prominence of sex endocrinological research in the preceding decades. For the additional materials in the second edition, see Zhu (1945) 1948, 335–80. See chapter 4 of this book on the development of endocrinological ideas in Republican China.
73. Zhu (1945) 1948, 5–16.
74. Pauwels 2006.
75. Lynch 1998, 223 (emphasis mine).
76. My arguments in this section build on the extensive body of literature on the important role visual images play in the production of scientific knowledge. See, e.g., Anderson and Dietrich 2012; Arnheim 1969; Baigrie 1996; Knorr-Cetina and Almann 1990; Latour 1990; Lynch and Woolgar 1988; Pauwels 2006; Taylor and Blum 1991.
77. On Fryer, see Wright 1996. On Kerr, see Szto 2002, 2014. The most comprehensive biography of Dudgeon to date is Gao 2009.
78. Heinrich 2013.
79. It should be noted that the contexts within which these images were circulated between the mid-nineteenth century and the early twentieth century differed significantly. For instance, Qing-era missionary anatomical treatises embodied the outlook of natural theology that was distinctively absent in the Republican-era illustrations.
80. On the rise of the commercial publishing houses and their impact in Shanghai, see Reed 2004.
81. Hüppauf and Weingart 2008, 18.
82. Whereas I adopt the more familiar term "sexual characteristics," the phrase "sexual character" seems to be the convention among this group of writers. See the quoted texts below.
83. Gao 1935, 12.
84. Gao 1935, 17.
85. Gao 1935, 17.

86. Wang 1926, 68.

87. Li Baoliang 1937, 39–41.

88. Li Baoliang 1937, 40–41.

89. Shen 1934, 33.

90. Wu 2008.

91. Ellis 1904. On Ellis's thinking on sexual differentiation in terms of secondary and tertiary orders, see Robinson 1973, 58; and Sengoopta 2000, 71.

92. Wang 1926, 70.

93. Wang 1926, 71.

94. See, e.g., Kwok 1965.

95. Lei 2014, 141–221.

96. Furth 1986.

97. Su 1935, 1.

98. Guo and Li 1930, 4.

99. Barlow 2004, 53.

100. Poovey 1990, 29.

101. Sinha 2006, 50.

102. Se 1924 (emphasis mine), cited in Wang 1999, 115.

103. Chen Xiefen, "Crisis in the Women's World," cited in Dooling and Torgeson 1998, 84.

104. Barlow 2004, 54.

105. Dikötter 1995.

106. Rocha 2010, 606.

107. Lanza 2010; Yeh 1990.

108. This is not to disregard the tremendous effort that Republican-era reformers of Chinese medicine had put in to stake new claims of being scientific in the process of modernizing their profession. See Andrews 2014; Lei 2014; Rogaski 2004.

109. Ren and Yi 1925, 1.

110. Ren and Yi 1925, 3.

111. Ren and Yi 1925, 4–5.

112. Kuriyama 2002, 111–51.

113. Ren and Yi 1925, 6.

114. Chai (1928) 1932, 28–29.

115. Chai (1928) 1932, 30.

116. Chai (1928) 1932, 31–32.

117. Cited in Dikötter 1995, 21.

118. Ko 2005, 9–37.

119. Hu 1935, 2.

120. Chai (1928) 1932, 51–52.

121. Shen 1934, 36–37.

122. Hu 1935, 122–23.

123. The most obvious contemporaneous example can be found in the writings of Patrick Geddes and John Arthur Thomson (1908). In fact, this rhetoric of active sperm / passive egg remains pervasive in biology textbooks today. See Martin 1991.

124. Daston and Lunbeck 2011, 1.

125. Bullock 1980; Yang Tsui-hua 1991.

126. Schneider 2003, 22. The most authoritative work on the history of the neo-Darwinian synthesis is Smocovitis 1996.

127. Schneider 2003, 23.

128. Ha 2010.

129. Chen 1924.

130. Late nineteenth-century and early twentieth-century European sexologists also entertained the notion of "psychical hermaphroditism." Two of the most reputable physicians advocating this idea were Richard von Krafft-Ebing ([1886] 1892) and Magnus Hirschfeld ([1914] 2000).

131. This is Stefan Helmreich and Sophia Roosth's (2010, 31) interpretation of *Lebensform*, a German equivalent of "living form" that first appeared in 1838 in *Jenaer Literatur-Zeitung*.

132. Zhu (1945) 1948, 224.

133. Zhu (1945) 1948, 224 (emphasis mine). An autosome is any chromosome that is not sex chromosome.

134. Zhu (1945) 1948, 225–26.

135. Zhu (1945) 1948, 226.

136. Zhu (1945) 1948, 223.

137. Zhu (1945) 1948, 243.

138. Zhu (1945) 1948, 243–44.

139. Zhu (1945) 1948, 250–51 (emphasis mine).

140. Zhu (1945) 1948, 312.

141. Daston and Galison 1992, 98.

142. Liu and Liu 1953.

143. Barlow 2004; Dikötter 1995; Hee 2013; Kang 2009; Leary 1994a; Peng 2001; Rocha 2010; Sang 2003; Shapiro 1998; Zhu 2015.

144. Weinbaum et al. 2008.

145. Barlow 2004, 53.

146. For an in-depth discussion of the place of scientific observation in studies of epistemology and ontology, see Daston 2008.

147. See, e.g., Dikötter 1992; Zarrow 2006, 2012.

3. Deciphering Desire

1. For a discussion of the authorship of the book, see Mao and Liu 1977, 90–95.
2. Brook 1999.
3. Ko 1995; Vitiello 2011; Volpp 1994, 2001.
4. The first proclaimed Chinese 3-D pornographic film came out in theater in early 2011 and is based on *The Carnal Prayer Mat*. Shiu et al. 2011. For an earlier version, see Lee 1991.
5. Zito 2001, 201.
6. Li 2011, 45.
7. Sedgwick 1985.
8. Li 2011, 71–73.
9. Sedgwick 1990, 5.
10. Sedgwick 1990, 1.
11. Barkey 2000, 56–71; Chung 2002, 107–19.
12. "Deguo xingxue boshi jianglai hu" 1931; "Xingxue boshi yanjiang suji" 1931a; "Xingxue boshi yanjiang suji" 1931b; Zhi 1931.
13. I have in mind, specifically, the notion of truthfulness used by Bernard Williams (2002). In this regard, I take cue from Ian Hacking (2008, 1–48) and use "truth" in this chapter as a formal (as opposed to a strictly realist) concept. Although scholars have pointed out that there is no fully adequate Chinese translation of the Western concept of "sexuality," the underlying premise of my study is that, at least on that level of epistemology, the emergence of homosexuality implies a broader emergence of the concept of sexuality itself. For a critical perspective on the Chinese translation of sexuality, see Ruan 1991. For a new collection of essays on the history of sexuality in China, see Chiang 2018.
14. Leary 1994a, 1994b; Lee 2006, 140–85; Peizhong Zhang 2008; Rocha 2010; Sakamoto 2004; Tsu 2005, 128–66; Xuefeng Wang 2006, 249–65.
15. Chung 2002; Dikötter 1992; Lee 2006, 186–217; Lü 2006; Rogaski 2004, 225–53; Sakamoto 2004; Tsu 2005, 98–166; Xuefeng Wang 2006, 197–232; Yanni Wang 2006.
16. For a brief analysis of Pan's sexological writings on homosexuality, see Sang 2003, 120–22. A more extended study can be found in Guo 2016; Kang 2009.
17. Dikötter 1995, 143–45.
18. McMillan 2006, 90.
19. Kang 2009; Sang 2003.
20. Chou 2000, 50; Dikötter 1995, 140–41; Kang 2009, 42–43; Sang 2003, 7. For an account that stresses the role of Western psychiatry and general political

trends but does not touch on the significance of the translation of "homosexuality," see Wu 2003.

21. Kang 2009, 42–43.
22. Sang 2003, 118.
23. Foucault 1990, 32.
24. Foucault 1990.
25. Altman 2001, 86–105; Farquhar 2002, 211–42; Rofel 2007, 85–110.
26. For important examples in the history of the French life sciences, see Appel 1987; Farley and Geison 1974.
27. Gulik 1974; Hinsch 1990; Xiaomingxiong 1984.
28. Sommer 2000. For earlier works that look at the legal construction of sodomy in China, see Meijer 1985; Ng 1989.
29. For an explanation of why homosexuality was not criminalized in the Republican period, see Kang 2010.
30. I distinguish "styles of argumentation" from "styles of reasoning" more carefully in the conclusion. On the epistemological applicability of "style," see also Chiang 2009.
31. Davidson 2001, 36, 37.
32. Halperin 2002, 131. On the historicism of homosexuality, see also Halperin 1990.
33. Andrews 2014; Lei 2014; Xu 1997.
34. Duara 2003, 3 and 6n6.
35. Barlow 1997, 6. By citing Barlow here, I am of course strategically endorsing her "colonial modernity" approach for the historical context under discussion, so this chapter must be seen as a continuation of the argument I posited in the earlier two chapters that "China" and "sex" have been bound by a mutually generative relationship.
36. Chen 2010, 244.
37. Rogaski 2004.
38. For recent reflections on the problem of Chinese self- or re-Orientalization, see Ang 2001; Chu 2008. For a similar argument on how certain formulations within queer studies could be complicit with homophobic reticent poetics by positing culturally essential "Chinese" nonhomophobic subjects and situations, see Liu and Ding 2005.
39. See, e.g., Arondekar 2009; Blackwood 2008; Chiang 2018; Chiang and Wong 2016, 2017; Garber 2005; Jackson 2009; Najmabadi 2005, 2014; Pflugfelder 1999.
40. Massad 2007, 41.
41. Massad 2007; Rofel 2007. The target of Rofel's critique is Altman 2001. Note the striking similarity in the titles of Massad's and Rofel's books.

42. For a recent set of essays that begin to push transnational Chinese queer studies in a more fruitful direction, see Chiang 2011; Chiang and Heinrich 2014.

43. Foucault 1972, 190.

44. Kwok 1965, 21.

45. Wu 1927b.

46. Li 1922.

47. Dikötter 1995, 2.

48. After the United States decided to remit its share of the Boxer indemnity to finance Chinese students' overseas education in the United States, the total number of students in America sent by the Chinese government was 847 by 1914 (Bailey 1990, 228). While the number of Chinese students who went to the United States for education increased, the number of those who went to Japan began to decline after 1906 with increasing governmental restrictions. The number had fallen to 1,400 by 1912 (Bailey 1990, 249n12).

49. Bailey 1990, 236. On the work-study movement, see also Levine 1993; Wang 1966.

50. Zhang (1924) 1998, (1925) 1998.

51. Peizhong Zhang 2008, 309.

52. Zhang (1926) 2005, 24–27.

53. Zhang (1926) 2005, 31.

54. On the historical relationship between science, medicine, and homosexuality in Western Europe and the United States, see, e.g., Bayer 1981; Drucker 2014; Irvine 1990; LeVay 1996; Minton 2002; Oosterhuis 2000; Rosario 1997; Terry 1999.

55. Zhang (1926) 2005, 31.

56. Zhang (1926) 2005, 27.

57. Peng 2002.

58. Vitiello 2011; Wu 2004.

59. Foucault 1990, 56.

60. I borrow the notion of "norms of scientific regularity" from Foucault: see Foucault 1990, 65. I borrow the phrase "facts of life" from the title of Roy Porter and Lesley Hall's (1995) book on the history of British sexuality.

61. See Zhang 1927a. Moreover, Zhang (1927a, 26; 1927d) claims that there is a "fourth kind of water" produced inside the womb.

62. Chai (1928) 1932, 42.

63. On "vaginal breathing," see Zhang 1927b, 1927c.

64. Zhang (1926) 2005, 110–11.

65. Foucault 1990, 61.

66. Kong 1927.

67. Zheng 1927.

68. Shi 1927.

69. He 1927.

70. Xu 1927.

71. On requests for a "male perspective," see Nan 1927; Zhi 1927. For examples of frustrations with the impracticality of Zhang's theory, see Chang 1927; Kuang 1927.

72. Su 1927.

73. Qin 1927, 63.

74. SSD 1927.

75. Foucault 1990, 24 (emphasis mine).

76. Tang 1927, 1.

77. Shan 1911.

78. Yan 1923.

79. Wei 1925.

80. Zhen 1946.

81. Jian 1934a.

82. Jian 1934b.

83. Jian 1934a, 12.

84. "Tongxing'ai jiehun zhi tongxun," 1934.

85. "Guanyu 'tongxing'ai,'" 1941.

86. For a discussion of the epistemic tension between Kinsey's statistical notion of sexual normality and American psychiatrists' framework of psychopathology around the mid-twentieth century, see Chiang 2008a.

87. Oosterhuis 2000, 212.

88. Foucault 1990, 63.

89. Rocha 2008.

90. Han 1927.

91. Qian 1927.

92. See, e.g., Yang 1930, 150 and 166; Zhang (1926) 1998; Zhou (1926) 1998. For a biography of Zhou in Chinese, see Xie 1991.

93. Qi 1929, 22.

94. Peizhong Zhang 2008, 347–61.

95. Pan 1927, 1.

96. Lü 2006, 46.

97. Kevles 1985, 45–46.

98. Pan (1941b) 1994. Pan has presented a similar view concerning Chinese musicians in an earlier article: see Pan (1932) 1994.

99. Kevles 1985, 60.

100. For Pan, individual and social health depended first and foremost in heredity and not behavior. He encouraged marriage and breeding among those deemed genetically superior, which would in turn strengthen the health of the nation.

Both Frank Dikötter's study on the Chinese conception of race (1992) and Ruth Rogaski's book on health and hygiene in Tianjin (2004, 225–53) have situated the significance of Pan's eugenic visions within the larger social and cultural expressions of modernization during the Republican period. See also Guldin 1994.

101. Dikötter 1992, 164–90.

102. Pan (1927) 1994, 402–3.

103. Pan (1927) 1994, 406.

104. Zhang 1927f, 126–27.

105. Zhang 1927e.

106. Zhang 1927g, 24.

107. Yang 1929, 403–4.

108. Kang 2009, 43–49.

109. Qiu 1929. This is a translation of Edward Carpenter's "The Homogenic Attachment," chapter 3 of *The Intermediate Sex* (1908).

110. Hu and Yang 1930.

111. Heike Bauer (2009) has distinguished British sexology from its continental European counterpart in terms of a strong literary tradition.

112. Yang 1929, 414.

113. On the development of sexology in modern Japan, see Früstück 2003; Pflugfelder 1999; Suzuki 2009, 2015.

114. For other analyses of the debates between Zhang Jingsheng and people like Zhou and Pan on the proper meaning of "sex science" and "sex education," see Chung 2002; Leary 1994b, 236–80; Xuefeng Wang 2006, 267–74. On the problem of scientism as a category of historical analysis, see my discussion in chapter 2 of this book.

115. Foucault 1990, 70–71.

116. Peizhong Zhang 2008, 354–56.

117. Zhou 1934c, 44–49 (the article was originally written in August 1933). On Zhou, see Daruvala 2000.

118. Yang 1929, 403.

119. Foucault 1990, 39.

120. Foucault 1990, 39.

121. Pan (1946) 1994. One should also note that, as the editor of *New Culture*, Zhang did publish several translated excerpts of Ellis's work (by himself or others) in the journal. One of these is an article on female homosexuality taken from Ellis's *Sexual Inversion*: see Xie 1927. This is a translation of Havelock Ellis's "The School Friendships of Girls" in *Studies in the Psychology of Sex*, vol. 2, *Sexual Inversion* (1897). But in general, Zhang's effort in translating Ellis's work was neither as comprehensive nor as extensive as Pan's.

122. For an extensive study of this appendix, see Guo 2016; Kang 2009.

123. Pan (1946) 1994, 701.

124. Qiu 1929. See also Shen 1923. This is a translation of Edward Carpenter's "Affection in Education," which was the fourth chapter of his *The Intermediate Sex* (1908).

125. Ellis and Symonds 1897.

126. See, e.g., Ba 1940; Hu 1933; Zhang 1925; Zhu 1928, esp. 108–13.

127. Zhou 1934c.

128. Zhao 1929.

129. My sources are replete with examples of this sort. See, e.g., Bin 1936; Cheng 1925, 148–53; Dongmin Zhang 1927, 46–47.

130. On American eugenicists' disregard for Freud, see Kevles 1985, 53.

131. Pan initially wrote a draft of this essay as a term paper for a history survey course taught by Liang Qichao at Qinghua University (Huang 2005, 181). He later revised it and published it as a book with additional materials after returning from the United States (Pan [1941a] 1994).

132. Lee 2006, 186–217; Tsu 2005, 98–166.

133. Lee 2006, 189.

134. Jing 1936, 231.

135. Pan (1934) 1994, 98.

136. Pan (1936) 1994, 375–76.

137. Pan (1946) 1994, 705–6.

138. Freud (1905) 2000, 10–11n1 (emphasis original).

139. Pan (1936) 1994, 376.

140. Gui 1936, 63–66. But it seems that Gui did not entirely agree with Freud on the interpretations of other types of psychopathology. This is most evident in her textbook (Gui 1932). In a 1903 interview with the Vienna newspaper *Die Zeit*, Freud indicated his "firm conviction that homosexuals must not be treated as sick people, for a perverse orientation is far from being a sickness." Cited in Abelove 1993, 382. Similarly, Arnold Davidson (2001, 79) has carefully historicized Freud's view of homosexuality in *Three Essays*: "what we ought to conclude, given the logic of Freud's argument and his radically new conceptualization . . . is precisely that cases of inversion can no longer be considered pathologically abnormal."

141. Wang 1944, 130.

142. Foucault 1990, 66–67.

143. Chang 1936, 4, 6. The article only indicates that this is a translation of a piece originally written by a German medical doctor.

144. Jian 1936, 6. The article only indicates that this is a translation of a piece originally written by an American medical doctor.

145. Mo 1936, 23–24.

146. See Hong 1936; Ping 1936. On sexuality in the prison environment, see also Xi 1937.
147. Yang 1936, 12.
148. Chai (1928) 1932, 117.
149. Mo 1936, 23.
150. Kong 1936, 3.
151. There is evidence that the readers of these sexological writings shared this view. See, e.g., Qin 1927, 64–66.
152. Gui 1932, 32.
153. Ding 1947.
154. Translated and cited in Sivin 1987, 98.
155. This is the phrase that Rogaski (2004, 37–40) uses to characterize discussions of sex in traditional Chinese medicine.
156. Furth 1988, 6 (emphasis mine). Cf. Goldin 2002.
157. Yan 1922.
158. Tang 1927; Wu 1927a.
159. "Kaichangbai," 1927.
160. Bu 1931.
161. Xue 1936.
162. Foucault 1990, 58. For a fuller articulation of this problem, see Chiang 2010.
163. Zhang 1950, 78.
164. Zhang 1950, 75.
165. Yang 1929, 436.
166. Pan (1946) 1994, 708–9. See also Pan (1941b) 1994, 255–58.
167. Wang 1935, 49.
168. Liu 1995, 1999.
169. On the association of male homosexual practice with national backwardness in the Republican period, see also Kang 2009, 115–44; Wu and Stevenson 2006.
170. Duara 1998. On the complicated historical layering of the *dan* figure, see Goldstein 2007; Tian 2000; Zou 2006.
171. For a classic discussion of the transformation from "culturalism" to "nationalism" in the Chinese political sphere, see Levenson 1965.
172. Zhang 1982, as translated (with my own modifications) and cited in Wu 2004, 42–43.
173. Evans 1997, 206.
174. Lu 1955, 53.
175. Lu 1955, 53–54.
176. I am being careful and specific when discussing "marriage to an opposite sex" because other scholars have unearthed the popularity of same-sex "marriages" in eighteenth-century China, especially in the region of Fujian (Szonyi 1998).

177. Heinrich 2008.

178. Kang 2009, 2010; Vitiello 2011, 13–14 and 200–201. For an excellent defense of the social constructionist approach, see Traub 2015.

179. Kang 2009, 21.

180. Sedgwick 1990.

181. Kang 2009, 21.

182. Kang 2010, 490.

183. Kang 2010, 492.

184. My disagreement with Kang in part can be viewed as the resurfacing of an earlier debate between Sedgwick (1990) and David Halperin (1990; 2002), with whom my analysis sides, on the genealogy of homosexuality in Western culture.

185. Halperin 2002, 117.

186. Wu and Stevenson 2010, 121. In this regard, Wu and Stevenson diverge from the view of Andrea Goldman (2008, 5), who discusses *huapu* authors' "awareness of the stigma that was associated with the sex trade in boy actors."

187. Vitiello 2011, 201.

4. Mercurial Matter

1. Clarke 1998; Fausto-Sterling 2000; Oudshoorn 1994; Sengoopta 2006.

2. Dikötter 1995, 14–101.

3. The term "hormone" was coined in 1905 by Ernest Henry Starling, professor of physiology at University College in London, who defined it as "chemicals that have to be carried from the organ where they are produced to the organ which they affect, by means of the blood stream." Cited in Fausto-Sterling 2000, 150.

4. Gu 1924, 23, 26.

5. Chai (1928) 1932, 42–43.

6. Clarke 1998; Oudshoorn 1994; Sengoopta 2006.

7. Wang 1927, 4.

8. Zhang 1936b, 327–29, 333.

9. Brown-Séquard 1889.

10. Yong 1937.

11. With the exception of this sentence, I employ the term "sex characteristics" (rather than the more conventional "sexual characteristics") throughout this chapter because this is the translation more commonly used by the writers discussed in this context.

12. Wang 1926, 137.

13. Li 1936, 4. On the sexing of the X and Y chromosomes, see Richardson 2013.
14. Li 1936, 6. Homosexuality was in fact a familiar topic to biologists, physicians, and other scientific and clinical researchers interested in glandular science around the turn of the twentieth century (Sengoopta 1998).
15. See, e.g., Rosario 1997; Terry 1999.
16. See, e.g., Minton 2002; Oosterhuis 2000.
17. Halberstam 1998; Meyerowitz 2002; Rosario 2002; Terry 1999.
18. Steinach 1940, 66.
19. Harms 1969; Meyerowitz 2002.
20. Cunningham 2008, 91. Johannes Meisenheimer's findings contradicted Steinach's, whose work has attracted more attention from historians of science and sexuality.
21. Fei 1924, 63.
22. The section titles echo the title of Steinach's article "Feminization of Males and Masculinization of Females" (1913).
23. Gu 1924, 42.
24. Gu 1924, 42–44.
25. Gu 1924, 44–45.
26. Wang 1926, 138.
27. Tan 1924, 30–31.
28. Xiang 1935a.
29. He 1928, 88, 90–91; Lillie 1917. As Anne Fausto-Sterling (2000, 64) notes, Lillie was aware of the similarity between the naturally occurring freemartins that he studied and the gonadally transplanted animals on which Steinach experimented, but he remained uncertain about the nature of male hormone activity.
30. "Electric Current Determines Sex," 1934, 58; Morange, 2011, 213.
31. "Electric Current Determines Sex," 1934, 58; Guo, 1934, 92.
32. "Electric Current Determines Sex," 1934, 58, 119.
33. Yun 1941.
34. Laqueur 1990; Sengoopta 2000.
35. Bailey 2007; Lutz 1971.
36. On feminism in modern China, see Barlow 2004; Ko and Wang 2007. On homosexuality, see chapter 3 of this book; Kang 2009; Sang 2003.
37. Meyerowitz 2002, 22.
38. "Xingde qubie shi chengdu de chayi," 1936.
39. On the role of Japan as an intermediary in the transmission of Western scientific knowledge, see, e.g., Rogaski 2004.
40. Wang 1926, 145–46.
41. Sengoopta 2000.

42. When Blair-Bell first published the book in 1916, his intention was to bring together a host of literature on the similar finding that reproductive functions are controlled by all the organs of internal secretion acting in concert.
43. Wang 1926, 150.
44. Wang 1926, 150.
45. Meyerowitz 2002, 29.
46. Zhou 1929.
47. Hu and Yang 1930.
48. Zhu 1928, 110–11.
49. Zhu 1928, 111.
50. Zhu 1928, 111.
51. Zhang 1936a, 140.
52. Yang 1929, 410.
53. Yang 1929, 411.
54. Wang 1944, 128.
55. A transgender activist, Virginia Prince founded the *Transvestia* magazine in 1960 and the first crossdressing organization in the United States, the Hose & Heels Club, in 1961.
56. Chen 1942, 101–2. For the coverage of this case in the American press, see Meyerowitz 2002, 39–41.
57. Rosario 2002, 193; Sengoopta 1998, 468.
58. Xing 1936, 104.
59. See chapter 1 of this book and see Jay 1993.
60. Gu 1924, 23.
61. Gu 1924, 24.
62. Gu 1924, 24.
63. Li 1936, 8.
64. Li 1936, 8. On the Skoptsy sect, see Engelstein 2003.
65. Li 1936, 8–9.
66. Li 1936, 9–11.
67. Consider the lingering effect of this trope as assumed in the work of Melissa Dale (2010) on Chinese eunuchs.
68. Zhang 1941, 40.
69. Zhang 1941, 40.
70. Chen 1937, 219.
71. Chen (1937) 1947, 242.
72. Wang 1926, 138.
73. Zhu 1928, 96.
74. He 1936, 41–43.
75. Li Shizhen, *Bencao gangmu*, juan 52, "人傀 (*renkui*)," as cited in Furth 1988, 5.
76. Furth 1988.

77. See chapter 2 of this book and see Chiang 2008b.
78. My analysis here excludes the section on "homosexuality," which is discussed in chapter 3 of this book.
79. Chai (1928) 1932, 115.
80. Chai (1928) 1932, 115–16.
81. The notion of "stone maidens" continued to hold some traction in the popular imagination well into the late 1940s, forming the basis of a complaint from a man in Guangzhou about his wife's "unpenetrable" condition. See "Shinü zhilei de qizi," 1948.
82. Chai (1928) 1932, 116.
83. Bridie Andrews (1997) has delineated a similar historical process for the reception of germ theory in early twentieth-century China.
84. Ding 1936. I thank Angela Leung for bringing this case to my attention. On Ding, see Andrews 2014, 122–33.
85. Liu (1928) 1935, 81–82. This distinction between true half *yin-yang* (真性半陰陽, *zhenxing banyinyang*) and pseudo-half *yin-yang* (假性半陰陽, *jiaxing banyinyang*) was also endorsed by the practicing gynecologist Gui Zhiliang (1936, 11).
86. Liu (1928) 1935, 82.
87. Interestingly, the biologist Zhu Xi also reproduced this image of Marie-Madeleine Lefort just after the title page of his book, *Changes in Biological Femaleness and Maleness* ([1945] 1948). On Lefort, see Dreger 1998, 54.
88. Liu (1928) 1935, 83.
89. Liu (1928) 1935, 84.
90. Liu (1928) 1935, 86.
91. See discussion of the term in Chiang 2014a; Kang 2009, 33–39; Zeitlin 1993, 104.
92. Liu (1928) 1935, 82.
93. Liu (1928) 1935, 88–89.
94. "Medicine: Girls into Boys," 1934.
95. "Nübiannan jiujing sheme daoli," 1934. On the coverage of Acces in the American press, see Meyerowitz 2002, 31–32.
96. Qu 1931.
97. Li 2012; Zhu 2010, 18.
98. Li 1934, 181–82.
99. Li 1934, 182.
100. In *How Sex Changed*, Joanne Meyerowitz questions Elbe's alleged hermaphroditism: "It seems highly unlikely that Elbe was intersexed" (2002, 30).
101. Meyerowitz 2002, 29–30.
102. Hoyer (1933) 2004. Niels Hoyer was a pseudonym for Ernst Ludwig Hathorn Jacobson. Some have suggested that the book was written by Elbe herself, and thus an autobiographical account.

103. Zhu 2015, 12.

104. Chen (1937) 1947, 245–47; Zhou 1934a, 185.

105. Xiang 1935b.

106. *Shenbao* 1935a.

107. See, e.g., Dan 1935; *Shenbao* 1935a; *Xinwenbao* 1935f.

108. *Shenbao* 1935a.

109. *Xinwenbao* 1935f.

110. *Shenbao* 1935b.

111. *Xinwenbao* 1935c.

112. *Shenbao* 1935b.

113. *Shenbao* 1935b.

114. *Shenbao* 1935d; *Xinwenbao* 1935d.

115. *Shenbao* 1930.

116. *Shenbao* 1935d.

117. *Shenbao* 1935e; *Xinwenbao* 1935g.

118. *Shenbao* 1935c.

119. *Xinwenbao* 1935e.

120. Han 1935.

121. See, e.g., "Nübian nanshen," 1934; "Nübiannan," 1934; Xu 1935.

122. The term "public passions" is adopted from Lean 2007.

123. *Xinwenbao* 1935b.

124. Fang 1935.

125. Fang 1935.

126. Xiao 1935a.

127. Xiao 1935a.

128. Bai 1935.

129. Xiao 1935b.

130. Gong 1935.

131. Yao 1935.

132. Mao 1935.

133. Meyer 2005.

134. "Nüxing biannan de jiepo," 1935, 17.

135. Tian 1935, preface.

136. Qi 1935.

137. Furth 1988, 18, 24.

138. "Xiaonü huanan," 1893.

139. Xin 1935.

140. Seven years before Yao's story, one writer claimed to have changed his sex in one evening by simply giving out some money to beggars on the street (Si 1928).

141. See, e.g., Hai 1935; *Xinwenbao* 1935a.

142. This was not entirely true. An article responding to Yao's incident, for example, documented numerous examples of male-to-female transformation as found in the historical record: Zhao 1938.
143. "Nühua nanshen," 1935, 11–12.
144. Lin 1997.
145. Gu 1940, 33.
146. Gu 1940, 93–97.
147. Gu 1940, 94. On the best-known authority on reproductive biology in twentieth-century China, see chapter 2 of this book and see Chen 2000; Chiang 2008b; Zhang Zhijie 2008.
148. Gu 1940, 94.
149. See Chiang 2008b, 408–17.
150. Gu 1940, 95.
151. Gu 1940, 96.
152. Lin 1997.
153. Gu 1940, 97–98.
154. Gu 1940, 97.
155. Gu 1940, 276–77.
156. My emphasis on "explicit" is important here. Cf. Dorothy Ko's work on the "implicitness" of female agency in the history of footbinding (2005).
157. Gu 1940, 216.
158. Gu 1940, 280.
159. Gu 1940, 335.
160. "Nanbiannü zhi qiwen," 1934; "Nühuanan," 1913; "Xiaonü huanan," 1893; "Youshi yige nübiannan," 1917.
161. Zhao 1938.
162. Zhang 1936. For an analysis of Koubek and Weston in the historical context of sex testing in international sport, see Heggie 2010.
163. On the press coverage of medicalized sex change in 1930s Britain, see Oram 2007, 109–28.
164. Xinyong Li 1937.
165. Mei 1947.
166. "Luoma qiwen," 1947.
167. Read by Chinese-speaking communities around the world, *West Wind* was a monthly magazine edited by Huang Chia-Yin (黃嘉音) that represented one of the most innovative ventures of Republican-era reformist elites. It popularized psychological knowledge, promoted mental hygiene, and drew attention to issues of family conflicts, emotional disturbances, and social pathologies in the Chinese context. On *West Wind* and the history of mental hygiene in China, see Blowers and Wang 2014; Wang 2011.

168. Coincidentally, in 1946, the Chinese press reported on an American story, billed as "Female Becoming Male," in which two twin brothers were raised as girls by their parents: Bi 1946.

169. Feng 1948. On Kinsey's view of cross-gender behavior, see Meyerowitz 2001.

5. Transsexual Taiwan

1. *Lianhebao* 1953f.

2. *Lianhebao* 1953a.

3. On Christine Jorgensen, see Meyerowitz 2002; Serlin 1995; Skidmore 2011.

4. On "Sinophone," I follow Shu-mei Shih's definition to refer to Sinitic-language cultures and communities outside of China or on the margins of the hegemonic productions of the Chinese nation-state or Chineseness. See Chiang and Heinrich 2014; Shih 2007, 2011; Shih, Tsai, and Bernards 2013.

5. Chen 2001, 165–71.

6. Liu 2017; Yang 2008.

7. The word "transexual" was first coined by the American sexologist David Cauldwell in 1949. Cauldwell (1949, 275) wrote: "When an individual who is unfavorably affected psychologically determines to live and appear as a member of the sex to which he or she does not belong, such an individual is what may be called a *psychopathic transexual*. This means, simply, that one is mentally unhealthy and because of this the person desires to live as a member of the opposite sex." In 1966 endocrinologist Harry Benjamin used the word "transsexual" in his magnum opus, *The Transsexual Phenomenon*. This book was the first large-scale work describing and explaining the kind of affirmative treatment for transsexuality that he had pioneered throughout his career. On the intellectual and social history of transsexuality in the United States, see Hausman 1995; Meyerowitz 2002; Stryker 2008.

8. In this chapter, I adopt the feminine pronoun when referring to Xie both generally and in the specific contexts after her first operation. I use the masculine pronoun to refer to Xie only in discussing her early media publicity, because she expressively refused sex reassignment before the first operation.

9. Stryker, Currah, and Moore 2008, 13.

10. Chiang 2014a; Chiang and Wang 2017.

11. On queer Sinophonicity, see Chiang and Heinrich 2014; Martin 2014.

12. *Lianhebao* 1953f.

13. *Taiwan xinshengbao* 1953a.

14. *Lianhebao* 1953f.

15. *Zhonghua ribao* 1953g.

16. *Lianhebao* 1953f; see also *Taiwan xinshengbao* 1953a; *Zhonghua ribao* 1953g, 1953i.
17. *Taiwan xinshengbao* 1953b.
18. *Lianhebao* 1953v.
19. *Lianhebao* 1953v.
20. *Lianhebao* 1953v.
21. *Lianhebao* 1953v.
22. *Taiwan xinshengbao* 1953f.
23. *Lianhebao* 1953v.
24. *Taiwan minsheng ribao* 1953f.
25. *Lianhebao* 1953v.
26. *Lianhebao* 1953v.
27. *Taiwan xinshengbao* 1953c; *Zhonghua ribao* 1953c.
28. *Lianhebao* 1953q.
29. *Lianhebao* 1953q.
30. *Lianhebao* 1953q; *Taiwan xinshengbao* 1953d; *Zhonghua ribao* 1953c.
31. *Lianhebao* 1953t.
32. *Taiwan xinshengbao* 1953e; *Zhonghua ribao* 1953j.
33. *Taiwan xinshengbao* 1953g; *Zhonghua ribao* 1953b.
34. *Lianhebao* 1953a.
35. *Lianhebao* 1953a.
36. *Lianhebao* 1953i.
37. *Lianhebao* 1953w.
38. On the significance of cultural concealment in American gay history, see, e.g., Chauncey 1994; D'Emilio 1983; Faderman 1991; Laughery 1998; Meeker 2006.
39. *Lianhebao* 1953m.
40. *Zhonghua ribao* 1953h.
41. *Lianhebao* 1953r.
42. *Lianhebao* 1953u; *Taiwan minsheng ribao* 1953c; *Zhonghua ribao* 1953a, 1953k.
43. Even after consenting to the sex reassignment, Xie still expressed strong desire to rejoin the army. See *Zhonghua ribao* 1953e.
44. *Lianhebao* 1953i.
45. *Lianhebao* 1953q.
46. Guan 1953.
47. Meyerowitz 1998, 173–74.
48. Guan 1953.
49. *Zhonghua ribao* 1953f.
50. Guan 1953.
51. *Lianhebao* 1953e, 1953j, 1953o, 1953z; *Taiwan minsheng ribao* 1953a.
52. See Fausto-Sterling 2000; Matta 2005; Meyerowitz 2002.

53. For theoretical considerations of the problem of civilizationalism, see Huntington 1993; and Chen 2006. For another example of the Xie-Jorgensen comparison, see *Lianhebao* 1954d.

54. *Lianhebao* 1953l.

55. Guan 1953.

56. Chen 2010, 177.

57. *Lianhebao* 1953k.

58. *Lianhebao* 1953k; *Taiwan minsheng ribao* 1953e.

59. *Lianhebao* 1953n.

60. *Lianhebao* 1953o.

61. *Lianhebao* 1953l.

62. *Lianhebao* 1953l. See also *Taiwan minsheng ribao* 1953g.

63. *Lianhebao* 1953j.

64. *Lianhebao* 1953aa. See also *Taiwan minsheng ribao* 1953d.

65. *Taiwan minsheng ribao* 1953b.

66. *Lianhebao* 1954g.

67. *Lianhebao* 1954g.

68. *Lianhebao* 1954i.

69. *Taiwan minsheng ribao* 1954; *Zhongyang ribao* 1954.

70. *Lianhebao* 1954i.

71. *Lianhebao* 1954h.

72. *Lianhebao* 1954h.

73. Lü was originally from Miaoli County.

74. *Lianhebao* 1953b.

75. On the historical meanings of "*renyao*" in Republican China, see Wenqing Kang 2009, 33–39; and Wong 2012. For the term's historical meaning in postwar Taiwan, see Huang 2011, 53–59. On the case of Zeng Qiuhuang, see Chiang 2014a, 208–13; Damm 2005, 69–70.

76. *Lianhebao* 1953b.

77. *Lianhebao* 1953p.

78. *Lianhebao* 1953h.

79. On "passing" in the history of gender and sexuality, see Garber 1992; Katz 1976, 317–422; San Francisco Lesbian and Gay History Project 1989; Sullivan 1990.

80. *Lianhebao* 1953d.

81. *Lianhebao* 1953c.

82. *Lianhebao* 1953c.

83. *Lianhebao* 1953g.

84. *Lianhebao* 1953g.

85. *Lianhebao* 1953g.

86. *Lianhebao* 1953g.

87. *Lianhebao* 1953g.

88. *Lianhebao* 1953g.

89. *Lianhebao* 1953s.

90. *Lianhebao* 1953s.

91. *Lianhebao* 1953y.

92. *Lianhebao* 1953x.

93. *Lianhebao* 1954f.

94. *Lianhebao* 1954j.

95. *Lianhebao* 1954a.

96. *Lianhebao* 1954b.

97. *Lianhebao* 1954c, 1954e.

98. *Lianhebao* 1954b.

99. *Lianhebao* 1955n.

100. *Zhongyang ribao* 1955f.

101. I put the words "third" and "final" in quotation marks here (and only here) because the official report released later in the year would contradict this count and indicate that this was actually Xie's fourth—and not final—operation. See discussion below.

102. *Lianhebao* 1955i.

103. *Zhongyang ribao* 1955d.

104. *Lianhebao* 1955k.

105. *Lianhebao* 1955k. See also *Zhongyang ribao* 1955d.

106. *Lianhebao* 1955l.

107. *Lianhebao* 1955l.

108. *Lianhebao* 1955l.

109. May 1988.

110. *Lianhebao* 1955l.

111. *Lianhebao* 1955j.

112. *Lianhebao* 1955m.

113. *Lianhebao* 1955a; *Taiwan minsheng ribao* 1955.

114. *Zhongguo shibao* 1956a; *Zhongyang ribao* 1955c.

115. *Lianhebao* 1955q.

116. *Lianhebao* 1955q.

117. *Lianhebao* 1955e, 1955o.

118. See Meyerowitz 1998, 2002, 81–97.

119. *Lianhebao* 1955b. On Reese, see Meyerowitz 2002, 84–85.

120. Quoted in Meyerowitz 2002, 85.

121. Greenson 1964, 218.

122. *Lianhebao* 1955p.

123. *Lianhebao* 1955g.

124. *Lianhebao* 1955f.

125. *Zhongguo shibao* 1955b.

126. *Zhongyang ribao* 1955b, 1955e.

127. *Gonglunbao* 1955; *Taiwan xinshengbao* 1955; *Zhonghua ribao* 1955b.

128. *Lianhebao* 1955h.

129. For voices that challenged the propriety and authority of the official report, pointing out that its explicit content was too invasive of Xie's privacy and that its "scientific" tone did not pay sufficient attention to Xie's post-op psychology, see *Lianhebao* 1955c; *Lianhebao* 1955d; *Zhongguo shibao* 1955a; *Zhonghua ribao* 1955a.

130. Yi 1955.

131. Quoted in Meyerowitz 2002, 65.

132. Jorgensen (1967) 2000; Small 1970.

133. *Lianhebao* 1956, 1958a, 1958b, 1959a, 1959b; *Zhongguo shibao* 1956b, 1959; *Zhongyang ribao* 1956, 1959.

134. Hsu 1998.

135. Personal interview with Jung-Kwang Wen on March 20, 2008.

136. Personal interview with Miss Lai on March 22, 2008. I thank Dr. Wen for introducing me to Miss Lai.

137. See, e.g., Chen 2016, 29–32; Ho 2003, 2006; Liu 1993.

138. *Lianhebao* 1955i.

139. Yu 2009.

140. On the discourse of intersexuality and gender variance in colonial Taiwan, see, for example, Chen 2013; Lin 2008. On the legacy of Japanese colonialism in the health care system of postwar Taiwan, see, for example, Fu 2005; Liu 2010.

141. For a detailed study of the Nationalist migration from mainland China to Taiwan, see Yang 2012.

142. Fan 2006; Liu 2009.

143. On the nationalization of Chinese medicine in early communist China, see Taylor 2005.

144. Carroll 2007, 140–66. For examples of queer cultural production in Hong Kong in the 1960s, see Weixing Shiguan Zhaizhu 1964, 1965. Scholars have begun to reconceptualize the history of love, intimacy, and sexuality in socialist China, but most revisionist readings are limited to discussions of heteronormative desires. See, e.g., Evans 1997; Honig 2003; Larson 1999; Yan 2003; Zhang 2015.

145. Chen 2010, 7.

146. Johnson and Chiu 2000, 1–2. See also Chiang 2017.

147. Chakrabarty 1992; Sakai 1988.

148. Nandy 1984.

149. Chen 2010, 216.

150. Chen 2010, 211–55.

151. Chen 2010, 197.

152. Chen 2010, 120. See also Chun 2017. On the "China-centered perspective," see Cohen 2010.

153. Chen 2010, 154.

154. Lionnet and Shih 2005, 5, 6.

155. Chen 2002.

156. Ng 2016.

157. Cumings 1999.

158. Chen 1998.

159. In October 1994, the *Daoyu bianyuan* 島嶼邊緣 (*Isle Margin*) magazine hosted a local workshop on queer and women's sexuality in Taipei, Taiwan. It was arguably the first sustained forum where scholars, authors, and activists debated on the proper translation and meaning of "queer" in Chinese-speaking communities. See Chi 1997; Chu 2003; Ho 1998, 47–87; Martin 2003, 2010. For a historical overview of the discourses and cultures of same-sex desire in Taiwan in the two decades preceding the lifting of martial law, see Damm 2005, 2017; Huang 2011.

160. Damm 2011, 172.

161. On the meaning, history, and politics of the term "*tongzhi*," see Chou 2000. On queer culture and politics in late twentieth-century Hong Kong, see Kong 2011; Leung 2008; Tang 2011.

162. Personal e-mail communication with Jens Damm on August 23, 2011.

163. This observation therefore challenges some of the conventional interpretations of Taiwanese intellectual history from the viewpoint of literature. These conventional readings tend to acknowledge the historical significance of gender and sexuality only with the rise of women's/feminist literature (女性文學, *nüxing wenxue*) and gay and lesbian literature (同志文學, *tongzhi wenxue*), along with the literatures of aborigines (原住民文學, *yuanzhumin wenxue*), military dependents' villages (眷村文學, *juancun wenxue*), and environmental groups (環保文學, *huanbao wenxue*), in the post-1987 era. Critics have called the 1980s in Taiwan's literary history the decade of "identity literature" (認同文學, *rentong wenxue*). See, e.g., Chen 1994, 235; 2002.

Bibliography

Abelove, Henry. 1993. "Freud, Male Homosexuality, and the Americans." In *The Lesbian and Gay Studies Reader*, ed. Henry Abelove, Michele Aina Barale, and David M. Halperin, 381–93. New York: Routledge.

Adas, Michael. 2004. "Contested Hegemony," *Journal of World History* 15, no. 1: 31–63.

Aisin Gioro Puyi 愛新覺羅溥儀. 1964. *Wo de qianbansheng* 我的前半生 [The first half of my life]. Hong Kong: Wentong shudian.

Altman, Dennis. 2001. *Global Sex*. Chicago: University of Chicago Press.

Anderson, Mary M. 1990. *Hidden Power: The Palace Eunuchs of Imperial China*. Buffalo, N.Y.: Prometheus.

Anderson, Nancy, and Michael R. Dietrich, eds. 2012. *The Educated Eye: Visual Culture and Pedagogy in the Life Sciences*. Hanover, N.H.: Dartmouth College Press.

Andrews, Bridie. 1997. "Tuberculosis and the Assimilation of Germ Theory in China, 1895–1937." *Journal of the History of Medicine and Allied Sciences* 52, no. 1:114–57.

——. 2014. *The Making of Modern Chinese Medicine, 1850–1960*. Vancouver: University of British Columbia Press.

Ang, Ien. 2001. *On Not Speaking Chinese: Living Between Asia and the West*. New York: Routledge.

Angelides, Steven. 2001. *A History of Bisexuality*. Chicago: University of Chicago Press.

Appel, Toby. 1987. *The Cuvier-Geoffroy Debate: French Biology in the Decades Before Darwin*. New York: Oxford University Press.

Arnheim, Rudolph. 1969. *Visual Thinking.* Berkeley: University of California Press.

Arondekar, Anjali. 2005. "Without a Trace: Sexuality and the Colonial Archive." *Journal of the History of Sexuality* 14: 10–27.

——. 2009. *For the Record: Sexuality and the Colonial Archive in India.* Durham, N.C.: Duke University Press.

Asen, Daniel. 2009. "'Manchu Anatomy': Anatomical Knowledge and the Jesuits in Seventeenth- and Eighteenth-Century China." *Social History of Medicine* 22, no. 1: 23–44.

——. 2016. *Death in Beijing: Murder and Forensic Science in Republican China.* Cambridge: Cambridge University Press.

Austin, John L. 1975. *How to Do Things with Words.* 2nd ed. Ed. by J. O. Urmson and Marina Sbisà. Cambridge, Mass.: Harvard University Press.

Ba Ni 巴尼, trans. 1940. *Ai de xingshenghuo* 爱的性生活 [Married love]. Shanghai: Shanghai Benliu shudian.

Bachelard, Gaston. (1938) 2002. *The Formation of the Scientific Method.* Trans. by Mary McAllester Jones. Manchester, U.K.: Clinamen Press.

Backhouse, Edmund. 1943. "Decadence Mandchoue." Memoirs of Sir Edmund Trelawny Backhouse Collection. University of Oxford Bodleian Library.

Bai Bi 白比. 1935. "Nühua nanshen" 女化男身 [Woman to man]. *Xinsheng zhoukan* 新生週刊 2, no 9: 182–83.

Baigrie, Brian S., ed. 1996. *Picturing Knowledge: Historical and Philosophical Problems Concerning the Use of Art in Science.* Toronto: University of Toronto Press.

Bailey, Paul. 1990. *Reform the People: Changing Attitudes Towards Popular Education in Early Twentieth Century China.* Edinburg: Edinburgh University Press.

——. 2007. *Gender and Education in China: Gender Discourses and Women's Schooling in the Early Twentieth Century.* London: Routledge.

Bao, Hongwei. 2011. "'Queer Comrades': Transnational Popular Culture, Queer Sociality, and Socialist Legacy." *English Language Notes* 49, no. 1: 131–37.

Barbier, Patrick. 1996. *The World of the Castrati: The History of an Extraordinary Operatic Phenomenon.* Trans. by Margaret Crosland. London: Souvenir Press.

Barkey, Cheryl Lynn. 2000. "Gender, Medicine and Modernity: The Politics of Reproduction in Republican China." PhD diss., University of California, Davis.

Barlow, Tani, ed. 1997. *Formations of Colonial Modernity in East Asia.* Durham, N.C.: Duke University Press.

——. 2004. *The Question of Women in Chinese Feminism.* Durham, N.C.: Duke University Press.

——. 2012. "Debates over Colonial Modernity in East Asia and Another Alternative." *Cultural Studies* 26, no. 5: 617–44.

Bauer, Heike. 2009. *English Literary Sexology: Translations of Inversion, 1860–1930.* New York: Palgrave Macmillan.

——. 2017. *The Hirschfeld Archives: Violence, Death, and Modern Queer Culture.* Philadelphia: Temple University Press.

Bayer, Ronald. 1981. *Homosexuality and American Psychiatry: The Politics of Diagnosis.* New York: Basic Books.

Benedict, Carol. 1996. *Bubonic Plague in Nineteenth-Century China.* Stanford, Calif.: Stanford University Press.

Benjamin, Harry. 1966. *The Transsexual Phenomenon.* New York: Julian Press.

Bhabha, Homi K. 1994. *The Location of Culture.* New York: Routledge.

Bi Jun 畢君. 1946. "Nübiannan" 女變男 [Female to male]. *Zhoubo* 週播 17: 8.

Bickers, Robert. 2004. "Backhouse, Sir Edmund Trelawny, second baronet (1873–1944)." In *Oxford Dictionary of National Biography.* Oxford: Oxford University Press, 2004. Online ed., http://www.oxforddnb.com/view/article/30513?docPos=2. Accessed July 31, 2014.

Bin 彬, trans. 1936. "Tongxing'ai" 同性愛 [Same-sex love]. *XKX* 1, no. 2: 92–94.

Blackwood, Evelyn. 2008. "Transnational Discourses and Circuits of Queer Knowledge in Indonesia." *GLQ* 14: 481–507.

Blair-Bell, William. 1916. *The Sex Complex: A Study of the Relationship of the Internal Secretions to the Female Characteristics and Functions in Health and Disease.* London: Baillière.

Blowers, Geoffrey and Shellen Xuelai Wang. 2014. "Gone with the *West Wind*: The Emergence and Disappearance of Psychotherapeutic Culture in China, 1936–68." In *Psychiatry and Chinese History*, ed. Howard Chiang, 143–60. London: Pickering and Chatto.

Bo Yang 柏楊. 1999. *Diyici huanguan shidai* 第一次宦官時代 [The first epoch of the rise of eunuchs]. Taipei: Yuan Liu.

——. 2001. *Dierci huanguan shidai* 第二次宦官時代 [The second epoch of the rise of eunuchs]. Taipei: Yuan Liu.

Bray, Francesca. 1995. "A Deathly Disorder: Understanding Women's Health in Late Imperial China." In *Knowledge and Scholarly Medical Traditions*, ed. Don Bates, 235–50. Cambridge: Cambridge University Press.

——. 1997. *Technology and Gender: Fabrics of Power in Late Imperial China.* Berkeley: University of California Press.

Brook, Timothy. 1999. *The Confusions of Pleasure: Commerce and Culture in Ming China.* Berkeley: University of California Press.

Brown-Séquard, Claude-Édouard. 1889. "The Effects Produced on Man by Subcutaneous Injection of a Liquid Obtained from the Testicles of Animals." *Lancet* 137: 105–7.

Brownell, Susan, and Jeffrey N. Wasserstrom, eds. 2002. *Chinese Femininities/Chinese Masculinities: A Reader.* Berkeley: University of California Press.

Bu Tong 不通. 1931. "Xingxuehui daxiezhen" 性學會大寫真 [A portrait of the Sexological Association]. *Qianyanbiao zazhi* 千嚴表雜誌 3: 1–2.

Bullock, Mary Brown. 1980. *An American Transplant: The Rockefeller Foundation and the Peking Union Medical College.* Berkeley: University of California Press.

Bullough, Vern. 1994. *Science in the Bedroom: A History of Sex Research.* New York: Basic Books.

Burton, Antoinette. 2003. *Dwelling in the Archive: Women Writing House, Home, and History in Late Colonial India.* New York: Oxford University Press.

Butler, Judith. 1997. *Excitable Speech: A Politics of the Performative.* New York: Routledge.

Cadden, Joan. 1993. *Meanings of Sex Difference in the Middle Ages: Medicine, Science, and Culture.* Cambridge: Cambridge University Press.

Cai Dengshan 蔡登山. 2011. "Zhongyijie caizi: Chen Cunren" 中醫界才子: 陳存仁 [A talent in Chinese medicine: Chen Cunren]. *Quanguo xinshu zixun yuekan* 全國新書資訊月刊 153: 37–42.

Canguilhem, Georges. (1943) 1991. *The Normal and the Pathological.* Trans. by Carolyn R. Fawcett. New York: Zone.

Carpenter, Edward. 1908. *The Intermediate Sex: A Study of Some Transitional Types of Men and Women.* London: George Allen & Unwin.

Carroll, John M. 2007. *A Concise History of Hong Kong.* Lanham, Md.: Rowman & Littlefield.

Cauldwell, David. 1949. "Psychopathia Transexualis." *Sexology* 16: 274–80.

Chai Fuyuan 柴福沅. (1928) 1932. *Xingxue ABC* 性學ABC [ABC of sexology]. Shanghai: Shijie shuju.

Chakrabarty, Dipesh. 1992. "Provincializing Europe: Postcoloniality and the Critique of History." *Cultural Studies* 6, no. 3: 337–57.

———. 2000. *Provincializing Europe: Postcolonial Thought and Historical Difference.* Princeton, N.J.: Princeton University Press.

Chan, Man Sing. 2012. "Sinicizing Western Science: The Case of *Quanti xinlun.*" *T'oung Pao* 98: 528–56.

Chang Hao. 1971. *Liang Ch'i-ch'ao and Intellectual Transition in China, 1890–1907.* Cambridge, Mass.: Harvard University Press.

Chang Hong 長虹, trans. 1936. "Biantai xingyu yu qi liaofa" 變態性欲與其療法 [Sexual perversion and its treatment]. *XKX* 2, no. 1: 3–7.

Chang Lu 昌瓐. 1927. "Xingyu tongxin" 性育通信 [Sex education letters]. *XWH* 1, no. 3: 69–70.

Chatterjee, Partha. 2004. *The Politics of the Governed: Popular Politics in Most of the World.* New York: Columbia University Press.

Chauncey, George. 1994. *Gay New York: Gender, Urban Culture, and the Making of the Gay Male World, 1890–1940.* New York: Basic Books.

Chen Cunren 陳存仁. 1990a. "Nanxing kuxing taijiankao" 男性酷刑太監考 [An investigation of male castration and eunuchs]. *ZJWX* 57, no. 3: 77–88.

——. 1990b. "Nanxing kuxing taijiankao." *ZJWX* 57, no. 4: 129–36.

——. 1990c. "Nanxing kuxing taijiankao." *ZJWX* 57, no. 5: 124–31.

——. 1990d. "Nanxing kuxing taijiankao." *ZJWX* 57, no. 6: 120–27.

——. 1991a. "Nanxing kuxing taijiankao." *ZJWX* 58, no. 1: 126–35.

——. 1991b. "Nanxing kuxing taijiankao." *ZJWX* 58, no. 2: 113–17.

——. 2008. *Beiyange de wenming: Xianhua Zhongguo gudai chanzu yu gongxing* 被閹割的文明: 閒話中國古代纏足與宮刑 [Castrated civilization: On footbinding and castration in ancient China]. Guilin: Guangxi Normal University Press.

Chen Fang-Ming 陳芳明. 1994. *Dianfan de zhuiqiu* 典範的追求 [The search for paradigm]. Taipei: Unitas.

——. 2002. *Houzhiming Taiwan: Wenxue shilun jiqi zhoubian* 後殖民台灣: 文學史論及其周邊 [Postcolonial Taiwan: Essays on Taiwanese literary history and beyond]. Taipei: Maitian.

Chen Fu 陳阜. 2000. *Zhu Xi* 朱洗. Hebei: Hebei Education Press.

Chen, Janet Y. 2012. *Guilty of Indigence: The Urban Poor in China, 1900–1952.* Princeton, N.J.: Princeton University Press.

Chen, Jian. 2001. *Mao's China and the Cold War.* Chapel Hill: University of North Carolina Press.

Chen, Kuan-Hsing. 2006. "Civilizationalism." *Theory, Culture & Society* 23, no. 2–3: 427–28.

——. 2010. *Asia as Method: Toward Deimperialization.* Durham, N.C.: Duke University Press.

Chen Pei-jean. 2013. "Xiandai 'xing' yu diguo 'ai:' Taihan zhimin shiqi tongxingai zaixian" 現代「性」與帝國「愛」: 台韓殖民時期同性愛再現 [Colonial modernity and the empire of love: The representation of same-sex love in colonial Taiwan and Korea]. *Taiwan wenxue xuebao* 台灣文學學報 23: 101–36.

Chen Shiyuan 陳始圜. 1942. "Nanbiannü" 男變女 [Male to female]. *Wanxiang* 萬象 1, no. 7: 101–3.

Chen Wei-Jhen. 2016. *Taiwan kuaxingbie qianshi: Yiliao, fengsuzhi yu yaji zaoyu* 台灣跨性別前史: 醫療、風俗誌與亞際遭逢 [Pre-transgender history in Taiwan: Medical treatment, hostess clubs and inter-Asia encounters]. New Taipei City: Transgender Punk Acitivist.

Chen Ying-zhen 陳映真. 1998. "Taiwan de meiguohua gaizao" 台灣的美國化改造 [Taiwan's Americanization]. In *Huigui de lütu* 回歸的旅途 [*The Trip of Return*], ed. Yang Dan 丹陽, 1–14. Taipei: Renjian.

Chen Yucang 陳雨蒼. 1937. *Renti de yanjiu* 人體的研究 [Research on the human body]. Shanghai: Zhengzhong shuju.

——. (1937) 1947. *Shenghuo yu shengli* 生活與生理 [Life and physiology]. Shanghai: Zhengzhong shuju.

Chen Zhen. 1924. *Putong shengwuxue* 普通生物學 [General biology]. Shanghai: Commercial Press.

Cheney, Victor T. 2006. *A Brief History of Castration*. 2nd ed. Bloomington, Ind.: AuthorHouse.

Cheng Hao 程浩. 1925. *Jiezhi shengyu de wenti* 節制生育的問題 [The problem of birth control]. Shanghai: Yadong tushuguan.

Chi Ta-wei 紀大偉. 1997. "Ku'er lun: Sikao dangdai Taiwan ku'er yu ku'er wensue" 酷兒論: 思考當代台灣酷兒與酷兒文學 [On *ku'er*: Thoughts on *ku'er* and *ku'er* Literature in Contemporary Taiwan]. In *Ku'er kuanghuan jie* 酷兒狂歡節 [Queer carnival], ed. Ta-wei Chi, 9–28. Taipei: Meta Media.

Chiang, Howard. 2008a. "Effecting Science, Affecting Medicine: Homosexuality, the Kinsey Reports, and the Contested Boundaries of Psychopathology in the United States, 1948–1965." *Journal of the History of the Behavioral Sciences* 44, no. 4: 300–18.

——. 2008b. "The Conceptual Contours of Sex in the Chinese Life Sciences: Zhu Xi (1899–1962), Hermaphroditism, and the Biological Discourse of *Ci* and *Xiong*, 1920–1950." *East Asian Science, Technology and Society: An International Journal* 2, no. 3: 401–30.

——. 2009. "Rethinking 'Style' for Historians and Philosophers of Science: Converging Lessons from Sexuality, Translation, and East Asian Studies." *Studies in History and Philosophy of Biological and Biomedical Sciences* 40: 109–18.

——. 2010. "Liberating Sex, Knowing Desire: *Scientia Sexualis* and Epistemic Turning Points in the History of Sexuality." *History of the Human Sciences* 23, no. 5: 42–69.

——, ed. 2011. "Queer Transnationalism in China." Topical cluster in *English Language Notes* 49, no. 1: 109–44.

——. 2014a. "Archiving Peripheral Taiwan: The Prodigy of the Human and Historical Narration." *Radical History Review*, no. 120: 204–25.

——, ed. 2014b. *Psychiatry and Chinese History*. London: Pickering and Chatoo.

——, ed. 2015a. *Historical Epistemology and the Making of Modern Chinese Medicine*. Manchester, U.K.: Manchester University Press.

——. 2015b. "Translating Culture and Psychiatry across the Pacific: How Koro Became Culture-Bound." *History of Science* 53, no. 1: 102–19.

——. 2017. "From Postcolonial to Subimperial Formations of Medicine: Superregional Perspectives from Taiwan and Korea." *East Asian Science, Technology and Society: An International Journal* 11, no. 4: 469–75.

——, ed. 2018. *Sexuality in China: Histories of Power and Pleasure*. Seattle: University of Washington Press.

Chiang, Howard, and Ari Larissa Heinrich, eds. 2014. *Queer Sinophone Cultures*. London: Routledge.

Chiang, Howard, and Yin Wang, eds. 2017. *Perverse Taiwan*. London: Routledge.

Chiang, Howard, and Alvin K. Wong. 2016. "Queering the Transnational Turn: Regionalism and Queer Asias." *Gender, Place and Culture: A Journal of Feminist Geography* 23, no. 11: 1643–56.

——. 2017. "Asia Is Burning: Queer Asia as Critique." *Culture, Theory and Critique* 58, no. 2: 121–26.

Chiang, Yung-chen. 2001. *Social Engineering and the Social Sciences in China, 1919–1949*. Cambridge: Cambridge University Press.

Chou, Min-Chih. 1978. "The Debate on Science and the Philosophy of Life in 1923." *Bulletin of the Institute of Modern History, Academia Sinica* 7: 557–81.

Chou, Wah-shan. 2000. *Tongzhi: Politics of Same-Sex Eroticism in Chinese Societies*. New York: Haworth.

Chow, Tse-Tsung. 1960. *The May Fourth Movement: Intellectual Revolution in Modern China*. Stanford, Calif.: Stanford University Press.

Chu Wei-cheng 朱偉誠. 2003. "Tongzhi•Taiwan: Xinggongmin, guozu jiangou huo gongmin shehui" 同志•台灣: 性公民、國族建構或公民社會 [Queer(ing) Taiwan: Sexual citizenship, nation building, or civil society]. *Nüxue xuezhi: Funü yu xingbie yanjiu* 女學學誌: 婦女與性別研究 15: 115–51.

Chu Yiu-Wai. 2008. "The Importance of Being Chinese: Orientalism Reconfigured in the Age of Global Modernity." *boundary 2* 35: 183–206.

Chun, Allen. 2017. *Forget Chineseness: On the Geopolitics of Cultural Identification*. Albany: State University of New York Press.

Chung, Yuehtsen Juliette. 2002. *Struggle for National Survival: Eugenics in Sino-Japanese Contexts, 1896–1945*. New York: Routledge.

Clarke, Adele. 1998. *Disciplining Reproduction: Modernity, American Life Sciences, and "The Problems of Sex."* Berkeley: University of California Press.

Cohen, Paul A. 2010. *Discovering History in China: American Historical Writing on the Recent Chinese Past*. 2nd ed. New York: Columbia University Press.

Cohen, William. 1996. *Sex Scandal: The Private Parts of Victorian Fiction*. Durham, N.C.: Duke University Press.

Cohn, Don J. 1987. *Vignettes from the Chinese: Lithographs from Shanghai in the Late Nineteenth Century*. Hong Kong: Chinese University of Hong Kong Press.

Coltman, Robert. 1891. *The Chinese, Their Present and Future: Medical, Political, and Social*. Philadelphia: F. A. Davis.

——. 1893. "Self-Made Eunuchs." *Universal Medical Journal*, November, 328–29.

——. 1894. "Peking Eunuchs." *China Medical Missionary Journal* 8: 28–29.

——. 1901. *Beleaguered in Peking: The Boxer's War Against the Foreigner*. Philadelphia: F. A. Davis.

Cook, Scott B. 1996. *Colonial Encounters in the Age of High Imperialism*. New York: Harper Collins.

Crossley, Pamela. 1999. *Translucent Mirror: History and Identity in Qing Imperial Ideology*. Berkeley: University of California Press.

Cumings, Bruce. 1999. *Parallax Visions: Making Sense of American-East Asian Relations at the End of the Century.* Durham, N.C.: Duke University Press.

Cunningham, J. T. 2008. *Hormones and Heredity.* Teddington, Middlesex, U.K.: Echo Library.

Curtis, Scott. 2012. "Photography and Medical Observation." In *The Educated Eye: Visual Culture and Pedagogy in the Life Sciences,* ed. Nancy Anderson and Michael R. Dietrich, 68–93. Hanover, N.H.: Dartmouth College Press.

Cvetkovich, Ann. 2003. *An Archive of Feelings: Trauma, Sexuality, and Lesbian Public Cultures.* Durham, N.C.: Duke University Press.

Dale, Melissa. 2000. "With the Cut of a Knife: A Social History of Eunuchs During the Qing Dynasty (1644–1911) and Republican Periods (1912–1949)." PhD diss., Georgetown University.

——. 2010. "Understanding Emasculation: Western Medical Perspectives on Chinese Eunuchs." *Social History of Medicine* 23, no. 1: 38–55.

Damm, Jens. 2005. "Same-Sex Desire and Society in Taiwan, 1970–1987." *China Quarterly* 181: 67–81.

——. 2011. "Discrimination and Backlash Against Homosexual Groups." In *Politics of Difference in Taiwan,* ed. Tak-Wing Ngo and Hong-zen Wang, 152–80. London: Routledge.

——. 2017. "From Psychoanalysis to AIDS: The Early Contradictory Approaches to Gender and Sexuality and the Recourse to American Discourses During Taiwan's Societal Transformation in the Early 1980s." In *Perverse Taiwan,* ed. Howard Chiang and Yin Wang, 64–85. London: Routledge.

Dan Weng 丹翁. 1935. "Nühuanan" 女化男 [Woman-to-man]. *JB,* March 20.

Daruvala, Susan. 2000. *Zhou Zuoren and an Alternative Chinese Response to Modernity.* Cambridge, Mass.: Harvard University Press.

Daston, Lorraine. 1994. "Historical Epistemology." In *Questions of Evidence: Proof, Practice, and Persuasion Across the Disciplines,* ed. James Chandler, Arnold Davidson, and Harry Harootunian, 282–89. Chicago: University of Chicago Press.

——. 2008. "On Scientific Observation." *Isis* 99, no. 1: 97–110.

Daston, Lorraine, and Peter Galison. 1992. "The Image of Objectivity." *Representations* 40: 81–128.

Daston, Lorraine, and Elizabeth Lunbeck. 2011. "Introduction: Observation Observed." In *Histories of Scientific Observation,* ed. Lorraine Daston and Elizabeth Lunbeck, 1–9. Chicago: University of Chicago Press.

Davidson, Arnold I. 2001. *The Emergence of Sexuality: Historical Epistemology and the Formation of Concepts.* Cambridge, Mass.: Harvard University Press.

"Deguo xingxue boshi jianglai hu" 德國性學博士將來滬 [German sexologist coming to Shanghai]. 1931. *Xinghua* 興華 28, no. 14: 43.

Delaporte, François. 1986. *Disease and Civilization: The Cholera in Paris.* Cambridge, Mass.: MIT Press.

D'Emilio, John. 1983. *Sexual Politics, Sexual Communities: The Making of a Homosexual Minority in the United States*. Chicago: University of Chicago Press.

Derrida, Jacques. 1994. *Specter of Marx: The State of the Debt, the Work of Mourning, and the New International*. Trans. by Peggy Kamuf. New York: Routledge.

Despeux, Catherine. 2005. "Visual Representations of the Body in Chinese Medical and Daoist Texts from the Song to the Qing period (Tenth to Nineteenth Century)." *Asian Medicine: Tradition and Modernity* 1, no. 1: 10–52.

Die Qi 蜨齊. 1940. "Xianhua taijian" 閒話太監 [On eunuchs]. *Sanliujiu huabao* 三六九畫報 6, 4: 22.

Dikötter, Frank. 1992. *The Discourse of Race in Modern China*. Stanford, Calif.: Stanford University Press.

——. 1995. *Sex, Culture, and Modernity in China: Medical Science and the Construction of Sexual Identities in the Early Republican Period*. Honolulu: University of Hawai'i Press.

——. 1998. *Imperfect Conceptions: Medical Knowledge, Birth Defects, and Eugenics in China*. New York: Columbia University Press.

——. 2008. *The Age of Openness: China before Mao*. Berkeley: University of California Press.

Ding Fubao 丁福保. 1936. "Ban yinyang yili" 半陰陽一例 [A case of half yin-yang]. *The Shin Yih Yaw* 新醫藥 [New medicine] 4, no. 9: 973–77.

Ding Zan 丁瓚. 1947. "Tantan tongxing'ai: Bie lanyong xinli bingtaixue de mingci" 談談同性愛: 別濫用心理病態學的名詞 [A discussion of homosexuality: Do not use psychopathological terms uncritically]. *Yichao yuekan* 醫潮月刊 [Medicine monthly] 1, 5: 14–15.

Dong Guo 東郭. 1985. *Taijian shengya* 太監生涯 [The life of eunuchs]. Yonghe City: Shishi chuban gongsi.

Dooling, Amy, and Kristina Torgeson, eds. 1998. *Writing Women in Modern China: An Anthology of Women's Literature from the Early Twentieth Century*. New York: Columbia University Press.

Dreger, Alice Domurat. 1998. *Hermaphrodites and the Medical Invention of Sex*. Cambridge, Mass.: Harvard University Press.

Drucker, Donna. 2014. *The Classification of Sex: Alfred Kinsey and the Organization of Knowledge*. Pittsburgh: University of Pittsburgh Press.

Du Wanyan 杜婉言. 1996. *Zhongguo huanguanshi* 中國宦官史 [History of Chinese eunuchs]. Taipei: Wenjin chubanshe.

Duara, Prasenjit. 1998. "The Regime of Authenticity: Timelessness, Gender, and National History in Modern China." *History and Theory* 37: 287–308.

——. 2003. *Sovereignty and Authenticity: Manchukuo and the East Asian Modern*. Lanham, Md.: Rowman & Littlefield.

Duhousset, Emile. 1877. "Moeurs orientales: De la circoncision des filles." *Bulletins de la Société d'anthropologie de Paris* 12, no. 2: 124–37.

——. 1896. "Discussion." In Jean-Jacques Matignon, "Les eunuques du Palais Impérial à Pékin." *Bulletins de la Société d'anthropologie de Paris* 7, no. 4: 325–36.

"Electric Current Determines Sex." 1934. *Popular Science Monthly* 124, no. 6: 58, 119.

Elliott, Mark C. 2001. *The Manchu Way: The Eight Banners and Ethnic Identity in Late Imperial China.* Stanford, Calif.: Stanford University Press, 2001.

Ellis, Havelock. 1904. *Man and Woman: A Study of Human Secondary Sexual Characters.* 4th ed. New York: Charles Scribner's Sons.

Ellis, Havelock, and John Addington Symonds. 1987. *Studies in the Psychology of Sex.* Vol. 2, *Sexual Inversion.* London: University of Watford Press.

Elman, Benjamin. 1991. "Political, Social, and Cultural Reproduction via Civil Service Examinations in Late Imperial China." *Journal of Asian Studies* 50, no. 1: 7–28.

——. 2004a. "From Pre-Modern Chinese Natural Studies to Modern Science in China." In *Mapping Meanings: The Field of New Learning in Late Qing China*, ed. by Michael Lackner and Natascha Vittinghoff, 25–72. Leiden: Brill.

——. 2004b. "Naval Warfare and the Refraction of China's Self-Strengthening Reforms into Scientific and Technological Failure, 1865–1895." *Modern Asian Studies* 38, no. 2: 283–326.

——. 2005. *On Their Own Terms: Science in China, 1550–1900.* Cambridge, Mass.: Harvard University Press.

——. 2006. *A Cultural History of Modern Science in China.* Cambridge, Mass.: Harvard University Press.

Elshakry, Marwa S. 2008. "Knowledge in Motion: The Cultural Politics of Modern Science Translations in Arabs." *Isis* 99, no. 4: 701–30.

Eng, David. 2001. *Racial Castration: Managing Masculinity in Asian America.* Durham, N.C.: Duke University Press.

Engelstein, Laura. 2003. *Castration and the Heavenly Kingdom: A Russian Folktale.* Ithaca, N.Y.: Cornell University Press.

Evans, Harriet. 1997. *Women and Sexuality in China: Female Sexuality and Gender Since 1949.* New York: Continuum.

Faderman, Lillian. 1991. *Odd Girls and Twilight Lovers: A History of Lesbian Life in Twentieth-Century America.* New York: Columbia University Press.

Fairbank, John K. 1977. "The Confidence of Man." Review of *The Hermit of Peking: The Hidden Life of Sir Edmund Backhouse*, by Hugh R. Trvor-Roper. *New York Review of Books*, April 14, pp. 3–5.

Faison, Seth. 1996. "The Death of the Last Emperor's Last Eunuch." *New York Times*, December 20.

Fan, Fa-ti. 2004. *British Naturalists in Qing China: Science, Empire, and Cultural Encounter.* Cambridge, Mass.: Harvard University Press.

Fan Yan-qiu 范燕秋. 2006. *Jibing, yixue yu zhimin xiandaixing: Rizhi Taiwan yi-xueshi* 疾病, 醫學與殖民現代性: 日治台灣醫學史 [*Diseases, medicine, and colonial modernity: History of medicine in Japan-ruled Taiwan*]. Taipei: Daw Shiang.

Fang Fei 芳菲. 1935. "Nüzhuan nanshen de yanjiu" 女轉男身的研究 [Research on female-to-male transformation]. *JB*, March 21.

Farley, John, and Gerald Geison. 1974. "Science, Politics and Spontaneous Generation in Nineteenth-Century France: The Pasteur-Pouchet Debate." *Bulletin of the History of Medicine* 48: 161–98.

Farquhar, Judith. 2002. *Appetites: Food and Sex in Post-Socialist China*. Durham, N.C.: Duke University Press.

Farrer, James. 2002. *Opening Up: Youth, Sex Culture, and Market Reform in Shanghai*. Chicago: University of Chicago Press.

Fausto-Sterling, Anne. 2000. *Sexing the Body: Gender Politics and the Construction of Sexuality*. New York: Basic Books.

Fei Hongnian 費鴻年. 1924. *Xin shengming lun* 新生命論 [New treatise on life]. Shanghai: Commercial Press.

Feng Fei 馮飛. 1920. *Nüxing lun* 女性論 [Treatise on womanhood]. Shanghai: Zhonghua shuju.

Feng Mingfang 馮明方. 1948. "Bubian nüzi shibuxiu" 不變女子誓不休 [Will not surrender until I become a woman]. *Xifeng* 西風 [West wind] 110: 180–82.

Fogel, Joshua A., and Peter G. Zarrow, eds. 1997. *Imagining the People: Chinese Intellectuals and the Concept of Citizenship, 1890–1920*. Armonk, N.Y.: M. E. Sharpe.

Forman, Ross G. 2013a. *China and the Victorian Imagination: Empires Entwined*. Cambridge: Cambridge University Press.

——. 2013b. "'Rivalling their Manchu compeers in aesthetic sexuality and unbridled exercise of unchaste instinct': The Chinese and Erotic Fictions of Empire in the Late Nineteenth and Early Twentieth Centuries." Paper presented at *Recovery: Memory, Corpus, Space*, Humanities Research Centre Staff Fellows Workshop, University of Warwick, Coventry, United Kingdom, May 8.

Foucault, Michel. 1972. *The Archeology of Knowledge and the Discourse on Language*. Trans. by Alan M. Sheridan Smith. New York: Pantheon.

——. 1973. *The Order of Things: An Archaeology of the Human Sciences*. New York: Vintage Books.

——. 1977. *Discipline and Punish: The Birth of the Prison*. Trans. by Alan Sheridan. New York: Pantheon Books.

——. 1980. "Nietzsche, Genealogy, History." In *Language, Counter-Memory, Practice: Selected Essays and Interviews*, ed. by D. F. Bouchard, 139–64. Ithaca, N.Y.: Cornell University Press.

——. 1990. *The History of Sexuality*. Vol. 1, *An Introduction*. Trans. by Robert Hurley. New York: Vintage Books.

———. 1994. *The Birth of the Clinic: An Archaeology of Medical Perception.* Trans. by Alan M. Sheridan Smith. New York: Vintage Books.

Fraser, Sarah. 2011. "Chinese as Subject: Photographic Genres in the Nineteenth Century." In *Brush and Shutter: Early Photography in China*, ed. Jeffrey W. Cody and Frances Terpak, 91–109. Los Angeles: Getty Research Institute.

Freedman, Estelle D. 1987. " 'Uncontrolled Desires': The Response to the Sexual Psychopath, 1920–1960." *Journal of American History* 74, no. 1: 83–106.

Freitas, Roger. 2009. *Portrait of a Castrato: Politics, Patronage, and Music in the Life of Atto Melani.* Cambridge: Cambridge University Press.

Freud, Sigmund. (1905) 2000. *Three Essays on the Theory of Sexuality.* Trans. and ed. by James Strachey. New York: Basic Books.

———. 1924. "The Passing of the Oedipus Complex." *International Journal of Psychoanalysis* 5: 419–24.

Friedman, David M. 2001. *A Mind of Its Own: A Cultural History of the Penis.* New York: Penguin Books.

Früstück, Sabine. 2003. *Colonizing Sex: Sexology and Social Control in Modern Japan.* Berkeley: University of California Press.

Fu Daiwie 傅大為. 2005. *Yaxiya de xinshenti: Xingbie, yiliao yu jindai Taiwan* 亞細亞 的新身體: 性別、醫療與近代台灣 [Assembling the new body: Gender/sexuality, medicine, andm Taiwan]. Taipei: Socio Publishing.

Fu, Louis. 2008. "A Forgotten Reformer of Anatomy in China: Wang Ch'ing-Jen." *ANZ Journal of Surgery* 78: 1052–58.

Furth, Charlotte. 1970. *Ting Wen-chiang: Science and China's New Culture.* Cambridge, Mass.: Harvard University Press.

———. 1986. "Blood, Body, and Gender: Medical Images of the Female Condition in China." *Chinese Science* 7: 43–66.

———. 1987. "Concepts of Pregnancy, Childbirth, and Infancy in Ch'ing Dynasty China." *Journal of Asian Studies* 46, no. 1: 7–35.

———. 1988. "Androgynous Males and Deficient Females: Biology and Gender Boundaries in Sixteenth- and Seventeenth-Century China." *Late Imperial China* 9, no. 2: 1–31.

———. 1999. *A Flourishing Yin: Gender in China's Medical History, 960–1665.* Berkeley: University of California Press.

Gao Xi 高晞. 2009. *Dezhen zhuan: Yige yingguo chuanjiaoshi yu wan Qing yixue jindaihua* 德貞傳: 一個英國傳教士與晚清醫學近代化 [A biography of John Dudgeon: A British missionary the modernization of late Qing medicine]. Shanghai: Fudan University Press.

Gao Xian 高銛, trans. 1935. *Xing ji shengzhi* 性及生殖 [Sex and reproduction]. Shanghai: Commercial Press.

Garber, Linda. 2005. "Where in the World Are the Lesbians?" *Journal of the History of Sexuality* 14: 28–50.

Garber, Marjorie. 1992. *Vested Interests: Cross-Dressing and Cultural Anxiety.* New York: Routledge.

Geddes, Patrick, and John Arthur Thomson. 1908. *The Evolution of Sex.* Rev. ed. New York: Walter Scott Publishing.

Glosser, Susan L. 2003. *Chinese Visions of Family and State, 1915–1953.* Berkeley: University of California Press.

Godard, Ernest. 1867. *Egypte et Palestine: Observations médicales et scientifiques.* Paris: Victor Masson et Fils.

Goldin, Paul. 2002. *The Culture of Sex in Ancient China.* Honolulu: University of Hawai'i Press.

Goldman, Andrea. 2008. "Actors and Aficionados in Qing Dynasty Texts of Theatrical Connoisseurship." *Harvard Journal of Asiatic Studies* 68, no. 1: 1–56.

Goldstein, Joshua. 2007. *Drama Kings: Players and Publics in the Re-creation of Peking Opera, 1870–1937.* Berkeley: University of California Press.

Gong Shou 恭壽. 1935. "Yu Songyun suojian zhi xiongfuci" 余松筠所見之雄孵雌 [Ya Songyun's witness of *xiong* becoming *ci*]. *JB*, March 27.

Gonglunbao 公論報. 1955. "Xie Jianshun shoushu yiwancheng" 謝尖順變性手術已完成 [Xie Jianshun's sex change operations completed]. October 28.

Greenson, Ralph R. 1964. "On Homosexuality and Gender Identity." *International Journal of Psycho-Analysis* 45: 216–18.

Gu Junzheng 顧均正. 1940. "Xingbian" 性變 [Sex change], *Kexue quwei* 科學趣味 [Scientific interest] 2, no. 1: 31–35; 2, no. 2: 93–98; 2, no. 4: 214–19; 2, no. 5: 276–80; 2, no. 6: 331–35.

Gu Mingsheng 顧鳴盛. 1913. "Taijian you qishi" 太監有妻室 [Eunuchs have wives]. *Yixue shijie* 醫學世界 [Medicine world] 26: 21–22.

Gu Rong 顧蓉, and Jinfang Ge 葛金芳. 1992. *Wuheng weiqiang—gudai huanguan qunti de wenhua kaocha* 霧橫帷牆—古代宦官群體的文化考察 [A study of the culture of ancient eunuchs]. Shanxi: Shanxi renmin jiaoyu chubanshe.

Gu Shoubai 顧壽白. 1924. *Neifenmi* 內分泌 [The internal secretions]. Shanghai: Commercial Press.

Guan Ming 管明. 1953. "Zhongguo Kelisiding" 中國克麗斯汀 [The Chinese Christine]. *LHB*, September 1.

"Guanyu 'tongxing'ai'" 關於「同性愛」 [About 'homosexuality']. 1941. *Liyan huabao* 立言畫報 [Liyan illustrated magazine] 137: 22.

Gui Zhiliang 桂質良. 1932. *Xiandai jingshen bingxue* 現代精神病學 [Modern psychopathology]. Shanghai: Xinyue shudian.

———. 1936. *Nüren zhi yisheng* 女人之一生 [The life of a woman]. Beijing: Zhengzhong shuju.

Guldin, Gregory E. 1994. *The Saga of Anthropology in China: From Malinowski to Moscow to Mao.* Armonk, N.Y.: M. E. Sharpe.

Gulik, Edward V. 1973. *Peter Parker and the Opening of China*. Cambridge, Mass.: Harvard University Press.

Gulik, Robert Hans van. 1974. *Sexual Life in Ancient China*. Leiden: Brill.

Guo Renyi 郭人驥, and Li Renling 酈人麟. 1930. *Nüxing weisheng* 女性衛生 [Women's hygiene]. Shanghai: Commercial Press.

Guo Shunping 郭舜平. 1934. "Dianliu jueding xingbie zhi jingren faxian" 電流決定性別之驚人發現 [The shocking discovery of sex-determining electric currents]. *Wenhua yuekan* 文化月刊 [Culture monthly] 1, no. 7: 92–94.

Guo, Ting. 2016. "Translating Homosexuality into Chinese: A Case Study of Pan Guangdan's Translation of Havelock Ellis' *Psychology of Sex: A Manual for Students* (1933)." *Asia Pacific Translation and Intercultural Studies* 3, no. 1: 47–61.

Ha, Nathan Q. 2010. "The Riddle of Sex: Biological Theories of Sexual Difference in the Early Twentieth Century." *Journal of the History of Biology* 44, no. 3: 505–46.

——. 2011. "Marking Bodies: A History of Genetic Sex in the Twentieth Century." PhD diss., Princeton University.

Hacking, Ian. 2002. *Historical Ontology*. Cambridge, Mass.: Harvard University Press.

——. 2009. *Scientific Reason*. Taipei: National Taiwan University Press.

Hai 害. 1935. "Jie Yao Jinping mu" 接姚錦屏幕 [Uncover Yao Jinping]. *JB*, March 26.

Halberstam, Judith. 1998. *Female Masculinity*. Durham, N.C.: Duke University Press.

——. 2011. *The Queer Art of Failure*. Durham, N.C.: Duke University Press.

Halperin, David. 1990. *One Hundred Years of Homosexuality and Other Essays on Greek Love*. New York: Routledge.

——. 2002. *How to Do the History of Homosexuality*. Chicago: University of Chicago Press.

Han 瀚. 1927. "Wuhu! Zhang Jingsheng de luanzhu" 嗚呼! 張競生的卵珠 [Woohoo! Zhang Jingsheng's Ovum]. *XZZ* 1, no. 1: 1–3.

Han Gong 憨公. 1935. "Yao Jinping qiangpo jianyan zhifalü yu daode wenti" 姚錦屏強迫檢驗之法律與道德問題 [The legal and ethical implications of Yao Jinping's forced examination]. *Yiyao pinglun* 醫藥評論 [Medicine review] 124: 43–44.

Han Suolin 韓索林. (1967) 1991. *Huanguan shanquan gailan* 宦官擅權概覽 [An overview of the power of eunuchs]. 1967. Reprint, Shenyang: Liaoning University Press.

Harms, Ernest. 1969. "Forty-Four Years of Correspondence between Eugen Steinach and Harry Benjamin: A Valuable Addition to the Manuscript Collection of the Library of the New York Academy of Medicine." *Bulletin of the New York Academy of Medicine* 45, no. 8: 761–66.

Harvey, Karen. 2004. *Reading Sex in the Eighteenth Century: Bodies and Gender in English Erotic Culture*. Cambridge: Cambridge University Press.

Hatzaki, Myrto. 2009. *Beauty and the Male Body in Byzantium*. New York: Palgrave Macmillan.

Hausman, Bernice. 1995. *Changing Sex: Transsexualism, Technology, and the Idea of Gender*. Durham, N.C.: Duke University Press.

He Qi 何琦. 1928. "Zhongjianxing yu neifenmi" 中間性與內分泌 [Intersexuality and hormones]. *Yanda yuekan* 燕大月刊 [Yanda monthly] 1, no. 4: 88–91.

He Shi 何士, trans. 1936. "Yin qushi ersheng de youqu xianxiang yu qi yuanli" 因去勢而生的有趣現象與其原理 [The effects of castration]. *XKX* 2, no. 1: 41–44.

He Zhifen 何芷分. 1927. "Xingyu tongxin" 性育通信 [Sex education letters]. *XWH* 1, no. 2: 100.

Hee, Wai Siam. 2013. "On Zhang Jing-sheng's Sexual Discourse: Women's Liberation and Translated Discourses on Sexual Differences in 1920s China." *Frontiers of Literary Studies in China* 7, no. 2: 235–70.

Heggie, Vanessa. 2010. "Testing Sex and Gender in Sports: Reinventing, Reimagining, and Reconstructing Histories." *Endeavour* 34, no. 4: 157–63.

Heinrich, Ari Larissa. 2008. *The Afterlife of Images: Translating the Pathological Body Between China and the West*. Durham, N.C.: Duke University Press.

——. 2013. "Dissection in China." In *Chinese Medicine and Healing: An Illustrated History*, edited by T. J. Hinrichs and Linda L. Barnes, 220–22. Cambridge, Mass.: Harvard University Press.

Helmreich, Stefan, and Sophia Roosth. 2010. "Life Forms: A Keyword Entry." *Representations* 112: 27–53.

Herdt, Gilbert, ed. 1993. *Third Sex, Third Gender: Beyond Sexual Dimorphism in Culture and History*. New York: Basic Books.

Hershatter, Gail. 1999. *Dangerous Pleasures: Prostitution and Modernity in Twentieth-Century Shanghai*. Berkeley: University of California Press.

Hevia, James. 2003. *English Lessons: The Pedagogy of Imperialism in Nineteenth-Century China*. Durham, N.C.: Duke University Press.

Hinsch, Bret. 1990. *Passions of the Cut Sleeve: The Male Homosexual Tradition in China*. Berkeley: University of California Press.

——. 2013. *Masculinities in Chinese History*. Lanham, Md.: Rowman and Littlefied.

Hirschfeld, Magnus. (1914) 2000. *The Homosexuality of Men and Women*. Trans. by Michael Lombardi-Nash. New York: Prometheus Books.

Ho, Josephine, ed. 1998. *Ku'er: Lilun yu zhengzhi* 酷兒: 理論與政治 [Queer politics and queer theory]. Zhongli, Taiwan: National Central University Center for the Study of Sexualities.

——, ed. 2003. *Kuaxingbie* 跨性別 [Trans]. Zhongli, Taiwan: National Central University Center for the Study of Sexualities.

———. 2006. "Embodying Gender: Transgender Body/Subject Formations in Taiwan." *Inter-Asia Cultural Studies* 7, no. 2: 228–42.

Ho, Loretta Wing Wah. 2010. *Gay and Lesbian Subculture in Urban China*. New York: Routledge.

Hobson, Benjamin. 1851. *Quanti xinlun* 全體新論 [A new treatise on anatomy]. Guangdong: Huiai yiguang 惠愛醫館.

———. (1851) 1857. *Zentai shinron* 全體新論 [A new treatise on anatomy]. Edo: Suharaya Mohē.

———. (1857) 1858. *Seii ryakuron* 西醫略論 [Outline of Western medicine]. Edo: Yorozuya Hyōshirō.

Hong 洪, trans. 1936. "Nüxing de tongxing'ai he xingde biantai" 女性的同性愛和性的變態 [Female homosexuality and sexual perversion]. *XKX* 2, no. 4: 13–15.

Honig, Emily. 2003. "Socialist Sex: The Cultural Revolution Revisited." *Modern China* 29, no. 2: 153–75.

Horesh, Niv. 2009. *Shanghai's Bund and Beyond: British Banks, Banknote Issuance, and Monetary Policy in China, 1842–1937*. New Haven, Conn.: Yale University Press.

Hou Jiafeng 侯甲峯. 1942a. "Taijian jingshen rugong I" 太監淨身入宮 I [Eunuchs enter the palace after castration I]. *Sanliujiu huabao* 三六九畫報 16, no. 6: 12.

———. 1942b. "Taijian jingshen rugong II." *Sanliujiu huabao* 16, no. 7: 12.

"How the Chinese Make Eunuchs." 1877. *Southern Medical Record: A Monthly Journal of Practical Medicine* 7: 301.

Howard, Patricia. 2014. *The Modern Castrato: Gaetano Guadagni and the Coming of a New Operatic Age*. Oxford: Oxford University Press.

Hoyer, Niels, ed. (1933) 2004. *Man into Woman: The First Sex Change*. London: Blue Boat.

Hsu Su-Ting 徐淑婷. 1998. "Bianxingyuzheng huanzhe bianxing shoushu hou de shenxin shehui shiying" 變性慾症患者變性手術後的身心社會適應 [The physical, psychological, and social adaptation among transsexuals after sex reassignment surgery: A study of six cases]. M.A. thesis, Kaohsiung Medical University.

Hu Boken 胡伯墾, trans. 1933. *Women de shenti* 我們的身體 [The human body]. Shanghai: Kaiming shudian.

———. 1935. *Renti gouzao yu shengli* 人體構造與生理 [The structure and physiology of the human body]. Shanghai: Yaxiya shuju.

Hu Qiuyuan 胡秋原, and Youtian Yang 楊尤天. 1930. *Tongxing'ai wenti taolun ji* 同性戀問題討論集 [Essays on the problem of homosexuality]. Shanghai: Beixin shuju.

Hua, Hsieh Bao. 2014. *Concubinage and Servitude in Late Imperial China*. Lanham, Md.: Lexington Books.

Huang, Hans Tao-Ming. 2011. *Queer Politics and Sexual Modernity in Taiwan*. Hong Kong: Hong Kong University Press.

Huang, Martin W. 2006. *Negotiating Masculinities in Late Imperial China*. Honolulu: University of Hawai'i Press.

Huang, Nicole. 2005. *Women, War, Domesticity: Shanghai Literature and Popular Culture of the 1940s*. Leiden: Brill.

Humana, Charles. 1973. *The Keeper of the Bed: A Study of the Eunuch*. London:
Arlington.

Hunt, Nancy Rose. 1999. *A Colonial Lexicon: Of Birth Ritual, Medicalization, and
Mobility in the Congo*. Durham, N.C.: Duke University Press.

Huntington, Samuel. 1993. "The Clash of Civilizations?" *Foreign Affairs* 72, no. 3:
22–49.

Hüppauf, Bernd, and Peter Weingart. 2008. "Images in and of Science." In
Science Images and Popular Images of the Sciences, ed. Bernd Hüppauf and Peter
Weingart, 3–31. New York: Routledge.

Hwang Jinlin 黃金麟. 2001. *Lishi, shenti, guojia: Jindai zhongguo de shenti xingcheng,
1895–1937* 歷史、身體、國家：近代中國的身體形成，1895–1937 [History, the body,
the nation: The formation of the body of modern China, 1895–1937]. Taipei:
Linking.

——. 2009. *Zhanzhen, shenti, xiandaixing: Xiandai Taiwan de junshi zhili yu shenti,
1895–2005* 戰爭、身體、現代性：近代台灣的軍事治理與身體，1895–2005 [War,
the body, and modernity: Military governmentality and the body in modern
Taiwan, 1895–2005]. Taipei: Linking.

Irvine, Janice M. 1990. *Disorders of Desire: Sex and Gender in Modern American
Sexology*. Philadelphia: Temple University Press.

Jackson, Peter. 2009. "Capitalism and Global Queering: National Markets, Parallels Among Sexual Cultures, and Multiple Queer Modernities." *GLQ* 15 (2009):
357–95.

James, Elizabeth, ed. 1997. *Women, Men and Eunuchs: Gender in Byzantium*. New
York: Routledge.

Jamieson, R. A. 1877. "Chinese Eunuchs." *Lancet* 110, no. 2813: 123–24.

Jay, Jennifer W. 1993. "Another Side of Chinese Eunuch History: Castration,
Adoption, Marriage, and Burial." *Canadian Journal of History* 28, no. 3:
459–78.

——. 1999. "Castration and Medical Images of Eunuchs in Traditional China."
In *Current Perspectives in the History of Science in East Asia*, ed. Yun-sik Kim and
Francesca Bray, 385–94. Seoul: Seoul National University Press.

Jefferys, W. Hamilton, and James L. Maxwell. 1910. *The Diseases of China: Including Formosa and Korea*. London: Bale & Danielsson.

Jeffreys, Elaine, and Haiqing Yu. 2015. *Sex in China*. Cambridge: Polity.

Jia Ben-ray. 1995. "*Zongguoren xing guan chutan* [A Preliminary Investigation of
the Chinese View on Sex]." *Si yu yan* [Thought and Language] 33: 27–75.

Jia Yinghua 賈英華. 1993. *Modai taijian miwen: Sun Yaoting zhuan* 末代太監秘聞: 孫耀庭傳 [The secret life of the last eunuch: A biography of Sun Yaoting]. Beijing: Zhishi chubanshe.

Jian 建, trans. 1936. "Zhenzheng de tongxing'ai keyi zhiliao ma?" 真正的同性愛可以治療嗎? [Can real homosexuality be cured?]. *XKX* 2, no. 4: 4–8.

Jian Yun 潤雲. 1934a. "Wo suo jianwen de nanguo nülang tongxing'ai I" 我所見聞的南國女郎同性愛 I [Female homosexuality in the south that I witnessed, part I]. *Choumou yuekan* 綢繆月刊 [Choumou monthly] 1, no. 1: 11–15.

———. 1934b. "Wo suo jianwen de nanguo nülang tongxing'ai II" 我所見聞的南國女郎同性愛 II [Female homosexuality in the south that I witnessed, part II]. *Choumou yuekan* 1, no. 2: 15–18.

Jiang, Lijing. 2016. "Retouching the Past with Living Things: Indigenous Species, Tradition, and Biological Research in Republican China, 1918–1937." *Historical Studies in the Natural Sciences* 46, no. 2: 154–206.

Jing Chun 靜淳, trans. 1936. "Yanjiu xingkexue de zhunbei" 研究性科學的準備 [Preparations for the scientific study of sex]. *XKX* 1, no. 4: 229–38.

Johnson, Marshall, and Fred Yen Liang Chiu. 2000. "Guest Editors' Introduction." *positions: east asia cultures critique* 8, no. 1: 1–7.

Jones, Andrew. 2011. *Developmental Fairy Tales: Evolutionary Thinking and Modern Chinese Culture*. Cambridge, Mass.: Harvard University Press.

Jordan-Young, Rebecca. 2010. *Brain Storm: The Flaws in the Science of Sex Differences*. Cambridge, Mass.: Harvard University Press.

Jorgensen, Christine. (1967) 2000. *Christine Jorgensen: A Personal Autobiography*. San Francisco: Cleis Press.

Judge, Joan. 1996. *Print and Politics: Shibao and the Culture of Reform in Late Qing China*. Stanford, Calif.: Stanford University Press, 1996.

"Kaichangbai" (開場白) [Prologue]. 1927. *XZZ* 1, no. 1:1–3.

Kang, Wenqing. 2009. *Obsession: Male Same-Sex Relations in China, 1900–1950*. Hong Kong: Hong Kong University Press.

———. 2010. "Male Same-Sex Relations in Modern China: Language, Media Representation, and Law, 1900–1949." *positions: east asia cultures critique* 18, no. 2: 489–510.

Katz, Jonathan, ed. 1976. *Gay American History: Lesbians and Gay Men in the U.S.A.* New York: Avon.

Keevak, Michael. 2011. *Becoming Yellow: A Short History of Racial Thinking*. Princeton, N.J.: Princeton University Press.

Kemp, Martin. 1996. "Temples of the Body and Temples of the Cosmos: Vision and Visualization in the Vesalian and Copernican Revolution." In *Picturing Knowledge: Historical and Philosophical Problems Concerning the Use of Art in Science*, ed. Brian Baigre, 40–85. Toronto: University of Toronto Press.

Kevles, Daniel. 1985. *In the Name of Eugenics: Genetics and the Uses of Human Heredity*. New York: Alfred Knopf.

King, Helen. 2013. *The One-Sex Body on Trial: The Classical and Early Modern Evidence*. Burlington, Vt.: Ashgate.

Knorr-Cetina, Karin, and Klaus Almann. 1990. "Images Dissection in Natural Scientific Inquiry." *Science, Technology, & Human Values* 15, no. 3: 259–83.

Ko, Dorothy. 1995. *Teachers of the Inner Chamber: Women and Culture in Seventeenth-Century China*. Stanford, Calif.: Stanford University Press.

——. 1997. "The Body as Attire: The Shifting Meanings of Footbinding in Seventeenth-Century China." *Journal of Women's History* 8, no. 4: 8–27.

——. 2005. *Cinderella's Sisters: A Revisionist History of Footbinding*. Berkeley: University of California Press.

Ko, Dorothy, and Zheng Wang, ed. 2007. *Translating Feminisms in China*. Malden, Mass.: Blackwell.

Kong Gu 空谷. 1927. "Tongxun" 通訊 [Letters]. *XWH* 1, no. 1: 49.

Kong Kongzhang 空空章, trans. 1936. "Xuesheng jian tongxing'ai yu fumu shizhang de jiaoyu" 學生間同性愛與父母師長的教育 [Homosexuality among students and the involvement of parents and teachers in education]. *XKX* 2, no. 4: 2–4.

Kong, Travis. 2011. *Chinese Male Homosexualities: Memba, Tongzhi, and Golden Boy*. New York: Routledge.

Korsakov, Vladimir Vikentevich. 1904. *V starom Pekine*. St. Petersburg, Russia: Trud.

Korsakow, V. 1989. "Die Eunuchen in Peking." *Deutsche Med Wochenschrift* 24: 338–40.

Koselleck, Rheinhart. 2002. *The Practice of Conceptual History: Timing History, Spacing Concepts*. Trans. by Todd Presner, Kerstin Behnke, and Jobst Welg. Stanford, Calif.: Stanford University Press.

Krafft-Ebing, Richard von. (1886) 1892. *Psychopathia Sexualis, with Especial Reference to Contrary Sexual Instinct: A Medico-Legal Study*. 7th ed. Trans. by Charles Gilbert Chaddock. Philadelphia: F. A. Davis.

Kuang Sheng 鄺生. 1927. "Xingyu tongxin" 性育通信 [Sex education letters]. *XWH* 1, no. 3: 71.

Kuo, Margaret. 2012. *Intolerable Cruelty: Marriage, Law, and Society in Early Twentieth-Century China*. Lanham, Md.: Rowman & Littlefield.

Kuo Wen-Hua 郭文華. 1998. "Meiyuan xia de weisheng zhengce: 1960 niandai Taiwan jiating jihua de tantao" 美援下的衛生政策：一九六〇年代臺灣家庭計畫的探討 [Politicizing family planning and medicalizing reproductive bodies: U.S. backed population control in 1960s Taiwan]. *Taiwan: A Radical Quarterly in Social Studies* 臺灣社會研究季刊 32: 39–82.

Kuriyama, Shigehisa. 2002. *The Expressiveness of the Body and the Divergence of Greek and Chinese Medicine.* New York: Zone Books.

Kutcher, Norman. 2010. "Unspoken Collusions: The Empowerment of Yuanming yuan Eunuchs in the Qianlong Period." *Harvard Journal of Asiatic Studies* 70, no. 2: 449–95.

Kwok, D. W. Y. 1965. *Scientism in Chinese Thought: 1900–1950.* New Haven, Conn.: Yale University Press.

Kwon, Nayoung Aimee. 2015. *Intimate Empire: Collaboration and Colonial Modernity in Korea and Japan.* Durham, N.C.: Duke University Press.

Lacan, Jacques. 1977. *Écrits: A Selection.* Trans. by Alan Sheridan. New York: Norton.

Lam, Tong. 2011. *A Passion for Facts: Social Surveys and the Construction of the Chinese Nation-State, 1900–1949.* Berkeley: University of California Press.

Lanza, Fabio. 2010. *Behind the Gate: Inventing Students in Beijing.* New York: Columbia University Press.

Laqueur, Thomas. 1990. *Making Sex: Body and Gender from Greeks to Freud.* Cambridge, Mass.: Harvard University Press.

Larson, Wendy. 1999. "Never So Wild: Sexing the Cultural Revolution." *Modern China* 25, no. 4: 423–50.

———. 2009. *From Ah Q to Lei Feng: Freud and the Revolutionary Spirit in 20th Century China.* Stanford, Calif.: Stanford University Press.

Lascaratos, J., and A. Kostakopoulus. 1997. "Operations on Hermaphrodites and Castration in Byzantine Times (324–1453 AD)." *Urologia internationalis* 58, no. 4: 232–35.

Latour, Bruno. 1990. "Drawing Things Together." In *Representation in Scientific Practice,* edited by Michael Lynch and Steve Woolgar, 19–67. Cambridge, Mass.: MIT Press.

Laughery, John. 1998. *The Other Side of Silence: Men's Lives and Gay Identities: A Twentieth-Century History.* New York: Henry Holt.

Lean, Eugenia. 2007. *Public Passions: The Trial of Shi Jianqiao and the Rise of Popular Sympathy in Republican China.* Berkeley: University of California Press.

Leary, Charles. 1994a. "Intellectual Orthodoxy, the Economy of Knowledge, and the Debate over Zhang Jingsheng's Sex Histories." *Republican China* 18: 99–137.

———. 1994b. "Sexual Modernism in China: Zhang Jingsheng and 1920s Urban Culture." PhD diss., Cornell University.

Lee, Haiyan. 2006. *Revolution of the Heart: A Genealogy of Love in China, 1900–1950.* Stanford, Calif.: Stanford University Press.

Lee, Jen-der. 1996. "Childbirth in Late Antiquity and Early Medieval China." *Bulletin of the Institute of History and Philology* 67, no. 3: 533–654.

———. 2005. "Childbirth in Early Imperial China." *Nan Nü* 7, no. 3: 216–86.

Lee, Leo Ou-fan. 1999. *Shanghai Modern: The Flowering of a New Urban Culture in China, 1930–1945*. Cambridge, Mass.: Harvard University Press.

Lee, Ying Kit. 1991. *Sex and Zen*. Dir. by Michael Mak. Hong Kong: Golden Harvest.

Legge, James. 1875. *The Chinese Classics: Translated into English, with Preliminary Essays and Explanatory Notes*, vol. 2. London: Trübner & Co.

Lei, Sean Hsiang-lin. 2014. *Neither Donkey nor Horse: Medicine in the Struggle over China's Modernity*. Chicago: University of Chicago Press.

Leung, Angela K. C., ed. 2006. *Medicine for Women in Imperial China*. Leiden: Brill.

——. 2009. *Leprosy in China: A History*. New York: Columbia University Press.

——. 2016. "Strategies of a Biomedical Hospital in 19th-Century Canton: Materiality Advertised in *Qizheng Lüeshu* 奇症略述 (Brief Account of Extraordinary Clinical Patterns), 1866." *Études chinoises* 35, no. 1: 175–96.

Leung, Helen Hok-Sze. 2008. *Undercurrents: Queer Culture and Postcolonial Hong Kong*. Vancouver: University of British Columbia Press.

LeVay, Simon. 1996. *Queer Science: The Use and Abuse of Research into Homosexuality*. Cambridge: MIT Press.

Levenson, Joseph. 1965. *Confucian China and Its Modern Fate: A Trilogy*. Berkeley: University of California Press.

Levine, Marilyn A. 1993. *The Found Generation: Chinese Communists in Europe During the Twenties*. Seattle: University of Washington Press.

Li 莉. 1941. "Lingjia, shule woba: Jishu wode tongxing'ai de shenghuo" 玲姐，恕了我吧：記述我的同性愛的生活 [Sister Lin, please forgive me: Narrating my homosexual life]. *Sanliujiu huabao* 三六九畫報 [369 illustrated magazine] 9, no. 10: 13.

Li Baoliang 李寶梁. 1937. *Xing de zhishi* 性的知識 [Sexual knowledge]. Shanghai: Zhonghua shuju.

Li, Bin. 2012. "The School of Mandarin Duck and Butterfly's Creative Push on Early Chinese Publishing Industry." *Asian Social Science* 8, no. 12: 164–70.

Li Peifen 李佩芬. 2008. "*Dianshizhai huabao* zhong de zhixuguan (1884–1898)" 《點石齋畫報》中的秩序觀 [1884–1898] [The perspectives on order in *Dianshizhai huabao*]. M.A. thesis, National Taiwan Normal University.

Li Shilian 李士璉. 1934. "Nanbiannü" 男變女 [Male to female]. *Libailiu* 禮拜六 [Saturday] 559: 181–83.

Li Xinyong 李心永. 1937. "Xingbian qitan" 性變奇談 [Marvelous stories of sex change]. *Xifeng* 西風 [West wind] 5: 557–64.

Li Yongnian 李永年. 1936. "Xingneifenmi kexue" 性內分泌科學 [Sex endocrinology]. *XKX* 2, 3: 2–34.

Li Yü. 2011. *Rouputuang miben* 肉普團密本 [The carnal prayer mat]. Taipei: Guojia Chubanshe.

Li Zongwu 李宗武. 1922. "Xingjiaoyu shang de yige zhongda wenti: Tongxing'ai zhi taorun" 性教育上的一個重大問題：同性愛之討論 [A paramount problem in

sex education: Some reflections on homosexuality]. *Juewu minguo ribao* 覺悟民
國日報, May 12, no. 4: 1–2.

Lianhebao 聯合報. 1953a. "Burang Kelisiding zhuanmei yuqian dabing jiang
biancheng xiaojie" 不讓克麗絲汀專美於前大兵將變成小姐 [Christine will not
be America's exclusive: Soldier destined to become a lady], August 21.

——. 1953b. "Dianyingjie zuowan chuxian renyao shaonian qiaozhuang modengnü
zikui shengwei nan'ersheng" 電影街昨晚出現人妖少年喬裝摩登女自愧生為男兒身
[Human prodigy appeared on movie street last night: A man dressing up like a
modern woman, loathing a natural male body], September 25.

——. 1953c. "Ji yinyangren hou you yi yiwen poufu qutai fang nanchan chuyun
shaofu liang zigong" 繼陰陽人後又一異聞剖腹取胎防難產初孕少婦兩子宮 [Another
strange news after the hermaphrodite: C-section performed on a woman
with two uteruses], September 28.

——. 1953d. "Jiananren youguai shaonü chongxian jietou bei juji" 假男人誘拐少
女重現街頭被拘拏 [Fake man abducts a young lady: Arrested when reappears
in public], September 28.

——. 1953e. "Nanbu junyou fenshe weiwen Xie Jianshun" 南部軍友分社慰問謝尖
順 [Soldiers from the southern station console Xie Jianshun],September 4.

——. 1953f. "Nanshi faxian yinyangren jiangdong shoushu bian nannü" 南市發
現陰陽人將動手術辨男女 [A hermaphrodite discovered in Tainan: Sex to be
determined after surgery], August 14.

——. 1953g. "Qingnian nanzi shen youyun yishi pouchu renxingliu" 青年男子身
有孕醫師剖出人形瘤 [Doctor excised a human-form growth from a pregnant
man], November 13.

——. 1953h. "Renyao zhanzhuan qijie nongde mianwu qunxie shangfeng baisu
juliu santian" 人妖輾轉起解弄得面污裙斜傷風敗俗拘留三天 [The human prod-
igy no longer appears fabulous: Detained for three days for offending public
morals], September 26.

——. 1953i. "Shoushu shunli wancheng gaizao juyou bawo" 手術順利完成改造具
有把握 [Surgery successfully completed: Alteration is feasible], August 21.

——. 1953j. "Wei qiuzheng shengli yi diding yinyang" 為求正生理易地定陰陽
[To validify physiology and yin or yang], December 5.

——. 1953k. "Xie Jianshun gaizao shoushu huo jianglai Taibei kaidao" 謝尖順改
造手術或將來台北開刀 [Xie Jianshun's alteration surgery might take place in
Taipei], September 24.

——. 1953l. "Xie Jianshun jue laibei kaidao xinan yinü nanding" 謝尖順決來北開
刀係男抑女難定 [Xie Jianshun has decided to relocate to Taipei for surgery: Sex
remains uncertain], November 27.

——. 1953m. "Xie Jianshun kaidaohou zuori qingkuang zhengchang" 謝尖順開
刀後昨日情況正常 [Xie Jianshun's operation proceeded normally yesterday],
August 22.

——. 1953n. "Xie Jianshun xiaojie de yintong" 謝尖順小姐的隱痛 [The pain of Miss Xie Jianshun], October 17.

——. 1953o. "Xie Jianshun xiaojie jiang beilai kaidao" 謝尖順小姐將北來開刀 [Miss Xie Jianshun coming to Taipei for surgery], October 29.

——. 1953p. "Yidu dianying dunzuo youtian" 一睹電影頓作憂天 [A trip to the movie: Fast becoming worries], September 26.

——. 1953q. "Yinyangren bianxing shoushu qian zhunbei hushi xiaojie qunyu tanxiao" 陰陽人變性手術前準備護士小姐群與談笑 [Before the hermaphrodite's sex change operation: Chatting with nurses], August 16.

——. 1953r. "Yinyangren daokou chaixian" 陰陽人刀口拆線 [The Hermaphrodite's stiches removed], August 28.

——. 1953s. "Yinyangren liulangji" 陰陽人流浪記 [The tale of a hermaphrodite], December 30.

——. 1953t. "Yinyangren Xie Jianshun jinkaidao biancixiong" 陰陽人謝尖順今開刀辨雌雄 [Hermaphrodite Xie Jianshun: Sex determined today through surgery], August 20.

——. 1953u. "Yinyangren Xie Jianshun tongyi gaizao nüxing" 陰陽人謝尖順同意改造女性 [Hermaphrodite Xiejianshun agreed to be turned into a woman], August 30.

——. 1953v. "Yinyangren xisu wangshi yuan cishen chengwei nan'er" 陰陽人細訴往事願此身成為男兒 [The hermaphrodite reveals his/her past: Hopes to remain a man], August 15.

——. 1953w. "Yinyangren yiyou zhiyin" 陰陽人已有知音 [Hermaphrodite already has an admirer], August 24.

——. 1953x. "You yi yinyangren linjia fangxin daluan wulu yindu yulang" 又一陰陽人臨嫁芳心大亂無路引渡漁郎 [Another hermaphrodite anxious prior to marriage but unable to be treated], December 31.

——. 1953y. "Yuanlizheng bingyi tijian faxian yige yinyangren" 苑裏鎮兵役體檢發現一個陰陽人 [A hermaphrodite discovered in Yuanli district], December 10.

——. 1953z. "Zenkuan Xie Jianshun buwang abingjie" 贈款謝尖順不忘阿兵姐 [Donating to Xiejianshun: Never forget the female soldier], September 21.

——. 1953aa. "Zhongguo Kelisiding zuori wei beilai" 中國克麗絲汀昨日未北來 [The Chinese Christine did not arrive at Taipei yesterday], December 6.

——. 1954a. "Benshu fujia shunü biancheng rongma zhangfu" 本屬富家淑女變成戎馬丈夫 [A well-to-do lady turned into a heroic warrior], December 10.

——. 1954b. "Bianxingren Liu Ming tulu zhenqing qianwei shengnü shijia yuanshi yifu meimei" 變性人劉敏吐露真情前謂生女是假原是異父妹妹 [Transsexual Liu Min reveals the truth: Her daughter is actually her stepsister], December 13.

——. 1954c. "Gechu nüxing qiguan keneng bianzuo nanxiang" 割除女性器官可能變作男相 [Possible masculinization by the removal of female genitals], December 14.

——. 1954d. "Hebi ruci feijing" 何必如此費勁 [Why go through so much trouble], March 26.

——. 1954e. "Liu Ming wuzui" 劉敏無罪 [Liu Min is innocent], December 20.

——. 1954f. "Rensheng guji dengxianguo touhuan yisi yinyangren" 人生孤寂等閒過投繯縊死陰陽人 [A hermaphrodite kills herself for loneliness], March 16.

——. 1954g. "Xie Jianshun bingfang shenju jingdai shoushu ding yiyang" 謝尖順病房深居靜待手術定陰陽 [Xie Jianshun residing in the hospital room: Waiting for sex-determination surgery], February 15.

——. 1954h. "Xie Jianshun jiju nüertai qinsi mantou fenbaimian youju mishi yizeng xiu" 謝尖順極具女兒態青絲滿頭粉白面幽居密室益增羞 [Xie Jianshun appears extremely feminine], June 25.

——. 1954i. "Xie Jianshun youju daibian yijue de titai jiaorou" 謝尖順幽居待變益覺得體態嬌柔 [Xie Jianshun secluding herself and becoming feminized], March 18.

——. 1954j. "Yinyangren gaizao chenggong qitai zhuidi nannü mobian yishu huitian mingzhu chenghui" 陰陽人改造成功奇胎墜地男女莫辨醫術回天明珠呈輝 [The sex of a hermaphrodite successfully transformed through surgery], April 12.

——. 1955a. "Banfa yingbianfei qianjin zeng hongzhuang" 頒發應變費千金贈紅妝 [Awarding Xie for building her new feminine look], September 21.

——. 1955b. "Daqian shijie mei bianxing sanbing zuo xinniang" 大千世界美變性傘兵作新娘 [American paratrooper becomes a bride], July 29.

——. 1955c. "Fabiao Xie Jianshun mimi weifan yishifa buwu shidangchu" 發表謝尖順祕密違反醫師法不無失當處 [Publicizing Xie Jianshun's secret goes against the legal regulation of medicine], October 29.

——. 1955d. "He yi wei Xie Jianshun" 何以慰謝尖順 [How to console Xie Jianshun], October 29.

——. 1955e. "Liuying fang yiren nanjie yimianyuan" 柳營訪異人難結一面緣 [Searching for a stranger at Liuying: Difficult to see a face], May 11.

——. 1955f. "Lujun diyi zongyiyuan xuanbu Xie Jianshun shoushu chenggong" 陸軍第一總醫院宣佈謝尖順變性手術成功 [No. 1 Hospital announces the completion and success of Xie Jianshun's sex change operation], October 28.

——. 1955g. "Ruoguan yinan jixing fayu shenju nannü liangxing" 弱冠役男畸形發育身具男女兩性 [Irregular development of dual-sexed genitalia on conscripted soldier], October 6.

——. 1955h. "Sici shoushu yibian erchai Xie Jianshun bianxing jingguo" 四次手術易弁而釵謝尖順變性經過 [Male to female transformation after four surgeries: The sex change experience of Xie Jianshun], October 28.

——. 1955i. "Woguo yixue shishang de chuangju Xie Jianshun bianxing shoushu chenggong" 我國醫學史上的創舉謝尖順變性手術成功 [A new chapter in the nation's medical history: The success of Xie Jianshun's sex change surgery], August 31.

——. 1955j. "Xie Jianshun bianxing shoushu hou yishi liangzhou hou kexue yan quansheng" 謝尖順變性手術後醫師兩週後科學驗全身 [Doctors will examine Xie Jianshun's body scientifically two weeks after sex change operation], September 3.

——. 1955k. "Xie Jianshun bianxing shoushu jingguo duanqi zhengshi gongbu" 謝尖順變性手術經過短期正式公佈 [The details of Xie Jianshun's sex change surgery to be publicized shortly], September 1.

——. 1955l. "Xie Jianshun de nü'erjing qi xumei busheng xiunao huai jilü jing-nian fangjie" 謝尖順的女兒經棄鬚眉不勝羞惱懷積慮經年方解 [Xie Jianshun's anxiety about menstruation problems finally resolved], September 2.

——. 1955m. "Xie Jianshun luoxiong zhaopian zhengshi xi gongpin" 謝尖順裸胸照片證實係贗品 [Xie Jianshun's half-nude photo: A hoax], September 10.

——. 1955n. "Xie Jianshun younan biannü shoushu yi jiejin chenggong" 謝尖順由男變女手術已接近成功 [The surgery of Xie Jianshun's male-to-female trans-formation almost complete], January 9.

——. 1955o. "Xu Zhenjie de mimi" 徐振傑的祕密 [Xu Zhenjie's secret], May 9.

——. 1955p. "Yinan Shi tijian jianshi yinyangren" 役男施體檢見是陰陽人 [Con-scripted soldier discovered to be a hermaphrodite], September 23.

——. 1955q. "Zilian feinan yi feinü huzhang yinzhong dong chunqing" 自憐非男亦非女虎帳隱衷動春情 [Self-loathing of sexual ambiguity], May 7.

——. 1956. "Xiri dabing tongzhi jianglai fulian huiyuan" 昔日大兵同志將來婦聯會員 [Former soldier comrade: A future member of women's association], October 10.

——. 1958a. "Xiri shachang zhanshi jinze jingru chuzi" 昔日沙場戰士今則靜如處子 [Former battle warrior: A present quiet virgin], September 17.

——. 1958b. "Yishi chuangzao nüren wunian duding qiankun" 醫師創造女人五年篤定乾坤 [Doctors building a woman: Sex determined in five years], September 4.

——. 1959a. "Xiaoyangnü Fu Xiuxia zhu Tatong Jiaoyangyuan" 小養女傅秀霞住大同教養院 [Little girl Fu Xiuxia boarding Tatong Relief Institute], June 7.

——. 1959b. "Xie Jianshun guanxiongfu" 謝尖順慣雄伏 [Xie Jianshun adjusting to feminine psyche], November 11.

Lillie, Frank R. 1917. "The Free-Martin: A Study of the Action of Sex Hormones in the Foetal Life of Cattle." *Journal of Experimental Zoology* 23, no. 2: 371–423.

Lin Jianqun 林健群. 1997. "Wan Qing kehuanxiaoshuo zhi yanjiu (1904–1911)" 晚清科幻小說之研: 1904–1911 [On science fictions in the late Qing, 1904–1911]. M.A. thesis, National Chung Cheng University.

Lin Shifang 林實芳. 2008. "Bainian duidui, zhihen kanbujian: Taiwan falü jiafengxia de nünü qinmi guanxi" 百年對對，只恨看不見: 台灣法律夾縫下的女女親密關係 [One hundred years of secrecy: Female-female intimacy in the fissures of Taiwanese law]. M.A. thesis, National Taiwan University.

Ling Haicheng 凌海成. 2003. *Zhongguo zuihou yige taijian* 中國最後一個太監 [The last eunuch in China]. Hong Kong: Heping tushu.

Lionnet, Françoise, and Shu-mei Shih. 2005. "Introduction: Thinking through the Minor, Transnationally." In *Minor Transnationalism*, ed. Françoise Lionnet and Shu-mei Shih, 1–23. Durham, N.C.: Duke University Press.

Liu Ben-Lih 劉本立 and Kai Liu 劉愷. 1953. "True Hermaphroditism: Report of Two Cases." *Chinese Medical Journal* 71: 148–54.

Liu Dao-Jie 劉道捷. 1993. "Biannan biannü bian bian bian" 變男變女變變變 [Transsexualism and sex-reassignment surgery in Taiwan]. M.A. thesis, National Taiwan University.

Liu, Jen-peng, and Ding Naifei. 2005. "Reticent Poetics, Queer Politics." *Inter-Asia Cultural Studies* 6, no. 1: 30–55.

Liu, Lydia. 1995. *Translingual Practice: Literature, National Culture, and Translated Modernity—China, 190–1937*. Stanford, Calif.: Stanford University Press.

——, ed. 1999. *Tokens of Exchange: The Problem of Translation in Global Circulations*. Durham, N.C.: Duke University Press.

——. 2004. *The Clash of Empires: The Invention of China in Modern World Making*. Cambridge, Mass.: Harvard University Press.

Liu, Lydia, Rebecca Karl, and Dorothy Ko, eds. 2013. *The Birth of Chinese Feminism: Essential Texts in Transnational Theory*. New York: Columbia University Press.

Liu, Petrus. 2015. *Queer Marxism in Two Chinas*. Durham, N.C.: Duke University Press.

Liu Piji 劉丕基. (1928) 1935. *Renjian wujie de shengwu* 人間誤解的生物 [Common misinterpretations of biology]. Shanghai: Commercial Press.

Liu, Shiyung 劉士永. 2009. *Prescribing Colonization: The Role of Medical Practice and Policy in Japan-Ruled Taiwan*. Ann Arbor, Mich.: Association for Asian Studies.

——. 2010. "Zhanhou Taiwan yiliao yu gongwei tizhi de bianqian" 戰後台灣醫療與公衛體制的變遷 [The transformation of medical care and public health regime in postwar Taiwan] *Huazhong shifan daxue xuebao* 華中師範大學學報 49, no. 4: 76–83.

——. 2017. "Transforming Medical Paradigms in 1950s Taiwan." *East Asian Science, Technology and Society: An International Journal* 11, no. 4: 477–97.

Liu Zhenqing 劉振卿. 1936. "Qingdai zhi taijian" 清代之太監 [The eunuchs of the Qing dynasty]. *Shibao banyuekan* 實報半月刊 14: 67–70.

Logan, Cheryl A. 2013. *Hormones, Heredity, and Race: Spectacular Failure in Interwar Vienna*. New Brunswick, N.J.: Rutgers University Press.

Loos, Tamara. 2006. *Subject Siam: Family, Law, and Colonial Modernity*. Ithaca, N.Y.: Cornell University Press.

Loshitzky, Yosefa, and Raya Meyuhas. 1992. " 'Ecstasy of Difference:' Bertolucci's *The Last Emperor*." *Cinema Journal* 31, no. 2: 26–44.

Louie, Kam. 2002. *Theorising Chinese Masculinity: Society and Gender in China*. Cambridge: Cambridge University Press.

Louie, Kam, and Morris Low, eds. 2003. *Asian Masculinities: The Meaning and Practice of Manhood in China and Japan*. London: Routledge.

Lu Huaxin 陸華信. 1955. *Shaonan shaonü xingzhishi* 少男少女性知識 [Sexual knowledge for young men and women]. Hong Kong: Xuewen shudian.

Lü Wenhao 呂文浩. 2006. *Pan Guangdan tuzhuan* 潘光旦圖傳. Wuhan: Hubei renmin chubanshe.

Luesink, David. 2015. "State Power, Governmentality, and the (Mis)Remembrance of Chinese Medicine." In *Historical Epistemology and the Making of Modern Chinese Medicine*, ed. Howard Chiang, 160–87. Manchester, U.K.: Manchester University Press.

"Luoma qiwen: Liangnüxing biannanzi" 羅馬奇聞: 兩女性變男子 [News from Rome: Two women become men]. 1947. *Xinfunü yuekan* 新婦女月刊 [New woman monthly] 14: 16.

Lutz, Jessie Gregory. 1971. *China and the Christian Colleges, 1850–1950*. Ithaca, N.Y.: Cornell University Press.

Lynch, Michael. 1998. "The Production of Scientific Images: Vision and Re-vision in the History, Philosophy, and Sociology of Science." *Communication & Cognition* 31, no. 2–3: 213–28.

Lynch, Michael, and Steve Woolgar. 1988. "Introduction: Sociological Orientations to Representational Practice in Science." *Human Studies* 11, no. 2–3: 259–83.

Ma Yongying 马永赢 and Gang Li 李岗. 2005. "Tantan Yangling chutu de 'huanzheyong'" 谈谈阳陵出土的"宦者俑 [Discussions of the eunuch statuettes excavated at Yangling]. *Kaogu yu wenwu* 考古与文物 4: 71–72.

Ma, Zhiying. 2014. "An 'Iron Case' of Civilization? Missionary Psychiatry, 'Chinese Family,' and a Colonial Dialect of Enlightenment." In *Psychiatry and Chinese History*, ed. Howard Chiang, 91–110. London: Pickering and Chatto.

Mao Dun 矛盾. 1935. "Yaojia nübiannan de gushi" 姚家女變男的故事 [The story of female-to-male transformation of Yao's]. *Manhua shenghuo* 漫畫生活 [Cartoon life] 8.

Mao, Nathan, and Ts'un-Yan Liu. 1977. *Li Yü*. Boston: Twayne.

Martin, Emily. 1991. "The Egg and the Sperm: How Science Has Constructed a Romance Based on Stereotypical Male-Female Roles." *Signs* 16, no. 3: 485–501.

Martin, Fran. 2003. *Situating Sexualities: Queer Representation in Taiwanese Fiction, Film and Public Culture*. Hong Kong: Hong Kong University Press.

——. 2010. *Backward Glances: Contemporary Chinese Culture and the Female Homoerotic Imaginary*. Durham, N.C.: Duke University Press.

———. 2014. "Transnational Queer Sinophone Cultures." In *Routledge Handbook of Sexuality Studies in East Asia*, ed. Mark McLelland and Vera Mackie, 35–48. London: Routledge.

Massad, Joseph. 2007. *Desiring Arabs*. Chicago: University of Chicago Press.

Matignon, Jean-Jacques. 1896. "Les Eunuches du Palais Impérial de Pékin." *Archives Cliniques de Bordeaux* 5: 193–204.

———. 1899. *Superstition, crime et misère en Chine: Souvenirs de biologie sociale*. Lyon: A. Storck & Cie.

Matta, Christina. 2005. "Ambiguous Bodies and Deviant Sexualities: Hermaphrodites, Homosexuality, and Surgery in the United States, 1850–1904." *Perspectives in Biology and Medicine* 48, no. 1: 74–83.

May, Elaine Tyler. 1988. *Homeward Bound: American Families in the Cold War Era*. New York: Basic Books.

McLaren, Angus. 2007. *Impotence: A Cultural History*. Chicago: University of Chicago Press.

McMillan, Joanna. 2006. *Sex, Science and Morality in China*. New York: Routledge.

"Medicine: Girls into Boys." 1934. *Time*, August 27.

Meeker, Martin. 2006. *Contacts Desired: Gay and Lesbian Communications and Community, 1940s–1970s*. Chicago: University of Chicago Press.

Mei Shi 梅拾. 1947. "Nübiannan" 女變男 [Female to male]. *Shengli* 勝利 [Victory] 125: 5.

Mei Xianmao 梅顯懋. 1997. *Luori wanzhong: Qingdai taijian zhidu* 落日晚鍾: 清代太監制度 [The system of Qing-dynasty eunuchs]. Shenyang: Liaohai chubanshe.

Meijer, Marinus J. 1985. "Homosexual Offences in Ch'ing Law." *T'oung Pao* 71: 109–33.

Meng Yue. 2006. *Shanghai and the Edges of Empires*. Minneapolis: University of Minnesota Press.

Métailé, Georges. 2001. "The *Bencao Gangmu* of Li Shizhen: An Innovation for Natural History?" In *Innovation in Chinese Medicine*, ed, Elisabeth Hsu, 221–61. Cambridge: Cambridge University Press.

Meyer, Richard J. 2005. *Ruan Ling-Yu: The Goddess of Shanghai*. Hong Kong: Hong Kong University Press.

Meyerowitz, Joanne. 1998. "Sex Change and the Popular Press: Historical Notes on Transsexuality in the United States, 1930–1955." *GLQ* 4, no. 2: 159–87.

———. 2001. "Sex Research at the Borders of Gender: Transvestites, Transsexuals, and Alfred C. Kinsey" *Bulletin of the History of Medicine* 75: 72–90.

———. 2002. *How Sex Changed: A History of Transsexuality in the United States*. Cambridge, Mass.: Harvard University Press.

Millant, Richard. 1908. *Les Eunuches à travers les ages*. Paris: Vigot.

Minton, Henry. 2002. *Departing from Deviance: A History of Homosexual Rights and Emancipatory Science in America*. Chicago: University of Chicago Press.

Mitamura, Taisuke. 1970. *Chinese Eunuchs: The Structure of Intimate Politics*. Trans. Charles A. Pomeroy. Tokyo: Charles E. Tuttle.

Miyasita, Saburo. 1967. "A Link in the Westward Transmission of Chinese Anatomy in the Later Middle Ages." *Isis* 58, no. 4: 486–90.

Mo 漠, trans. 1936. "Tongxing'ai de yanjiu he fangzhi" 同性愛的研究何防止 [The study and prevention of homosexuality]. *XKX* 2, no. 4: 15–26.

Morache, Georges. 1869 *Pékin et ses habitants: Étude d'hygiene*. Paris: J.-B. Baillière.

Morange, Michel. 2011. "What History Tells Us XXIV: The Attempt of Nikolai Koltzoff (Koltsov) to Link Genetics, Embryology and Physical Chemistry." *Journal of Biosciences* 36, no. 2: 211–14.

Najmabadi, Afsaneh. 2005. *Women with Mustaches and Men Without Beards: Gender and Sexual Anxieties in Iranian Modernity*. Berkeley: University of California Press.

——. 2014. *Professing Selves: Transsexuality and Same-Sex Desire in Contemporary Iran*. Durham, N.C.: Duke University Press.

Nan Xi 南溪. 1927. "Xingyu tongxin" 性育通信 [Sex education letters]. *XWH* 1, no. 3: 66–67.

"Nanbiannü zhi qiwen" 男變女之奇聞 [News of male to female]. 1934. *Sheying huabao* 攝影畫報 [Photography pictorial] 10, no. 18: 5.

Nandy, Ashis. 1984. *The Intimate Enemy: Loss and Recovery of Self Under Colonialism*. New York: Oxford University Press.

Nappi, Carla. 2009. *The Monkey and the Inkpot: Natural History and Its Transformations in Early Modern China*. Cambridge, Mass.: Harvard University Press.

Nedostup, Rebecca. 2009. *Superstitious Regimes: Religion and the Politics of Chinese Modernity*. Cambridge, Mass.: Harvard University Press.

New York Daily News. 1952. "Ex-GI Becomes Blonde Beauty." December 1.

New York Times. 1931. "Dr. Robert Coltman, Royalty's Friend, Dies; Was Physician to the Former Imperial Family of China, Where He Lived for Forty Years." November 5.

Ng, Jason. 2016. *Umbrellas in Bloom: Hong Kong's Occupancy Movement Uncovered*. Hong Kong: Blacksmith Books.

Ng, Vivian. 1989. "Homosexuality and State in Late Imperial China." In *Hidden from History: Reclaiming the Gay and Lesbian Past*, ed. Martin B. Duberman, Martha Vicinus, and George Chauncey, 76–89. New York: New American Library.

"Nübian nanshen" 女變男身 [Woman to man]. 1934. *Juxing* 聚星 2, no. 3: 10.

"Nübiannan" 女變男 [Female to male]. 1934. *Xinzhonghua zazhi* 新中華雜誌 [New China magazine] 2, no. 6: 52.

"Nübiannan jiujing sheme daoli" 女變男究竟什麼道理 [What is the explanation for female-to-male sex transformation]. 1934. *Kexue de Zhongguo* 科學的中國 [Scientific China] 4,no. 9: 398.

"Nühua nanshen" 女化男身 [Woman to man]. 1935. *Jianzheng zhoukan* 監政週刊 110: 11–12.

"Nühuanan" 女化男 [Female to male]. 1913. *Yixue shijie* 醫學世界 [Medical world] 25: 8.

"Nüxing biannan de jiepo" 女性變男的解剖 [The anatomy of female-to-male transformation]. 1935. *Xinghua zhoukan* 興華週刊 32, no. 12: 16–18.

"Obituary: J. J. Matignon." 1900. *British Medical Journal* 2, no. 2065: 268.

Ochiai Naobumi and Haga Yaichi. 1927. *Fountain of Words: Comprehensive Japanese Dictionary (Gensen nihon dai jiten)*. Tokyo: Nihon Tosho Senta.

Oosterhuis, Harry. 2000. *Stepchildren of Nature: Krafft-Ebing, Psychiatry, and the Making of Sexual Identity*. Chicago: University of Chicago Press.

Oram, Alison. 2007. *Her Husband Was a Women! Women's Gender-Crossing in Modern British Culture*. London: Routledge.

Oudshoorn, Nelly. 1994. *Beyond the Natural Body: An Archeology of Sex Hormones*. New York: Routledge.

Pan Guangdan 潘光旦. (1927) 1994. "Jinri zhi xingjiaoyu yu xingjiaoyuzhe" 今日之性教育與性教育者 [Today's sex education and sex educator]. *Shishi xinbao xuedeng* 時事新報學燈 [Current events newsletter], May 5, June 24, and June 14. Reprinted in *PGDWJ*, vol. 1, 401–12.

——. 1927. "'Xin wenhua' yu jiakexue" 『新文化』與假科學 ['New Culture' and fake science]. *XZZ* 1, no. 2: 1–8.

——. (1932) 1994. "Lujiang Huangshi de yinyuecai" 鷺江黃氏的音樂才 [The musical talent of Huang's in Lujiang]. *Yousheng yuekan* 優生月刊 [Eugenics monthly] 2, no. 2. Reprinted in *PGDWJ*, vol. 8, 409–40.

——. (1934) 1994. *Xing de Jiaoyu* 性的教育 [Sexual education]. Shanghai: Qinnian xiehue shuju. Reprinted in *PGDWJ*, vol. 12, 1–99.

——. (1936) 1994. "Xing'ai zai jinri" 性愛在今日 [Sexuality today]. *Huanian* 華年 5 (45, 49, and 50), November 21, December 19, and December 26. Reprinted in *PGDWJ*, vol. 9, 370–87.

——. (1941a) 1994. *Feng Xiaoqin: Yijian yinglian zhi yanjiu* 馮小青: 一件影戀之研究 [Feng Xiaoqin: A study of unconscious desire]. Shanghai: Commercial Press. Reprinted in *PGDWJ*, vol. 1, 1–66.

——. (1941b) 1994. *Zhongguo lingren xieyuan zhi yanjiu* 中國伶人血緣之研究 [Research on the Pedigrees of Chinese Actors]. Shanghai: Commercial Press. Reprinted in *PGDWJ*, vol. 2, 73–303.

——, trans. (1946) 1994. *Xing xinlixue* 性心理學 [Psychology of Sex]. Reprinted in *PGDWJ*, vol. 12, 197–714.

Park, Katharine. 2006. *Secrets of Women: Gender, Generation, and the Origins of Human Dissection*. New York: Zone Books.

Pauwels, Luc. 2006. "Introduction: The Role of Visual Representation in the Production of Scientific Reality." In *Visual Cultures of Science: Rethinking Representational Practices in Knowledge Building and Science Communication*, ed. Luc Pauwels, vii–xix. Hanover, N.H.: Dartmouth College Press.

Peng Hsiao-yen. 2001. "Wusi de xin xingdaode: Nüxing qingyu lunshu yu jian-gou minzu guojia" 五四的新性道德: 女性情慾論述與建構民族國家 [The new sexual ethics of May Fourth: The discourse of female sexuality and nation-building]. In *Haishang shou qingyu* 海上說情慾 [Talking about love and desire on the sea], 1–26. Taipei: Institute of Chinese Literature & Philosophy, Academia Sinica.

——. 2002. "*Sex Histories*: Zhang Jingsheng's Sexual Revolution." In *Critical Studies: Feminism/Femininity in Chinese Literature*, ed. Peng-hsiang Chen and Whitney Crothers Dilley, 159–77. Amsterdam: Rodopi.

Peschel, Enid Rhodes, and Richard E. Peschel. 1987. "Medical Insights into the Castrati in Opera." *American Scientist* 75: 581–82.

Pflugfelder, Gregory M. 1999. *Cartographies of Desire: Male-Male Sexuality in Japanese Discourse*. Berkeley: University of California Press.

Ping 平, trans. 1936. "Jiachong huo xide de tongxing'ai de tezhi" 假充或習得的同性愛的特質 [The characteristics of fake or acquired homosexuality]. *XKX* 2, no. 4: 9–11.

Poovey, Mary. 1990. "Speaking of the Body: Mid-Victorian Constructions of Female Desire." In *Body/Politics: Women and the Discourses of Science*, ed. Mary Jacobus, Evelyn Fox Keller, and Sally Shuttleworth, 29–46. New York: Routledge.

Porter, Roy, and Lesley Hall. 1995. *The Facts of Life: The Creation of Sexual Knowledge in Britain, 1650–1950*. New Haven, Conn.: Yale University Press.

Pujia 溥佳 and Pujie 溥傑. 1984. *Wan Qing gongting shenghuo jianwen* 晚清宮廷生活見聞 [Life in Late-Qing Imperial Palace]. Taipei: Juzhen shuwu.

Pusey, James R. 1983. *China and Charles Darwin*. Cambridge, Mass.: Harvard University Press.

——. 1998. *Lu Xun and Evolution*. Albany: State University of New York Press.

Qi Fan 豈凡. 1935. "Nübiannan yu nübannan" 女變男與女扮男 [Female-to-male and female transvestism]. *Renyan zhoukan* 人言週刊 2, no. 9: 171.

Qi Ren 齊人. 1929. "Yu youren lun xingxue ji xingjiaoyu shu" 與友人論性學及性教育書 [Discussing sexology and sex education books with friends]. *Fujian pinglun* 福建評論 3: 19–23.

Qian Qian 倩倩. 1927. "Disan zhongshui de yanjiu" 第三種水的研究 [Research on the third kind of water]. *XZZ* 1, no. 2: 1–12.

Qin Xin 芹心. 1927. "Tongxing lian'ai taolun" 同性戀愛討論 [Discussion on homosexuality]. *XWH* 1, no. 3: 63–66.

Qing Gui 慶桂 et al., ed. 1994. *Guochao gongshi xubian* 國朝宮史續編 [A supplemental history of the palace of the reigning dynasty]. Repr. ed. Beijing: Beijing guji chubanshe.

"Qinghuangshi quzhu taijian" 清皇室驅逐太監 [The Qing expels eunuchs]. 1923. *Xinghua* 興華 20, no. 29: 17–20.

Qiu Jun 丘畯. 1926. "Dongwu de 'tongxing lianai'" 動物的『同性戀愛』 ['Homosexuality' in animals]. *Shengwuxue zazhi* 生物學雜誌 [Biology magazine] 1: 69–75.

Qiu Shi 裘湜, ed. 1990. "Biekai shengdemian taijian zuotanhui" 別開生的面太監座談會 [A forum with eunuchs]. *ZJWX* 57, no. 4: 124–28.

Qiu Yuan 秋原 (possibly Hu Qiuyuan, 胡秋原), trans. 1929. "Tongxinglian'ai lun" 同性戀愛論 [On same-sex romantic love]. *Xin nüxing* 新女性 [New woman] 4, no. 4: 513–34; and no. 5: 605–28.

Qu Xianguai 曲綫怪. 1931. "Xingbian" 性變 [Sex change]. *Beiyang huabao* 北洋畫報 [Beiyang pictorial] 13, no. 648: 3.

Rawski, Evelyn. 2001. *The Last Emperors: A Social History of Qing Imperial Institutions.* Berkeley: University of California Press.

Reed, Christopher. 2004. *Gutenberg in Shanghai: Chinese Print Capitalism, 1876–1937.* Vancouver: University of British Columbia Press.

Ren Baitao 任白濤, and Jianxu Yi 易家鉞. 1925. *Qingnian zhi xing de weisheng ji daode* 青年之性的衛生及道德 [The sexual hygiene and morals of adolescents]. Shanghai: Commercial Press.

Rheinberger, Hans-Jörg. 1997. *Toward a History of Epistemic Things: Synthesizing Proteins in the Test Tube.* Stanford, Calif.: Stanford University Press.

——. 2010a. *An Epistemology of the Concrete: Twentieth-Century Histories of Life.* Durham, N.C.: Duke University Press.

——. 2010b. *On Historicizing Epistemology: An Essay.* Stanford, N.C.: Stanford University Press.

Rhoads, Edward J. M. 2000. *Manchus & Han: Ethnic Relations and Political Power in Late Qing and Early Republican China, 1861–1928.* Seattle: University of Washington Press, 2000.

Richardson, Sarah S. 2013. *Sex Itself: The Search for Male and Female in the Human Genome.* Chicago: University of Chicago Press.

Ringrose, Kathryn M. 2003. *The Perfect Servant: Eunuchs and the Social Construction of Gender in Byzantium.* Chicago: University of Chicago Press.

——. 2007. "Eunuchs in Historical Perspective." *History Compass* 5, no. 2: 495–506.

Robinson, David. 2001. *Bandits, Eunuchs and the Son of Heaven: Rebellion and the Economy of Violence in Mid-Ming China.* Honolulu: University of Hawai'i Press.

Robinson, Paul. 1973. "Havelock Ellis and Modern Sexual Theory." *Salmagundi* 21: 27–62.

Rocha, Leon Antonio. 2008. "Zhang Jingsheng's Utopian Project." Paper presented at the 81st Annual Meeting of the American Association for the History of Medicine, Rochester, New York, April 11–13.

——. 2010. "*Xing*: The Discourse of Sex and Human Nature in Modern China." *Gender and History* 22, no. 3: 603–28.

Rofel, Lisa. 2007. *Desiring China: Experiments in Neoliberalism, Sexuality, and Public Culture.* Durham, N.C.: Duke University Press.

Rogaski, Ruth. 2004. *Hygienic Modernity: Meanings of Health and Disease in Treaty-Port China.* Berkeley: University of California Press.

Rojas, Carlos. 2015. *Homesickness: Culture, Contagion, and National Reform in Modern China.* Cambridge, Mass.: Harvard University Press.

Rosario, Vernon, ed., 1997. *Science and Homosexualities.* New York: Routledge.

——. 2002. *Homosexuality and Science: A Guide to the Debates.* Santa Barbara, Calif.: ABC-CLIO.

Ruan, Fang Fu. 1991. *Sex in China: Studies in Sexology in Chinese Culture.* New York: Plenum.

Sakai, Naoki. 1988. "Modernity and Its Critique: The Problem of Universalism and Particularism." *South Atlantic Quarterly* 87, no. 3: 475–504.

Sakamoto, Hiroko. 2004. "The Cult of 'Love and Eugenics' in May Fourth Movement Discourse." *positions: east asia cultures and critique* 12: 329–76.

San Francisco Lesbian and Gay History Project. 1989. " 'She Even Chewed Tobacco:' A Pictorial Narrative of Passing Women in America." In *Hidden from History: Reclaiming the Gay and Lesbian Past,* ed. Martin Duberman, Martha Vicinus, and George Chauncey, 183–94. New York: New American Library.

Sandhaus, Derek. 2011. Introduction to *Décadence Mandchoue: The China Memoirs of Edmund Trelawny Backhouse,* by Edmund Trelawny Backhouse, ix–xxv. Ed. Derek Sandhaus. Chicago: Earnshaw Books.

Sang, Tze-lan Deborah. 2003. *The Emerging Lesbian: Female Same-Sex Desire in Modern China.* Chicago: University of Chicago Press.

Schiebinger, Londa. 1989. *The Mind Has No Sex? Women in the Origin of Modern Science.* Cambridge, Mass.: Harvard University Press.

Schmalzer, Sigrid. 2008. *The People's Peking Man: Popular Science and Human Identity in Twentieth-Century China.* Chicago: University of Chicago Press.

Schneider, Laurence A. 2003. *Biology and Revolution in Twentieth-Century China.* Lanham, Md.: Rowman & Littlefield.

Scholz, Piotr O. 2001. *Eunuchs and Castrati: A Cultural History.* Trans. John A. Broadwin and Shelley L. Frisch. Princeton, N.C.: Markus Wiener.

Schwartz, Benjamin. 1964. *In Search of Wealth and Power: Yan Fu and the West.* Cambridge, Mass.: Harvard University Press.

Scott, Charles, and Trent Holmberg. 2003. "Castration of Sex Offenders: Prisoners' Rights Versus Public Safety." *Journal of the American Academy of Psychiatry and Law* 31: 502–9.

Scott, Joan. 1991. "The Evidence of Experience." *Critical Inquiry* 17, no. 4: 773–97.

Se Lu. 1924. "Zuijin shiniannei funüjie de huigu" [A review of women in the past ten years]. *FNZZ* 10, no. 1: 21–22.

Sedgwick, Eve Kosofsky. 1985. *Between Men: English Literature and Male Homosocial Desire*. New York: Columbia University Press.

——. 1990. *Epistemology of the Closet*. Berkeley: University of California Press.

Sengoopta, Chandak. 1998. "Glandular Politics: Experimental Biology, Clinical Medicine, and Homosexual Emancipation in Fin-de-Siècle Central Europe." *Isis* 89, no. 3: 445–73.

——. 2000. *Otto Weininger: Sex, Science, and Self in Imperial Vienna*. Chicago: University of Chicago Press.

——. 2006. *The Most Secret Quintessence of Life: Sex, Glands, and Hormones, 1850–1950*. Chicago: University of Chicago Press.

Serlin, David. 1995. "Christine Jorgensen and the Cold War Closet." *Radical History Review*, no.62: 136–65.

Shan Zai 善哉. 1911. "Funü tongxing zhi aiqing" 婦女同性之愛情 [Same-sex erotic love between women]. *Funü shibao* 婦女時報 [Women's times] 1, no. 7: 36–38.

Shapiro, Hugh. 1998. "The Puzzle of Spermatorrhea in Republican China." *positions: east asia cultures critique* 6, no. 3: 551–96.

Shen Chichun 沈霽春. 1934. *Xing de shenghuo* 性的生活 [The life of sex]. Shanghai: Shijie shuju.

Shen Zemin 沈澤民, trans. 1923. "Tongxing'ai yu jiaoyu" 同性愛與教育 [Same-sex love and education]. *Jiaoyu zazhi* 教育雜誌 [Education magazine] 15, no. 8: 22115–124.

Shenbao 申報. 1930. "Ershiyi sui nanzi hubian er nüzi" 二十一歲男子乎變而女子 [21-year-old man suddenly becomes a woman], October 29.

——.1935a. "Nühua nanshen zhi Yao Jinping yeyi di Hu" 女化男身之姚錦屏業已抵滬 [The woman-to-man Yao Jinping has arrived in Shanghai], March 17.

——. 1935b. "Yao Jinping huanan fangwen ji" 姚錦屏化男訪問記 [Interview with Yao Jinping about her sex transformation], March 18.

——. 1935c. "Yao Jinping wanquan nüershen" 姚錦屏完全女兒身 [Yao Jinping is a complete woman], March 21.

——. 1935d. "Yao Jinping xi fushen huananti" 姚錦屏係副腎化男體 [Yao Jinping's masculinization of the adrenal gland], March 19.

——. 1935e. "Yao Jinping zuoru yiyuan jianyan" 姚錦屏昨入醫院檢驗 [Yao Jinping entered the hospital yesterday for examination], March 20.

Shi Fen 世芬. 1927. "Tongxun" 通訊 [Letters]. *XWH* 1, no. 1: 51.

Shi Kekuan 施克寬. 1988. *Zhongguo huanguan mishi: renzao de disanxing* 中國宦官祕史: 人造的第三性 [The secret history of Chinese eunuchs: The man-made third sex]. Beijing: Zhongguo xiju chubanshe.

Shih, Shu-mei. 2001. *The Lure of the Modern: Writing Modernism in Semicolonial China, 1917–1937*. Berkeley: University of California Press.

——. 2007. *Visuality and Identity: Sinophone Articulations Across the Pacific*. Berkeley: University of California Press.

——. 2011. "The Concept of the Sinophone." *PMLA* 126, no. 3: 709–18.

Shih, Shu-mei, Chien-hsin Tsai, and Brian Bernards, ed. 2013. *Sinophone Studies: A Critical Reader*. New York: Columbia University Press.

Shin, Gi-wook, and Michael Robinson, eds. 2001. *Colonial Modernity in Korea*. Cambridge, Mass.: Harvard University Press.

Shinian Kanchai 十年砍柴. 2007. *Huangdi, wenchen he taijian: Mingchao zhengju de "sanjiaolian"* 皇帝、文臣和太監: 明朝政局的"三角戀" [The emperor, scholar officials, and eunuchs: The triangular relationship of the political situation in the Ming dynasty]. Nanning: Guangxi Normal University Press.

"Shinü zhilei de qizi" 石女之類的妻子 [A stone maiden wife]. 1948. *Xifeng* 西風 [West wind] 112: 365.

Shiu, Stephen, et al. 2011. *3-D Sex and Zen: Extreme Ecstasy*. Dir. Christopher Suen. Hong Kong: One Dollar Production.

Si Ke 司克. 1928. "Bianxing" 變性 [Sex change]. *Bairi xinwen fukan* 白日新聞副刊 [Day newspaper supplement] 30: 20.

Sinha, Mrinalini. 2006. *Specters of Mother India: The Global Restructuring of Empire*. Durham, N.C.: Duke University Press.

Sivin, Nathan. 1987. *Traditional Medicine in Contemporary China*. Ann Arbor: Center for Chinese Studies, University of Michigan.

Skidmore, Emily. 2011. "Constructing the 'Good Transsexual': Christine Jorgensen, Whiteness, and Heteronormativity in the Mid-Twentieth Century Press." *Feminist Studies* 37, no. 2: 270–300.

Small, Edward. 1970. *The Christine Jorgensen Story*. Dir. Irving Rapper. Los Angeles: United Artists.

Smith-Rosenberg, Carroll. 1985. *Disorderly Conduct: Visions of Gender in Victorian America*. New York: Alfred Knopf.

Smocovitis, Betty. 1996. *Unifying Biology: The Evolutionary Synthesis and Evolutionary Biology*. Princeton, N.J.: Princeton University Press.

Sommer, Matthew H. 1997. "The Penetrated Male in Late Imperial China: Judicial Constructions and Social Stigma." *Modern China* 23, no. 1: 140–80.

——. 2000. *Sex, Law, and Society in Late Imperial China*. Stanford, Calif.: Stanford University Press.

Soothill, William Edward, and Lewis Hodous. 1937. *A Dictionary of Chinese Buddhist Terms, with Sanskrit and English Equivalents and a Sanskrit-Pali Index.* London: Kegan Paul.

Spiegel, Gabrielle M. 2009. "The Task of the Historian." *American Historical Review* 114, no. 1: 1–15.

SSD. 1927. "Xingyu tongxin" 性育通信 [Sex education letters]. *XWH* 1, no. 3: 71–73.

Steinach, Eugen. 1912. "Willkürliche Umwandlung von Säugetiermännchen in Tiere mit ausgeprägt weiblichen Geschlechtscharacteren und weiblicher Psyche." *Pflügers Archiv* 144: 71.

——. 1913. "Feminierung von Männchen und Maskulierung von Weibchen." *Zentralblatt für Physiologie* 27: 717.

——. 1940. *Sex and Life: Forty Years of Biological and Medical Experiments.* New York: Viking.

Stent, G. Carter. 1877. "Chinese Eunuchs." *Journal of the North China Branch of the Royal Asiatic Society*, no. 11: 143–84.

Stern, Alexandra Minna. 2005. *Eugenic Nation: Faults and Frontiers of Better Breeding in Modern America.* Berkeley: University of California Press.

Stolberg, Michael. 2003. "The Anatomy of Sexual Difference in the Sixteenth and Early Seventeenth Centuries." *Isis* 94: 274–99.

Stryker, Susan. 2008. *Transgender History.* Berkeley, Calif.: Seal Press.

Stryker, Susan, Paisley Currah, and Lisa Jean Moore. 2008. "Introduction: Trans-, Trans, or Transgender?" *Women's Studies Quarterly* 36, no. 3–4: 11–22.

Su Ya 素雅. 1927. "'Xing' zhishi pupian le jiu meiyo 'qiangjian'" '性' 智識普遍了就沒有 '強姦' [Rape will be gone after sex education has become common]. *XWH* 1, no. 2: 104.

Su Yizhen 蘇儀貞. 1935. *Nüxing weisheng changshi* 女性衛生常識 [Hygiene manual for women]. Shanghai: Zhonghua shuju.

Sullivan, Louis. 1990. *From Female to Male: The Life of Jack Bee Garland.* Boston: Alyson.

Sun Yaoting 孫耀廷. 1990a. "Wozai mingguo zuo taijian" 我在民國作太監 [Being a eunuch in the Republican period]. *ZJWX* 57, no. 2: 113–33.

——. 1990b. "Wozai mingguo zuo taijian." *ZJWX* 57, no. 3: 115–34.

Suryadinata, Leo, ed. 2005. *Admiral Zheng He and Southeast Asia.* Singapore: Institute of Southeast Asian Studies.

Suzuki, Michiko. 2009. *Becoming Modern Women: Love and Female Identity in Prewar Japanese Literature and Culture.* Stanford, Calif.: Stanford University Press.

——. 2015. "The Translation of Edward Carpenter's *Intermediate Sex* in Early Twentieth Century Japanese Feminist Fiction and Politics." In *Sexology and*

Translation: Cultural and Scientific Encounters across the Modern World, 1880–1930, ed. Heike Bauer, 197–215. Philadelphia: Temple University Press.

Szonyi, Michael. 1998. "The Cult of Hu Tianbao and the Eighteenth-Century Discourse of Homosexuality." *Late Imperial China* 19, no. 1: 1–25.

Szto, Peter Paul. 2002. "The Accommodation of Insanity in Canton, China: 1857–1935." PhD diss., University of Pennsylvania.

——. 2014. "Psychiatric Space and Design Antecedents: The John G. Kerr Refuge for the Insane." In *Psychiatry and Chinese History*, ed. Howard Chiang, 71–90. London: Pickering and Chatto.

Taiwan minsheng ribao 台灣民聲日報. 1953a. "Gejun youshe xuanwei zhongxingren Xie Jianshun" 各軍友社宣慰中性人謝尖順 [Soldiers console intersex Xie Jianshun], September 24.

——. 1953b. "Xie Jianshun pazuo shiyanpin jujue beishang jiaozhi" 謝尖順怕做試驗品拒絕北上矯治 [Xie Jianshun afraid of becoming an experimental subject: Refuses to relocate to Taipei for treatment], December 9.

——. 1953c. "Xie Jianshun tianju zhiyuanshu yuanyi bianwei xiaojie" 謝尖順填具志願書願意變為小姐 [Xie Jianshun consents to becoming a lady], September 4.

——. 1953d. "Xie Jianshun yanqi beishang" 謝尖順延期北上 [Xie Jianshun delays moving to Taipei], December 6.

——. 1953e. "Yinyangren Xie Jianshun fangxin liulian Tainan" 陰陽人謝尖順芳心留戀台南 [Hermaphrodite Xie Jianshun prefers to remain in Tainan], September 27.

——. 1953f. "Zhongguo Kelisiding mingri diyici dongshoushu" 中國克麗絲汀明日第一次動手術 [Chinese Christine's first operation tomorrow], August 19.

——. 1953g. "Zhongxingren Xie Jianshun fengling dingqi beishang" 中性人謝尖順奉令定期北上 [Intersex Xie Jianshun obeys order to relocate to Taipei], November 28.

——. 1954. "Shuyi bingta baochi jimi" 數易病塌保持機密 [Maintains secretive to avoid failed treatment], February 2.

——. 1955. "Dabing bian 'guniang'" 大兵變「姑娘」 [Soldier becomes a 'lady'], September 21.

Taiwan xinshengbao 台灣新生報. 1953a. "Pushuo mili bubian cixiong" 撲朔迷離不便雌雄 [Bewildering: Sex difficult to distinguish], August 14.

——. 1953b. "Tainan yinyangren weiwei tanshenshi" 台南陰陽人微微談身世 [Tainan hermaphrodite discusses his life], August 15.

——. 1953c. "Yinyangren danyan junzhongshi bushi chuxiacai" 陰陽人但言軍中事不是廚下才 [Hermaphrodite discusses military life: Not a cook], August 18.

——. 1953d. "Yinyangren jingshenti" 陰陽人淨身體 [Hermaphrodite cleans body], August 19.

———. 1953e. "Yinyangren jinpofu qiankun keding" 陰陽人今剖腹乾坤可定 [Hermaphrodite's sex will be determined today after operation], August 20.

———. 1953f. "Yinyangren ruobiannü zhongshen keyoukao" 陰陽人若變女終身可有靠 [If hermaphrodite becomes a woman: A life partner confirmed], August 16.

———. 1953g. "Yinyangren yinsheng yangshuai" 陰陽人陰盛陽衰 [Hermaphrodite's yin flourishes and yang depletes], August 21.

———. 1955. "Guifu shengong jidao fenkai yinyanglu" 鬼斧神工幾刀分開陰陽錄 [Magical axe rewrites the fate of a hermaphrodite], October 28.

Takao Club. "George Carter Stent." http://www.takaoclub.com/personalities /Stent/. Accessed July 21, 2014.

Tan Xihong 譚錫鴻, trans. 1924. "Ciji zhi bianxing" 雌雞之變性 [Sex change in hen]. Lingnan nongke daxue nongshi yuekan 嶺南農科大學農事月刊 2, no. 12: 29–32.

Tang, Denise. 2011. Conditional Spaces: Hong Kong Lesbian Desires and Everyday Life. Hong Kong: Hong Kong University Press.

Tang Hao 唐豪. 1927. "Lian'ai yu xing de jiqiao zhi meishuhua" 戀愛與性的技巧之美術化 [The art of the technique of love and sex]. XZZ 1, no. 2: 1–6.

Tang Yinian 唐益年. 1993. Qing gong taidian 清宮太監 [Qing palace eunuchs]. Shenyang: Liaoning University Press.

Taylor, Gary. 2000. Castration: An Abbreviated History of Western Manhood. New York: Routledge.

Taylor, Kim. 2005. Chinese Medicine in Early Communist China, 1945–1963: A Medicine of Revolution. New York: Routledge.

Taylor, Peter, and Ann S. Blum. 1991. "Pictoral Representation in Biology." Biology and Philosophy 6: 125–34.

Terry, Jennifer. 1999. An American Obsession: Science, Medicine, and Homosexuality in Modern Society. Chicago: University of Chicago Press.

Tian Junzuo 田軍作. 1935. Bayue de xiangcun 八月的鄉村 [August village]. Shanghai: Rongguang shuju.

Tian, Min. 2000. "Male Dan: The Paradox of Sex, Acting, and Perception of Female Impersonation in Traditional Chinese Theatre." Asian Theatre Journal 17, no. 1: 78–97.

"Tongxing'ai jiehun zhi tongxun" 同性愛結婚之通訊 [Communication on same-sex marriage]. 1934. Sheyiing huabao 攝影畫報 [Illustrated magazine] 10, no. 26: 10.

Tougher, Shaun, ed. 2002. Eunuchs in Antiquity and Beyond. Swansea, U.K.: Classical Press of Wales.

Traub, Valerie. 2015. Thinking Sex with the Early Moderns. Philadelphia: University of Pennsylvania Press.

Trevor-Roper, Hugh Redwald. 1976. The Hermit of Peking: The Hidden Life of Sir Edmund Backhouse. New York: Alfred Knopf.

Tsai, Hung, and Lien-teh Wu. 1929. "The Practice of Surgery and Anesthetics in Ancient China." In *Proceedings of the First Pan-Pacific Surgical Conference*, 477–482. Honolulu: Pan Pacific Union.

Tsai, Shih-shan Henry. 1996. *The Eunuchs in the Ming Dynasty*. Albany: State University of New York Press.

——. 2002. "Eunuch Power in Imperial China." In *Eunuchs in Antiquity and Beyond*, ed. Shaun Tougher, 221–33. London: Classical Press of Wales and Duckworth.

Tsu, Jing. 2005. *Failure, Nationalism, and Literature: The Making of Modern Chinese Identity, 1895–1937*. Stanford, Calif.: Stanford University Press.

Tsu, Jing, and Benjamin Elman, eds. 2014. *Science and Technology in Modern China, 1880s–1940s*. Leiden: Brill.

Unshuld, Paul. 1986. *Nan-Ching: The Classic of Difficult Issues*. Berkeley: University of California Press.

Valussi, Elena. 2008. "Blood, Tigers, Dragons: The Physiology of Transcendence of Women." *Asian Medicine* 4, no. 1: 46–85.

Veith, Ilza, trans. 2002. *The Yellow Emperor's Classic of Internal Medicine*. Berkeley: University of California Press.

Visker, Rudi. 1995. *Michel Foucault: Genealogy as Critique*. London: Verso.

Vitiello, Giovanni. 2011. *The Libertine's Friend: Homosexuality and Masculinity in Late Imperial China*. Chicago: University of Chicago Press.

Volpp, Sophie. 1994. "The Discourse on Male Marriage: Li Yu's 'A Male Mencius's Mother.'" *positions: east asia cultures critique* 2: 113–32.

——. 2001. "Classifying Lust: The Seventeenth-Century Vogue for Male Love." *Harvard Journal of Asiatic Studies* 61: 77–117.

Wagenseil, F. 1933. "Chinesische Eunuchen (Zugleich ein Beitrag zur Kenntnis der Kastrationsfolgen und der rassialen und körperbaulichen Bedeutung der anthropologischen Merkmale)." *Zeitschrift für Morphologie und Anthropologie* 32: 415–68.

Wang Erh-min 王爾敏. 1990. "Zhongguo jindai zhishi pujihua chuanbo zhi tushuo xingshi—Dianshizhai huabaoli" 中國近代知識普及化傳播之圖說形式──點石齋畫報例 [The illustrated form of news for the diffusion of the modern world knowledge in nineteenth-century China: The Tien-Shih-Chai pictorial newspaper, 1884–1900]. *Bulletin of the Institute of Modern History Academia Sinica* 19: 135–72.

Wang, Hsiu-yun. 2017. "Postcolonial Knowledge from Empires: The Beginnings of Menstrual Educaiton in Taiwan, 1950s–1980s." *East Asian Science, Technology and Society: An International Journal* 11 (4): 519–40.

Wang Hui. 1997. "The Fate of 'Mr. Science' in China: The Concept of Science and Its Application in Modern Chinese Thought." In *Formations of Colonial*

Modernity in East Asia, edited by Tani Barlow, 21–81. Durham, N.C.: Duke University Press.

——. 2008. "Scientific Worldview, Cultural Debates, and the Reclassification of Knowledge in Twentieth-Century China." *boundary 2* 35, no. 2: 125–55.

Wang Jingzhong 汪靖中. 2009. *Wugen zhi gen: Zhongguo huanguan shihua* 無根之根: 中國宦官史話 [The roots of the rootless: History of Chinese eunuchs]. Beijing: Dongfang chubanshe.

Wang Jueming 汪厥明, trans. 1926. *Xing zhi yuanli* 性之原理 [The principle of sex]. Shanghai: Commercial Press.

Wang Shounan 王壽南. 2004. *Tangdai de huanguan* 唐代的宦官 [Tang-dynasty eunuchs]. Taipei: Commercial Press.

Wang, Wen-Ji. 2011. "*West Wind Monthly* and the Popular Mental Hygiene Discourse in Republican China." *Taiwanese Journal for Studies of Science, Technology and Medicine* 13: 15–88.

Wang Xianli 王顯理. 1944. "Tongxinglian huanzhe" 同性戀患者 [Homosexual persons]. *Wanxiang* 萬象 4, no. 1: 128–31.

Wang Xuefeng 王雪峰. 2006. *Jiaoyu zhuanxing zhijing: Ershi shiji shangbanshi Zhongguo de xingjiaoyu sixiang yu shijian* 教育轉型之鏡: 20世紀上半時中國的性教育思想與實踐 [Mirror of the education paradigm shift: Theories and practices of sex education in China in early twentieth century]. Beijing: Social Sciences Academic Press.

Wang, Y. C. 1966. *Chinese Intellectuals and the West, 1872–1949*. Chapel Hill: University of North Carolina Press.

Wang Yang 汪洋. 1927. "Renlei faqingqi zhi xingdetezheng" 人類發情期之性的特徵 [The sexual characteristics of the human estrus period]. *XZZ* 1, no. 2: 1–10.

——. 1935. *Fufu xingweisheng* 夫婦性衛生 [The sexual hygiene of married couples]. Shanghai: Zhongyang shudian.

Wang Yanni 王燕妮. 2006. *Guangdan zhihua* 光旦之華. Wuhan: Changjiang wenyi chubanshe.

Wang Yude 王玉德. 1995. *Shenmi de disanxing—Zhongguo taijian* 神秘的第三性—中國太監 [The mysterious third sex: Zhongguo taijian]. Hong Kong: Minchuang chubanshe.

Wang, Zheng. 1999. *Women in the Chinese Enlightenment: Oral and Textual Histories*. Berkeley: University of California Press.

Watson, Burton. 1958. *Ssu-ma Ch'ien: Grand Historian of China*. New York: Columbia University Press.

Weeks, Jeffrey. 1981. *Sex, Politics and Society: The Regulation of Sexuality Since 1800*. London: Longman.

Wei Sheng 薇生, trans. 1925. "Tongxing'ai zai nüzi jiaoyu shang de xin'yiyi" 同性愛在女子教育上的新意義 [The new meaning of same-sex love in women's education]. *FNZZ* 11, no. 6: 1064–69.

Weinbaum, Alys, Lynn M. Thomas, Priti Ramamurthy, Uta G. Poiger, Madeleine Yue Dong, and Tani E. Barlow, eds. 2008. *The Modern Girl Around the World: Consumption, Modernity, Globalization.* Durham, N.C.: Duke University Press.

Weixing Shiguan Zhaizhu 唯性史觀齋主. 1964 and 1965. *Zhongguo tongxinglian mishi* 中國同性戀秘史 [*The secret history of Chinese homosexuality*]. 2 vols. Hong Kong: Yuzhou chuban.

Williams, Bernard. 2002. *Truth and Truthfulness: An Essay in Genealogy.* Princeton, N.J.: Princeton University Press.

Williams, Raymond. 1976. *Keywords: A Vocabulary of Culture and Society.* Oxford: Oxford University Press.

Wilson, Jean D., and Claus Roehrborn. 1999. "Long-Term Consequences of Castration in Men: Lessons from the Skoptzy and the Eunuchs of the Chinese and Ottoman Courts." *Journal of Clinical Endocrinology and Metabolism* 84, no. 12: 4324–31.

Winichakul, Thongchai. 1994. *Siam Mapped: A History of the Geo-Body of a Nation.* Honolulu: University of Hawai'i Press.

"Women weisheme qiaobuqi heshang he taijian" 我們為什麼瞧不起和尚和太監 [Why we look down on monks and eunuchs]. 1941. *Beihua yuekan* 北華月刊 1,no.4: 69–71.

Wong, Alvin K. 2012. "Transgenderism as a Heuristic Device: On the Cross-Historical and Transnational Adaptations of the *Legend of the White Snake*." In *Transgender China*, ed. Howard Chiang, 127–58. New York: Palgrave Macmillan.

Wong, K. Chimin, and Lien-teh Wu. 1936. *History of Chinese Medicine: Being a Chronicle of Medical Happenings in China from Ancient Times to the Present Period.* 2nd ed. Shanghai: National Quarantine Service.

Wong, Young-tsu. 1989. *Search for Modern Nationalism: Zhang Binglin and Revolutionary China, 1869–1936.* Hong Kong: Oxford University Press.

Woolgar, Steve. 1988. *Science: The Very Idea.* London: Travistock.

Wright, David. 1996. "John Fryer and the Shanghai Polytechnic: Making Space for Science in Nineteenth-Century China." *British Journal for the History of Science* 29, no. 1: 1–16.

——. 2000. *Translating Science: The Transmission of Western Chemistry into Late Imperial China, 1840–1900.* Leiden: Brill.

Wu Chieh Ping, and Gu Fang-Liu. 1990. "The Prostate in Eunuchs." In *EORTC Genitourinary Group Monograph 10: Eurological Oncology: Reconstructive Surgery, Organ Conservation, and Restoration of Function*, ed. Philip H. Smith and Michele Pavone-Macaluso, 249–55. New York: Wiley-Liss.

Wu, Cuncun. 2004. *Homoerotic Sensibilities in Late Imperial China.* London: Routledge.

Wu, Cuncun, and Mark Stevenson. 2006. "Male Love Lost: The Fate of Male Same-Sex Prostitution in Beijing in the Late Nineteenth and Early Twentieth Centuries." In *Embodied Modernities: Corporeality, Representation, and Chinese Cultures*, ed. Fran Martin and Ari Larissa Heinrich, 42–59. Honolulu: University of Hawai'i Press.

——. 2010. "Speaking of Flowers: Theatre, Public Culture, and Homoerotic Writing in Nineteenth-Century Beijing." *Asian Theatre Journal* 27, no. 1: 100–129.

Wu Guozhang 吳國璋. 1999. *Beiyange de wenming: Zhongguo taijian wenhualun* 被閹割的文明: 中國太監文化論 [Castrated civilization: On the culture of Chinese eunuchs]. Beijing: Zhishi chubanshe.

Wu, Jie-ping, and Gu Fang-liu. 1987. "The Prostate 41–56 Years Post Castration: An Analysis of 26 Eunuchs." *Chinese Medical Journal* 100, no. 4: 271–72.

Wu, Jin. 2003. "From 'Long Yang' and 'Dui Shi' to Tongzhi: Homosexuality in China." In *The Mental Health Professions and Homosexuality: International Perspectives*, ed. Vittorio Lingiardi and Jack Drescher, 117–43. New York: Haworth.

Wu Ruishu 吳瑞書. 1927a. "Lian'ai yu xingyu zhi qubie" 戀愛與性慾之區別 [The differences between love and sexual desire]. *XZZ*1, no. 1: 1–6.

——. 1927b. "Xingjiaoyu zhi biyao" 性教育之必要 [The necessity of sex education]. *XZZ*1, 1: 1–4.

Wu Xiaocong 吳晓丛, and Baoping Wang 王保平. 2003. "Beiyange de lingyurou" 被阉割的灵与肉 [Castrated soul and flesh]. *Shoucangjie* 收藏界 11: 20–23.

Wu, Yi-Li. 2008. "The Gendered Medical Iconography of the *Golden Mirror* (*Yuzuan yizong jinjian* 御纂醫宗金鑑, 1742)." *Asian Medicine: Tradition and Modernity* 4, no. 2: 452–91.

——. 2010. *Reproducing Women: Medicine, Metaphor, and Childbirth in Late Imperial China*. Berkeley: University of California Press.

——. 2011. "Body, Gender, and Disease: The Female Breast in Late Imperial Chinese Medicine." *Late Imperial China* 32, no. 1: 83–128.

——. 2013. "The Qing Period." In *Chinese Medicine and Healing: An Illustrated History*, ed. T. J. Hinrichs and Linda Barnes, 161–207. Cambridge, Mass.: Harvard University Press.

——. 2016. "The Menstruating Womb: A Cross-Cultural Analysis of Body and Gender in Hŏ Chun's *Precious Mirror of Eastern Medicine* (1613)." *Asian Medicine: Tradition and Modernity* 11, no. 1–2: 21–60.

Wu Youru 吳有如 et al., ed. 2008. *Qingmuo fushihui*: Dianshizhai huabao *jingxuanji* 清末浮世繪: 《點石齋畫報》精選集 [Late Qing Lithographs: Best Collections of *Dianshizhai huabao*]. Taipei: Yuanliu.

Xi Tuo 西拓, trans. 1937. "Meiguo qiufan de xingshenghuo" 美國囚犯的性生活 [The sexual lives of American prisoners]. *XKX* 4, no. 1: 51–57.

Xiang Ru 湘茹. 1935a. "Cong ciji bianxiong shuodao nühua nanshen" 從雌雞變雄說到女化男身 [From the sex change of hens to human female-to-male transformation]. *Beiyang huabao* 北洋畫報 [Beiyang pictorial] 25, no. 1225: 2.

——. 1935b. "Cong ciji bianxiong shuodao nühua nanshen" 從雌雞變雄說到女化男身 [From the sex change of hens to human female-to-male transformation]. *Beiyang huabao* 25, no. 11228: 3.

Xiao Ying 小英. 1935a. "Yin Yao Jinping huixiang A'nidu" 因姚錦屏迴想阿尼度 [Reminded of A'nidu because of Yao Jinping]. *JB*, March 20.

——. 1935b. "Yin Yao Jinping nüshen xiangqi taijian" 因姚錦屏女身想起太監 [Reminded of eunuchs because of Yao Jinping's female body]. *JB*, March 24.

Xiaomingxiong 小明雄. 1984. *Zhongguo tongxing'ai shilu* 中國同性愛史錄 [The history of homosexual love in China]. Hong Kong: Fenhong sanjiao chubanshe.

"Xiaonü huanan" 孝女化男 [A filial daughter becomes a man]. 1893. *Dianshizhai huabao* 點石齋畫報 [Dianshizhai pictorial] 339.

Xie Dexian 謝德銑. 1991. *Zhou Jianren pingzhuan* 周建人評傳 [Biography of Zhou Jianren]. Chongqing: Chongqing chubanshe.

Xie Se 謝瑟, trans. 1927. "Nüxuesheng de tongxing'ai" 女學生的同性愛 [Same-sex love among female students]. *XWH* 1, no. 6: 57–74.

Xin Sheng 辛生. 1935. "Nühuanan xinwen changshi" 女化男新聞常識 [Common knowledge about news of woman-to-man transformation]. *JB*, March 23.

Xinhua zidian 新華字典 [New China character dictionary]. 10th ed. 2004. Beijing: Commercial Press.

Xinwenbao 新聞報. 1935a. "Jinping qu" 錦屏曲 [Jinping qu], March 23.

——. 1935b. "Nühua nanshen" 女化男身 [Woman-to-man bodily transformation], March 18.

——. 1935c. "Yao Jinping de zizhu hou jiang qiuxue" 姚錦屏得資助後將求學 [Yao Jinping will return to school after receiving financial support], March 18.

——. 1935d. "Yao Jinping nühua nanshen" 姚錦屏女化男身 [Yao Jinping's body transformed into a man], March 19.

——. 1935e. "Yao Jinping yanxi wanquan nüshen" 姚錦屏驗係完全女身 [Yao Jinping has been confirmed to be a complete woman], March 22.

——. 1935f. "Yao Jinping yeyi di Hu" 姚錦屏業已抵滬 [Yao Jinping has arrived in Shanghai], March 17.

——. 1935g. "Zuo song Shanghai yiyuan jianyan" 昨送上海醫學院檢驗 [Entered the Shanghai Hospital for examination], March 20.

Xing 邢, trans. 1936. "Yizhi xingxian neng gaibian xingbie" 移植性腺能改變性別 [Transplanting the sex glands can change sex]. *XKX* 2, no. 2: 102–4.

"Xingde qubie shi chengdu de chayi" 性的區別是程度的差異 [Sex difference is a matter of relative difference]. 1936. *XKX* 1, no. 3: 110.

"Xingxue boshi yanjiang suji" 性學博士演講速記 [Notes on the lecture of a sexology doctor]. 1931a. *Linglong* 玲瓏 1, no. 10: 331.

——. 1931b. *Linglong* 1, no. 11: 370.

Xu Jingzai 徐敬仔. 1927. "Xingyu tongxin" 性育通信 [Sex education letters]. *XWH* 1, no. 3: 59–63.

Xu, Xiaoqun. 1997. "'National Essence' vs 'Science:' Chinese Native Physicians' Fight for Legitimacy, 1912–37." *Modern Asian Studies* 31, no. 4: 847–77.

Xu Youwen 徐友文. 1935. "Nübiannan qiwen" 女變男奇聞 [News of female to male]. *Libailiu* 禮拜六 [Saturday] 605: 97.

Xue 雪, trans. 1936. "Duiyu 'xing' kexue de taidu" 對於「性」科學的態度 [Attitudes toward 'sex' science]. *XKX* 1, no. 6: 305–7.

Yan Dongmei 閻東梅 and Cunfa Dong 董存發. 1995. *Ren zhong yao—wan Qing quanjian zhi mi* 人中妖——晚清權監之謎 [Monsters among humans: The riddle of late-Qing powerful eunuchs]. Beijing: China Renmin University Press.

Yan Shi 晏始. 1922. "Funü wenti yu xing de yanjiu" 婦女問題與性的研究 [The woman question and research on sex]. *FNZZ* 8, no. 2: 18–19.

——. 1923. "Nannü de geli yu tongxing'ai" 男女的隔離與同性愛 [The segregation between the sexes and same-sex love]. *FNZZ* 9, no. 5: 14–15.

Yan Xinzhi 晏新志, Yusheng Liu 刘宇生, and Huajun Yan 闫华军. 2009. "Han Jingdi Yangling yanjiu de huigu yu zhanwang" 汉景帝阳陵研究的回顾与展望 [Review and prospect of the Yang mausoleum of Emperor Jingdi, Han dynasty]. *Wenbo* 文博 [Relics and museology] 1: 25–33.

Yan, Yunxiang. 2003. *Private Life Under Socialism: Love, Intimacy, Family Change in a Chinese Village, 1949–1999*. Stanford, Calif.: Stanford University Press.

Yang Guanxiong 楊冠雄. 1930. *Xing jiaoyufa* 性教育法 [Doctrines of sex education]. Shanghai: Liming shuju.

Yang Kai 楊開. 1936. "Xing de diandao'zheng—tongxing'ai" 性的顛倒症——同性愛 [Sexual inversion—homosexuality]. *XKX* 2, no. 4: 11–13.

Yang, Linren 杨林仁. 2003. *Han Yangling* 汉阳陵. Xi'an: Sanqin chubanshe.

Yang, Meng-Hsuan. 2012. "The Great Exodus: Sojourn, Nostalgia, Return, and Identity Formation of Chinese Mainlanders in Taiwan, 1940s–2000s." PhD diss., University of British Columbia.

Yang, Ruisong 楊瑞松. 2010. *Bingfu, huanghuo yu shuishi: "Xifang" shiye de Zhongguo xingxiang yu jindai Zhongguo guozu lunshu xiangxiang* 病夫、黃禍與睡獅：「西方」視野的中國形象與近代中國國族論述想像 [Sick man, yellow peril, and sleeping lion: The images of China from the Western perspectives and the discourses and imagination of Chinese national identity]. Taipei: Chengchi University Press.

Yang Tsui-hua 楊翠華. 1991. *Zhongjihui dui kexue de zanzhu* 中基會對科學的贊助 [Patronage of science: The China Foundation for the Promotion of Education and Culture]. Taipei: Institute of Modern History, Academia Sinica.

——. 2008. "Meiyuan dui Taiwan de weishen jihua yu yiliao tizhi de xingsu" 美援對台灣的衛生計畫與醫療體制的形塑 [U.S. aid in the formation of health planning and the medical system in Taiwan]. *Bulletin of the Institute of Modern History, Academia Sinica* 62: 91–139.

Yang Youtian 楊憂天. 1929. "Tongxing'ai de wenti" 同性愛的問題 [The problem of same-sex love]. *Beixin* 北新 3, no. 2: 403–39.

Yang Zhengguang 楊爭光. 1991. *Zhongguo zuihou yige taijian* 中國最後一個太監 [The last eunuch in China]. Beijing: Qunzhong chubanshe.

Yao Wu 耀五. 1935. "Nübiannan jiqita" 女變男及其他 [Female to male and others]. *Dushu shenghuo* 讀書生活 [Reading life] 1, no. 11: 15–16.

Yates, Robin. 2005. "Medicine for Women in Early China: A Preliminary Survey." *Nan Nü* 7, no.3: 127–81.

Yeh, Wen-hsin. 1990. *The Alienated Academy: Culture and Politics in Republican China, 1919–1937*. Cambridge, Mass.: Harvard University Press.

——. 2007. *Shanghai Splendor: Economic Sentiments and the Making of Modern China, 1843–1949*. Berkeley: University of California Press.

Yi Yi 憶漪. 1955. "Xie Jianshun xiaojie de gushi" 謝尖順小姐的故事 [The story of Miss Xie Jianshun]. *LHB*, October 13–November 18.

Yong Jia 永嘉. 1937. "Zhipei xingyu de neifenmi yaosu" 支配性慾的內分泌要素 [Hormones that determine sexual desire]. *XKX* 3, no. 6: 15–18.

"Youshi yige nübiannan" 又是一個女變男 [Another woman becoming man]. 1917. *Xiaoyi* 小鐸, February 20.

Yu Hsin-ting 余欣庭. 2009. "Taiwan zhanhou yiduanxing/shenti de guansu lishi: Yi tongxinglian han yinyangren weilie, 1950s–2008" 臺灣戰後異端性/身體的管束歷史:以同性戀和陰陽人為例, 1950s–2008 [Regulating deviant sexualities and bodies in Taiwan, 1950s–2008: The cases of homosexuality and hermaphrodites]. M.A. thesis, Kaohsiung Medical University.

Yu Minzhong 于敏中, ed. 1965. *Guochao gongshi* 國朝宮史 [A history of the palace during the Qing period]. 5 vols. Taipei: Taiwan xuesheng shuju.

Yuan Yuan 袁媛. 2010. *Jindai shenglixue zai Zhongguo* 近代生理學在中國 [Modern physiology in China]. Shanghai: Shanghai renmin chubanshe.

Yun 雲. 1941. "Yu de cixiongxing bianhuan" 魚的雌雄性變換 [Sex change in fish]. *Xin kexue* 新科學 [New science] 4, no. 3: 284.

Zarrow, Peter. 2006. "Liang Qichao and the Conceptualization of 'Race' in Late Qing China." *Bulletin of The Institute of Modern History, Academia Sinica* 52: 113–64.

——. 2012. *After Empire: The Conceptual Transformation of the Chinese State, 1885–1924*. Stanford, Calif.: Stanford University Press.

Zeitlin, Judith. 1993. *Historian of the Strange: Pu Songling and the Chinese Classical Tale*. Stanford, Calif.: Stanford University Press.

Zhang 章, trans. 1936a. "Xingbie zhi benyi" 性別之本意 [The meaning of sex difference]. *XKX* 1, no. 3: 136–40.

——, trans. 1936b. "Xingde neifenmi yaosu" 性的內分泌要素 [Sex hormones]. *XKX* 1, no. 6: 326–34.

Zhang Dai 張岱. 1982. *Tao'an mengyi* 陶庵夢憶 [Dream reminiscences of Tao'an]. Shanghai: Shanghai shudian.

Zhang Dongmin 張東民. 1927. *Xing de chongbai* 性的崇拜 [The worship of sex]. Shanghai: Beixin shuju.

Zhang, Everett Yuehong. 2015. *The Impotence Epidemic: Men's Medicine and Sexual Desire in Contemporary China*. Durham, N.C.: Duke University Press.

Zhang Jingsheng 張競生. (1924) 1998. *Meide renshengguan* 美的人生觀 [A way of life based on beauty]. Reprinted in *ZJSWJ*, vol. 1, 24–38.

——. (1925) 1998. *Meide shehui zuzhifa* 美的社會組織法 [Organizational principles of a society based on beauty]. Reprinted in *ZJSWJ*, vol. 1, 139–264.

——. (1926) 1998. "Da Zhou Jianren xiansheng 'Guanyu *Xingshi* de jijü hua'" 答周建人先生《關於〈性史〉的幾句話》 [A response to Mr. Zhou Jianren's "Few Words on *Sex Histories*"]. *Yiban* 一般 [Ordinary magazine], November. Reprinted in *ZJSWJ*, vol. 2, 420–22.

——. 1927a. "Disanzhongshui yu luanzhu ji shengji de dian he yosheng de guanxi" 第三種水與卵珠及生機的電和優生的關係 [The relationship between the third kind of water and the ovum, and, the electric vital moment and eugenics]. *XWH* 1, no. 2: 23–48.

——. 1927b. "Xingbu huxi" 性部呼吸 [The breathing of sex anatomy]. *XWH* 1, no. 4: 21–32.

——. 1927c. "Xingbu yu dantian huxi" 性部與丹田呼吸 [The breathing of sex anatomy and the diaphragm]. *XWH* 1, no. 5: 1–23.

——. 1927d. "Xingyu tongxin" 性育通信 [Sex education letters]. *XWH* 1, no. 2: 111.

——. 1927e. "'Yimang' yu xingxue" 醫氓與性學 ["Quack doctors" and sexology]. *XWH* 1, 3: 113–16.

——. 1927f. "Youchu yige guaitou" 又出一個怪頭 [The appearance of another weirdo]. *XWH* 1, no. 4: 126–28.

——. 1927g. "Yu Jingbao lunjin yinshu erchang xingxue de fangfa" 與晶報論禁淫書而倡性學的方法 [Discussing with *Crystal* methods for banning pornography and promoting sexology]. *XWH* 1, no. 6: 23–25.

——. (1926) 2005. *Xingshi 1926* 性史 1926 [Sex histories 1926]. Taipei: Dala.

Zhang, Jingyuan. 1992. *Psychoanalysis in China: Literary Transformations, 1919–1949*. Ithaca, N.Y.: Cornell University East Asia Program.

Zhang Luqi 張祿祺. 1941. "Shengzhixian de neifenmi" 生殖腺的內分泌 [The internal secretions of the sex glands]. *Xuesheng zhiyou* 學生之友 3, no. 4–5: 39–43.

Zhang Minyun 張敏筠. 1950. *Xing kexue* 性科學 [Sexological science]. Shanghai: Shidai Shuju.

Zhang Peizhong 張培忠. 2008. *Wenyao yu xianzhi: Zhang Jingsheng zhuan* 文妖與先知: 張竟生傳 [Literary freak and prophet: A Biography of Zhang Jingsheng]. Beijing: Sanlian shudian.

Zhang Ruogu 張若谷. 1936. "Yingnü yundongyuan huashen weinan" 英女運動員化身為男 [British woman athlete becomes male]. *XKX* 2, no. 2: 105.

Zhang Xichen 章錫琛, ed. 1925. *Xindaode taolunji* (性道德討論集) [On sexual morality]. Shanghai: Liangxi tushuguan.

Zhang Yaoming 張躍銘. 1996. *Zhanggong huanguan quanshu: Lidai taijian mishi* 掌宮宦官全書: 歷代太監密史 [The secret histories of eunuchs throughout the dynasties]. Harbin: Heilongjiang renmin chubanshe.

Zhang Yunfeng 張雲風. 2004. *Zhongguo huanguan shilui* (中國宦官事略) [Matters regarding Chinese eunuchs]. Taipei: Dadi.

Zhang Zhijie 張之傑. 2008. "Zhu Xi yu wu zhengfu zhuyi: Wei shengwuxuejia Zhu Xi zhuanji buyi" 朱洗與無政府主義: 為生物學家朱洗傳記補遺 [Zhu Xi and anarchy: A biographical supplement for the biologist Zhu Xi]. *Zhonghua kejishixue huikan* 中華科技史學會刊 [Journal of the Chinese Association for the History of Science] 12: 25–38.

——. 2011. "Gongxing gongnali?" 宮刑宮哪裡? [What is being castrated?]. In *Kexue shihua* 科學史話 [History of science], 188–192. Taipei: Shangwu yingshuguan.

Zhang, Zhongmin. 2009. "Publishing and Cultural Politics: The Books on Reproductive Medicine and Their Readership in Late Qing." *Academic Monthly* 41, no. 1: 128–42.

Zhao Jingshen 趙景深. 1929. "Tongxing lian'ai xiaoshuo de chajin" 同性戀愛小說的查禁 [The banning of a homosexual novel]. *Xiaoshuo yuebao* 小說月報 [Novel monthly] 20, no. 3: 611–12.

Zhao Pingqi 趙蘋起. 1938. "Nühuanan nanhuanü" 女化男男化女 [Female-to-male and male-to-female]. *Beining tielu yuekan* 北寧鐵路月刊 [Beining railway monthly] 8, no. 4: 20–21.

Zhen Ni 珍妮. 1946. "Tongxing lian'ai" 同性戀愛 [Homosexuality]. *Xin funü yuekan* 新婦女月刊 [New women's monthly] 5: 13–14.

Zheng Renjia 鄭仁家. 1991. "Diaonian Chen Cunren zhongyishi" 掉念陳存仁中醫師 [An obituary of the Chinese medicine physician Chen Cunren]. *ZJWX* 58, no. 2: 118–22.

Zheng Yi 正誼. 1927. "Tongxun" 通訊 [Letters]. *XWH* 1, no. 1: 47.

Zhi Jun 芝君. 1927. "Xingyu tongxin" 性育通信 [Sex education letters]. *XWH* 1, no. 3: 73.

Zhi Qin 志勤. 1931. "Diyipao: Zhongguo funü zhuzhong xingxue: Shanghai funühui de yongwei" 第一砲: 中國婦女注重性學: 上海婦女會的勇為 [The first shot: Chinese women value sexology: The courage of the Shanghai women's association]. *Zhongguo sheying xuehui huabao* 中國攝影學會畫報 6: 298.

Zhi Yan 隻眼. 1919. "Taijian yu chanzu" 太監與纏足 [Eunuchs and footbinding]. *Meizhou pinglun* 每週評論 [Weekly review], no. 3 (April 6, 1919).

Zhong, Xueping. 2000. *Masculinity Besieged: Issues of Modernity and Male Subjectivity in Chinese Literature of the Late Twentieth Century.* Durham, N.C.: Duke University Press.

Zhongguo renmin zhengzhi xieshang huiyi quanguo weiyuanhui wenshi ziliao yanjiu weiyuanhui 中國人民政治協商會議全國委員會文史資料研究委員會, ed. 1982. *Wan Qing gongting shenghuo jianwen* 晚清宮廷生活見聞 [Life in Late-Qing Imperial Palace]. Beijing: Wenshi ziliao chubanshe.

Zhongguo shibao 中國時報. 1955a. "Lujun zongyiyuan fabiao Xie Jianshun jingguo" 陸軍總醫院發表謝尖順變性經過 [No. 1 Hospital publicized Xie Jianshun's sex change story], October 29.

——. 1955b. "Xie Jianshun bianxingji Lujun diyi zongyiyuan zuofabiao gongbao" 謝尖順變性記陸軍第一總醫院昨發表公報 [No. 1 hospital yesterday published report on Xie Jianshun's sex change], October 28.

——. 1956a. "Dabing xiaojie Xie Jianshun zhongri chifan shuijiao" 大兵小姐謝尖順終日吃飯睡覺 [Miss soldier Xie Jianshun is preoccupied with eating and sleeping daily], October 10.

——. 1956b. "Ding Hao qufang Xie Jianshun" 丁皓昨趨訪謝尖順 [Ding Hao visited Xie Jianshun yesterday], November 7.

——. 1959. "Dangnian yingqi haomai jinri fumei wenrou" 當年英氣豪邁今日撫媚溫柔 [Heroic previously, gentle today], November 1.

Zhonghua ribao 中華日報. 1953a. "Dabing bian xiaojie Xie Jianshun buzai fandui" 大兵變小姐謝尖順不再反對 [Soldier becomes a lady: Xie Jianshun no longer refuses]. August 30.

——. 1953b. "Nanshi yinyangren dongshoushu shengli jiego quanshu nüxing" 南市陰陽人動手術生理結構全屬女性 [Tainan hermaphrodite's female biology revealed after surgery], August 21.

——. 1953c. "Nanshi yinyangren jingchubu jianyan" 南市陰陽人經初步檢驗 [Tainan hermaphrodite after preliminary examination], August 14.

——. 1953d. "Nanshi yinyangren mingtian shishoushu" 南市陰陽人明天施手術 [Tainan hermaphrodite to be operated tomorrow], August 19.

——. 1953e. "Ruguo biancheng nüxing renyuan shangqianxian shadi" 如果變成女性仍願上前線殺敵 [If becomes a woman: Still willing to join the army], September 8.

——. 1953f. " 'Ta' heyi beiai" 「她」何以悲哀 [Why is 'she' disappointed], August 23.

——. 1953g. "Tainanshi faxian jiaxing yinyangren" 台南市發現假性陰陽人 [A pseudo-hermaphrodite discovered in Tainan city], August 14.

——. 1953h. "Xie Jianshun kaidaohou shenghuo qingxing zhengchang" 謝尖順開刀後生活情形正常 [Xie Jianshun resumes normal life after surgery], August 22.

——. 1953i. "Yinyangren" 陰陽人 [Hermaphrodite], August 17.

——. 1953j. "Yinyangren shinan shinü jintian fenxiao" 陰陽人是男是女今天分曉 [Heramphrodite's sex determined today], August 20.

——. 1953k. "Yinyangren Xie Juanshun qianzi tongyi bianwei nüxing" 陰陽人謝尖順簽字同意變為女性 [Hermaphrodite Xie Jianshun consents to becoming female], September 4.

——. 1955a. "Xielou bingren mimi weifan yiwu daode" 洩漏病人秘密違反醫務道德 [Leaking medical secrets of a patient goes against medical ethics], October 29.

——. 1955b. "Yibian erchai Xie Jianshun bianxingji" 易弁而釵謝尖順變性記 [Man to woman: Xie Jianshun's sex change story], October 28.

Zhongyang ribao 中央日報. 1954. "Xie Jianshun yiyou nü'er tai" 謝尖順已有女兒態 [Xie Jianshun appears feminine already], March 18.

——. 1955a. "Dabing Xie Jianshun zuoxing sanci xingbian shoushu" 大兵謝尖順昨行三次性變手術 [Soldier Xie Jianshun had a third sex change surgery yesterday], August 31.

——. 1955b. "Lishi liangzai shoushu sici" 歷時兩載手術四次 [Four operations in two years], October 28.

——. 1955c. "Xie Jianshun bianxing chenggong yi guanyu funü shenghuo" 謝尖順變性成功已慣於婦女生活 [Xie Jianshun's sex change successful: Already used to the lifestyle of a woman], September 9.

——. 1955d. "Xie Jianshun bianxing Lujun diyi zongyiyuan duanqinei gongbu xiangqing" 謝尖順變性陸軍第一總醫院短期內公佈詳情 [No. 1 Hospital soon releases details regarding Xie Jianshun's sex change], September 1.

——. 1955e. "Xie Jianshun bianxing jingguo" 謝尖順性變經過 [Xie Jianshun's sex change experience], October 28.

——. 1955f. "Xie Jianshun xingbian jinkuang" 謝尖順性變近況 [Recent updates on Xie Jianshun's sex change], January 9.

——. 1956. "Guanyu Xie Jianshun xiaojie"關於謝尖順小姐 [Regarding Miss Xie Jianshun], June 18.

——. 1959. "Datong furu jiaoyangyuan zuoqing chengli liuzhounian" 大同婦孺教養院昨慶成立六週年 [Tatong Relief Institute celebrated sixth anniversary yesterday], November 1.

Zhou Jianren 周健人. (1926) 1998. "Da Zhang Jingsheng xiansheng" 答張競生先生 [Response to Mr. Zhang Jingsheng] *Yiban*, November. Reprinted in *ZJSWJ*, vol. 2, 423–26.

——. (1927) 1928. *Xing yu rensheng* 性與人生 [Sex and human life]. Shanghai: Kaiming shudian.

——. 1931. *Xingjiaoyu* 性教育 [Sex education]. Shanghai: Commercial Press.

Zhou Shoujuan 周瘦鵑. 1929. "Diandao xingbie zhi guainü qinan" 顛倒性別之怪女奇男 [Deviant women and men of sexual inversion]. *Zi ruo lan* 紫羅蘭 4, no. 5: 1–6.

Zhou Zuoren 周作人. 1934a. "Nanhuanü" 男化女 [Man into woman]. In *Yeduchao* 夜讀抄 [Notes from night readings], 184–92. Shanghai: Beixin shuju.

——. 1934b. "Taijian" 太監 [Eunuchs]. In *Yeduchao* 夜讀抄 [Notes from night readings], 261–68. Shanghai: Beixin shuju.

——. 1934c. "Xing de xinli" 性的心理 [The psychology of sex]. In *Yeduchao* 夜讀抄 [Notes from night readings], 43–52. Shanghai: Beixin shuju.

——. 1935. "Guanyu gongxing" 關於宮刑 [On castration]. In *Kucha suibi* 苦茶隨筆 [Jottings from drinking bitter tea], 251–54. Shanghai: Beixin shuju.

Zhu Jianxia 朱劍霞, trans. 1928. *Xing zhi shengli* 性之生理 [The physiology and psychology of sex]. Shanghai: Commercial Press.

Zhu Jihuang 朱季潢, ed. 1980a "Taijian tan wanglu" 太監談往錄. *Zijincheng* 紫禁城 [Forbidden city] 1: 42–44.

——. 1980b. "Taijian tan wanglu." *Zijincheng* 2: 40–41.

Zhu, Ping. 2015. *Gender and Subjectivities in Early Twentieth-Century Literature and Culture*. New York: Palgrave Macmillan.

Zhu Rutong 朱汝曈. 2010. *Zhongguo xiandai wenxue liupai mantan* 中國現代文學流派漫談 [Modern Chinese literary schools]. Taipei: Showwe.

Zhu Xi 朱洗. 1923. "Sishiyi ge yue de qingong jianxue shenghuo" 四十一個月的勤功兼學生活 [My forty-one months of life in the Work-Study Movement]. *Minguo ribao* 民國日報, March–May.

——. (1939) 1946. *Danshengren yu renshengdan* 蛋生人與人生蛋 [Human from eggs and egg from humans]. Rev. ed. Shanghai: Wenhua shenghuo chubanshe.

——. (1940) 1948. *Women de zuxian* 我們的祖先 [Our ancestors]. Rev. ed. Shanghai: Wenhua shenghuo chubanshe.

——. (1941) 1948. *Zhognü qingnan* 重女輕男 [Women over men]. Rev. ed. Shanghai: Wenhua shenghuo chubanshe.

——. (1945) 1948. *Cixiong zhi bian* 雌雄之變 [Changes in biological femaleness and maleness]. Rev. ed. Shanghai: Wenhua shenghuo chubanshe.

——. 1946. *Aiqing de laiyuan* 愛情的來源 [The origins of love]. Rev. ed. Shanghai: Wenhua shenghuo chubanshe.

——. (1946) 1948. *Zhishi de laiyuan* 智識的來源 [The origins of intellectual knowledge]. Rev. ed. Shanghai: Wenhua shenghuo chubanshe.

Zito, Angela. 2001. "Queering Filiality, Raising the Dead." *Journal of the History of Sexuality* 10, no. 2: 195–201.

——. 2006. "Bound to be Represented: Theorizing/Fetishizing Footbinding." In *Embodied Modernities: Corporeality, Representation, and Chinese Culture*, ed. Fran Martin and Ari Larissa Heinrich, 21–41. Honolulu: University of Hawai'i Press.

——. 2007. "Secularizing the Pain of Footbinding in China: Missionary and Medical Stagings of the Universal Body." *Journal of the American Academy of Religion* 75, no. 1: 1–24.

Zou, John. 2006. "Cross-Dressed Nation: Mei Lanfang and the Clothing of Modern Chinese Men." In *Embodied Modernities: Corporeality, Representation, and Chinese Cultures*, ed. Fran Martin and Ari Larissa Heinrich, 79–97. Honolulu: University of Hawai'i Press.

Zou Lü 鄒律. 1988. *Lidai mingtaijian miwen* 歷代名太監祕聞 [The secrets of famous eunuchs]. Tianjin: Tianjin renming chubanshe.

Index

518 Hospital, 239, 242, 245, 250

A'nidu, 218–20
Acces, Alice Henriette, 205–6
Adanson, Michel, 11
Albinus, Bernhard, 82
Altman, Dennis, 132
anatomy, 1, 4, 10, 90, 105, 108, 116, 119,
    128, 173, 174, 178, 179, 199, 200;
    anatomical aesthetic, 72–86, 88, 97,
    100, 101, 104, 111–12, 113, 121, 283,
    299n7; anatomical drawings or
    illustrations, 72, 79–82, 86, 93, 97,
    98–99, 100, 101, 103, 107, 123–24, 134,
    283, 300n30; anatomical text, 71,
    74–75, 198, 302n.79; Anatomy Law of
    1913, 97; and castration, 27, 31, 32, 40,
    50, 54; reproductive, 3, 75–78, 82, 84,
    95, 97, 100, 102, 109, 144, 147, 161,
    301n41; style of reasoning, 134; of Xie
    Jianshun, 243, 247, 248, 260; of Yao
    Jinping, 219. See also dissection
animal sex change, 5, 27–28, 178–79,
    183–88, 193, 195, 197, 204, 208, 227, 284

archive, 3, 20–22, 39, 45, 57, 68, 125, 136,
    143, 160, 239, 278, 294n27; Qing
    palace, 29, 48–49, 58
Arondekar, Anjali, 21
Art Life (*Meishu shenghuo*), 208
Articles of Favorable Treatment of the
    Emperor of Great Qing After His
    Abdication, 63
asylum, 3, 71, 290n17, 299n6. *See also*
    psychiatry

Bachelard, Gaston, 13
Backhouse, Edmund, 36–39
Baldwin, E. M., 188
banlu chujia, 55
bao, 74, 85. *See also* womb
Barlow, Tani, 105–6, 121, 135
Bataillon, J. E., 94
Beijing (Beiping, 1928–1949), 20, 29, 31,
    56, 59, 94, 125, 129, 159, 176, 184,
    213, 234, 262–63
Beixin, 156
Beixin shuju, 156
Beiyang Army, 63

Broster, Lennox Ross, 233–34

Brown-Séquard, Claude-Édouard, 181–82

Bushell, Stephen W., 22

Cai Yuanpei, 141

Canguilhem, Georges, 13

Capital Newspaper (Jingbao), 142

Carnal Prayer Mat, The (Rouputuan), 125–28, 177

carnality, 4, 14, 125–29, 177, 178, 234, 284, 286

Carpenter, Edward, 147, 156–57, 160, 190–91

castrated civilization, 4, 18, 69, 70–71, 124, 173, 293n20

castration, 5, 70, 75, 124, 126, 164, 185, 203, 207, 219–20, 228–29, 235, 278, 284–86; archiving of, 3, 20–22; critiques of, 1, 4; and discrepant definitions of masculinity, 29–39; domestic accounts of, 66–67; and feminization of male sex, 193–98; *gongxing* (castration for punishment), 16, 32–33; medical imaging of 39–49; natural, 199; parasitic, 227; as social and cultural reproduction, 49–55; Westerners' accounts of, 22–29. *See also* eunuch; *tianyan*

Cauldwell, David, 318n7

Central Daily News (Zhongyang ribao), 252–53, 264, 268, 271

Chai Fuyuan, 88, 97, 99, 108–10, 144, 164, 180–81, 199–201, 203, 208, 218

Chakrabarty, Dipesh, 276

Chang Hong, 163

Changes in Biological Femaleness and Maleness (Cixiong zhi bian), 95, 114

Chaozhou, Guangdong, 239

Chen Cunren, 15–16, 152

Chen Duxiu, 11

Chen, Fang-Ming, 279

Chen, Jian, 237

Chen, Kuan-Hsing, 135, 275–76

Chen Xiefen, 106

Chen Yucang, 97, 110, 196–97, 208

Chen Zhen, 94, 112

Chengdu, 141

Chi Huanqing, 56–57

Chiang Kai-Shek, 266, 272

Chiayi County, 267

childbirth, 1, 73, 104, 238, 269. See also *fuke*; pregnancy

China Daily News (Zhonghua ribao), 240, 244, 248, 252, 271

China Times (Zhongguo shibao), 268, 271

Chinese Academy of Health, 168

Chinese Academy of Sciences: Experimental Biology Institute, 94

"Chinese Christine," 6, 236–37, 244, 248–50. *See* Xie Jianshun

Chinese Communist Party, 94, 237

Chinese Imperial Maritime Customs Service, 24

Chinese medicine, 2, 16, 27, 28, 74, 75, 80, 104, 107, 134, 166, 169, 200, 201, 275, 303n108, 311n155, 322n143

Chiu, Fred Yen Liang, 276

Choumou Monthly, 148

chouren, 10–11

chromosome, 3–4, 103, 107, 112, 114, 116–19, 121, 128, 182, 188, 198, 227, 304n113

Chu En-ming, 38

ci, 4, 72, 88–92, 95–97, 100–101, 103, 107, 113–14, 116–18, 140, 201–2, 205, 211–12, 219–20, 228, 234, 257, 284

ci-xiong human, 201–2, 205

civility (*wen*), 19, 21

cixiong tongti, 114. *See* hermaphroditism

Classic of Difficult Issues, The (Nanjing), 84–85

transsexuals, 282, 285–86; and
Confucian norms, 51; critique and
condemnation of, 1, 16, 18–19, 65;
employment, 1, 57–58, 61, 62, 65;
eunuchism, 3–4, 7, 16, 19–23, 25, 37,
49–51, 61–63, 65–67, 70, 194, 218,
285; gender identity, 4, 19, 22, 36–37,
51, 194; illiteracy, 58; natural, 1–2,
58; self-narration, 56–62; sexual
identity, 1, 4, 71; as third sex, 18. *See*
castration
evidence, 3, 7, 21, 24, 31, 42, 44–46,
56–57, 61, 64, 68, 70, 71–72, 83, 101,
108, 113, 127, 134, 154, 155, 157,
175–76, 181, 184–86, 189, 196,
209–10, 216–18, 222, 238, 289n5

Fairbank, John K., 36
Fang Fei, 217, 219
fangzu (letting foot out), 67–68
Farquhar, Judith, 132
Fei Hongnian, 184, 186
feinan (nonmale), 2, 256
feinü (nonfemale), 2, 256
"Female Becoming Male," 224
feminism, 3, 19, 106, 107, 129, 168, 181,
188, 221, 235, 274
Feng Fei, 90
Feng Mingfang, 234–35
Feng Xiaoqing, 161
Feng Yuxiang, 20
Fine Arts Research Society, 146
First Epoch of the Rise of Eunuchs,
16, 33
First Sino-Japanese War (1894–1895),
134
footbinding, 3, 4, 5, 16, 19, 21, 25, 50,
60, 61, 63, 67, 107, 110, 194, 230, 231
Forbidden City, 20, 36, 63
Forel, Auguste, 160
Forman, Ross, 36

Foucault, Michel, 8, 13, 131–32, 137, 146,
150, 158, 169, 174, 177
Fraser, Sarah, 45
fraudulent science, 151
Friedman, David, 29–30
Freud, Sigmund, 5, 32, 149, 161–63,
166–67, 189, 191
Fryer, John, 97
Fu Chun, 251
Fujian, 141
fuke (gynecology or women's medicine),
105; development, 84; diagnosis of
gender, 3, 73–74, 79; emergence, 2,
72–73; experts, 2, 82
funü, 105–6
Furth, Charlotte, 2, 73, 84, 89, 167,
198, 222

Galen, 79
Galton, Francis, 154–55
Gao Rimei, 214
Gao Xian, 100
Gaozi, 8
Gangshan District, 270
gate of life, 85–86
gay, 132, 137, 175, 246, 274, 280–81
gaze; subscellular, 112–21, 123
Geddes, Patrick, 111
gender, 5, 8, 13, 126, 133, 136, 196, 217,
221, 229–32, 235, 239, 242, 247,
255–56, 260, 267, 269–70, 273, 275,
279, 281, 284; biologization, 7,
70–124, 190; differentiation, 1;
equality, 181–82, 188–89; of eunuchs,
15–69; expression, 3, 183; hierarchy,
2; identity, 4, 183, 238, 267, 269, 274;
inversion, 130, 157; liminal, 1;
malleability, 263; naturalization, 3;
segregation, 140, 147; social division,
1; transgression, 174, 176, 222, 238,
254, 257, 263, 281

Huang Huiming, 225
Huang Jiede, 258
Hundred Days' Reform, 65
Huiyaolou, 64
Humans from Eggs and Eggs from Humans, 95, 227
human/nonhuman divide, 3, 72, 88–89, 113–14
Humana, Charles, 28
Hunan, 141, 232
Hunt, Nancy Rose, 50
Hunter, John, 101
Hunter, William, 82
Huwei District, 257

Imperial Chinese Railways, 40
Imperial Maritime Customs, 22, 24, 27, 40
Imperial School of Combined Learning (Tongwen Guan), 40
Imperial University of Peking, 40
impersonation, 257
intersexuality, 113–20, 124, 187, 196, 198, 204–5, 208–9, 224, 233–34, 236–39, 244, 247, 249, 260–61, 263, 271, 275. *See also* hermaphroditism
Ishimpo, 82

Jai Ben-ray, 9
Jamieson, R. A., 27–28
Japan, 22, 33, 87, 110, 125, 141, 142, 180, 186, 196, 209, 216, 260, 275–76, 280–81, 285
Jay, Jennifer, 51
Jefferys, W. Hamilton, 23
Jia Yinghua, 30
Jian Yun, 148
Jianfu Palace, 63–65
Jiang Peizhen, 274
Jiang Xizheng, 252
Jiangnan Arsenal, 10

Jiangping Children's Home, 220
Jichangtang Hospital, 258
Jicuiting, 64
jing ("essence"), 75–76, 78, 84–85, 167
Jinghai County, Hebei, 59
Jingyixuan, 64
jiulong huishengsan, 67
Jiyunlou, 64
Johnson, Marshall, 276
Johnston, Reginald, 65
Jorgensen, Christine, 6, 236–37, 239, 248–50, 260, 269, 272–73
Journal of Sexual Science, 168

kagaku, 10
Kaiming Bookstore, 225
kang ("depository"), 26, 59–60
Kang, Wen-Qing, 130–32, 156, 174–75
Kangde Emperor, 20
Kaohsiung County, 270, 274
Katholieke Universiteit Leuven, 213
Kerr, John G., 97
kexue (science), 10–12, 151–52, 155
kexuehua (scientization), 104
kidney: 83–85; inner, 85; outer, 54, 75–76, 85
Kinsey, Alfred, 149, 235
knifer, 26–29, 54, 57, 60–61
Ko, Dorothy, 19, 109
Koltzoff, Nicholas K., 188
Korea, 237, 275
Korean War (1950–1953), 237, 275
Korsakov, Vladimir Vikentevich, 40
Koubková, Zdeňka (Zdeněk Koubek), 233
Koyasu Takashi, 9
Krafft-Ebing, Richard von, 149, 157
Krieg, Paul, 31
Kung Sheung Daily News, 267